5G Multimedia Communication

5G Multimedia Communication

Communication
Technology, Multiservices, and Deployment

Edited by
Zoran S. Bojkovic, Dragorad A. Milovanovic,
and Tulsi Pawan Fowdur

CRC Press
Taylor & Francis Group
Boca Raton London New York

CRC Press is an imprint of the
Taylor & Francis Group, an **informa** business

First edition published 2020
by CRC Press
6000 Broken Sound Parkway NW, Suite 300, Boca Raton, FL 33487-2742

and by CRC Press
2 Park Square, Milton Park, Abingdon, Oxon, OX14 4RN

First issued in paperback 2022

Library of Congress Cataloging-in-Publication Data
Names: Bojkovic, Z. S., editor. | Milovanovic, Dragorad A., editor. |
Fowdur, Tulsi Pawan, editor.
Title: 5G multimedia communication : technology, multiservices, and
deployment / edited by, Zoran Bojkovic, Dragorad Milovanovic, Tulsi
Pawan Fowdur.
Other titles: Five G multimedia communication
Description: First edition. | Boca Raton, FL : CRC Press, 2020. |
Includes bibliographical references and index.
Identifiers: LCCN 2020030689 (print) | LCCN 2020030690 (ebook) |
ISBN 9780367178505 (hbk) | ISBN 9781003096450 (ebk)
Subjects: LCSH: 5G mobile communication systems.
Classification: LCC TK5103.25 .A146 2020 (print) | LCC TK5103.25 (ebook) |
DDC 621.3845/6—dc23
LC record available at https://lccn.loc.gov/2020030689
LC ebook record available at https://lccn.loc.gov/2020030690

ISBN 13: 978-0-367-56115-4 (pbk)
ISBN 13: 978-0-367-17850-5 (hbk)
ISBN 13: 978-1-003-09645-0 (ebk)

Typeset in Palatino
by codeMantra

Contents

Section I Critical Enabling Technology

Section II Multiservices Network

Section III Deployment Scenarios

Foreword

The deployment of devices such as IoT and drones in everyday life, the ongoing explosion of the media services and the continuously increasing number of connected users and devices pose a significant challenge to the current ICT's architectural and operational paradigms, especially when considering the need to handle and accommodate the colossal amount of data generated.

In this context, 5G is viewed as the key network technology that will meet the aforementioned challenge and allow for the realization of a "hyperconnected society" where billions of users, via their respective end devices (e.g., wearables, smart home appliances, connected cars, smart phones, laptops) and machines (e.g., IoT devices, drones, etc.), will be able to exchange data and offer/receive services at a higher quality of service (QoS) level. 5G mobile communication networks are expected to enhance not only the data rates but also the scalability, connectivity and energy efficiency of the networks. It is expected that by 2030, 50 billion devices will be connected to the global network, which would appear to present a challenge. Various standardized fora have defined 5G key properties, which are the following: enhanced mobile broadband (eMBB), ultrareliable low-latency communication (URLLC) and massive machine-type communications (mMTCs). Therefore, 5G has considered to support applications in various verticals such as smart cities, health care, digital manufacturing and creative industries.

There is an evolution of immersive media to support new type of applications and services. Such evolution includes higher resolution (4K, 8K), new type of capturing (360°, light field, volumetric), new type of immersive perception (use of augmented/virtual/mixed/extreme reality) and new type of devices in terms of PPI, FoV, tracking and immersive consumption capabilities. Such evolution to immersive media technologies leads stakeholders to cope with new issues in terms of interaction metaphors, perception of distances, coexistence and presence.

As an effect, there are new challenges associated with multimedia functionalities across 5G networks in terms of compression, encoding, transmission, manipulation, synchronization, storage and contextual. Besides capabilities in terms of network bandwidth to accommodate new types of immersive media, 5G can introduce distributed cloud capabilities to manipulate, transcode and cache immersive media. Furthermore, the use of artificial intelligence (AI) and machine learning (ML) can provide personalize immersive media and analytics in different environments to end users.

Professor Tasos Dagiuklas
London, March 2020

Preface

By bringing this book to the readers, the aim is to present current work and directions on a challenging subject with theoretical and practical roots of the fifth-generation (5G) mobile wireless systems. The book provides the reader with an overview of current 5G progress and achievements, including feature functions and underlying technologies for new multimedia services that accommodate a massive number of connected devices while maintaining high quality of user experience. Besides a detailed introduction describing the representative modules and services, the critical enabling technologies are identified in the current situation along with future directions and challenges.

The past two decades have witnessed an extremely fast evolution of mobile cellular network technologies. In general, in 5G development, it is all about a question of how to optimize the degrees of freedom (DoFs) of time, frequency and spatial resources to achieve the varied and refined requirements of society. Based on this perspective, the potential trends of utilizing these resources are discussed. Recently, the fifth generation of mobile wireless systems has achieved the first milestones toward finalization and deployment by 2020. It is also important to consider the performance of 5G technology by analyzing network capabilities of the operators and of the end users in terms of data rates, capacity, coverage, energy efficiency, connectivity and latency. New 5G networks need to fulfill a number of new performance targets of extreme mobile broadband, massive IoT and critical communications.

This book is divided into three major parts, and each part contains four to seven chapters. Part 1 discusses the critical enabling technologies, such as green communications, fast statistical channel modeling, massive MIMO and ML-based networks. In Part 2, different methodologies and standards for multimedia communications have been discussed, and exclusive chapters have been dedicated on each of the open research challenges such as multimedia operating over a 5G environment, network slicing optimization, mobile cloud computing (MCC), mobile edge computing (MEC) as well as immersive media delivery, mobile video multicast/broadcast, QoE (quality of experience)/QoS management, integrated satellite and drone 5G communications. Part 3 of this book opens up the major deployments scenarios for different services having diverging requirements of enhanced mobile broadband (eMBB), massive machine-type communications (mMTC) and ultrareliable and low-latency critical communication (uRLLC). It includes chapters on MES integration in smart cities, challenges of 5G connectivity for IoT, implementations of innovative services, intelligent transportation use cases and innovative services.

The book is written by experts in the field from the United Kingdom, China, Serbia, Mauritius, India, Pakistan and Russia, who bring out the intrinsic challenges of 5G multimedia communication. The authors have introduced blends of scientific and engineering concepts, covering this emerging wireless communication area. The book can be read cover to cover or selectively in the areas of the interest for readers. It will be a very good reference book for undergraduate students, young researchers and practitioners in the field of wireless multimedia communications.

Since this book project would have been impossible without them, we would at first like to thank all the authors who answered to our call for invited chapters. With valuable suggestions and knowledge, they contributed to further improvement of this book project.

Zoran S. Bojkovic
Dragorad A. Milovanovic
Tulsi Pawan Fowdur
Belgrade-Mauritius, January 2020

Editors

Zoran S. Bojkovic is a full professor of Electrical Engineering University of Belgrade, Serbia, Life Senior Member of IEEE, full member of Engineering Academy of Serbia, member of Scientific Society of Serbia, member of Athens Institute for Education and Research ATINER. He was and still is a visiting professor worldwide. He is author/coauthor of more than 470 publications: monographies, books (Prentice Hall, Wiley, McGraw-Hill, Springer, CRC Press/Taylor & Francis Group, IGI GLOBAL, WSEAS Press), book chapters, peer-reviewed journals and conference and symposium papers. Some of the books have been translated in China, India, Canada and Singapore. His research focuses on computer networks, multimedia communications, 3D video coding, smart grid, green communications, 5G and beyond. He is a highly regarded expert in the IEEE, contributing to the growth of communication industry and society reviewing process in many books and journals as well as organizing special session workshops and being General Chair and TPC member at numerous conferences all over the world. The other activities include serving as editor-in-chief and associate editor in international journals WSEAS, NAUN and IARAS.

Dragorad A. Milovanovic received his Diploma in Electrical Engineering and magistar degree from the University of Belgrade, Serbia. From 1987 to 1991, he was a research assistant and PhD researcher from 1991 to 2001 at the Department of Electrical Engineering, where his interest includes simulation and analysis of digital communications systems. He has been working as R&D engineer for DSP software development in digital television industry. Also, he is serving as an ICT lecturer and consultant in digital television and medicine/sports informatics for implementation standard-based solutions. He participated in research/innovation projects and published more than 250 papers in international journals and conference proceedings. He also coauthored several textbooks/chapters in the field of multimedia communications. Present projects include adaptive coding of 3D Video, QoE immersive media, IoMT and 5G/6G wireless technology.

Tulsi Pawan Fowdur received his BEng (Hons) degree in Electronics and Communication Engineering with first class honors from the University of Mauritius in 2004. He was the recipient of a gold medal for having produced the best degree project at the Faculty of Engineering in 2004. In 2005, he obtained a full-time PhD scholarship from the Tertiary Education Commission of Mauritius and was awarded his PhD degree in Electrical and Electronic Engineering in 2010 by the University of Mauritius. He is also a registered chartered engineer of the Engineering Council of the United Kingdom, member of the Institute of Telecommunications Professionals of the United Kingdom and the IEEE. He joined the University of Mauritius as an academic in June 2009 and is presently an associate professor at the Department of Electrical and Electronic Engineering of the University of Mauritius. His research interests include mobile and wireless communications, multimedia communications, networking and security, telecommunications applications development, Internet of things and AI. He has published several papers in these areas, is actively involved in research supervision and reviewing of papers and has been the general co-chair of two international conferences.

Contributors

Juan Carlos Augusto
Research Group on Intelligent Environments
Middlesex University
The Burroughs, Hendon, London,
    United Kingdom

Bojan Bakmaz
Faculty of Transport and Traffic
    Engineering
University of Belgrade
Belgrade, Republic of Serbia

Miodrag Bakmaz
Faculty of Transport and Traffic Engineering
University of Belgrade
Belgrade, Republic of Serbia

V. Bassoo
Faculty of Engineering, Department of
    Electrical and Electronic Engineering
University of Mauritius
Réduit, Republic of Mauritius

Girish Bekaroo
Middlesex University Mauritius
Uniciti, Flic-en-Flac, Mauritius

Y. Beeharry
Faculty of Engineering, Department of
    Electrical and Electronic Engineering
University of Mauritius
Réduit, Republic of Mauritius

Robin Singh Bhadoria
Department of Computer Science and
    Engineering
Birla Institute of Applied Sciences (BIAS)
Bhimtal, Uttarakhand, India

Natasa Bojkovic
Faculty of Transport and Traffic
    Engineering
University of Belgrade
Belgrade, Republic of Serbia

Zoran S. Bojkovic
University of Belgrade
Belgrade, Republic of Serbia

Yigang Cen
School of Computer and Information
    Technology, Institute of
    Information Communication
    and Networks
Beijing Jiaotong University
Beijing, China

Dragan Cetenovic
Power Systems Engineering,
    Electrical Energy and
    Power Systems Group,
    Department of Electrical and
    Electronic Engineering
University of Manchester
Manchester, United Kingdom

Danijel Djosic
Faculty of Sciences
University of Pristina
Kosovska Mitrovica,
    Republic of Serbia

Tulsi Pawan Fowdur
Faculty of Engineering,
    Department of Electrical and
    Electronic Engineering
University of Mauritius
Réduit, Republic of Mauritius

Mussawir Ahmad Hosany
University of Mauritius
Réduit, Republic of Mauritius

V. Hurbungs
SIS Department, Faculty of Information,
    Communication and Digital
    Technologies
University of Mauritius
Réduit, Republic of Mauritius

Rachit Jain
Department of Electronics and
    Communication Engineering
ITM Group of Institutions (ITM GOI)
Gwalior, Madhya Pradesh, India

Asutosh Kar
Department of Electronics and
    Communication Engineering
Indian Institute of Information Technology,
    Design and Manufacturing
Kancheepuram, Tamil Nadu, India

Jahangir Khan
Faculty of Computer Science and
    Information Technology
Sarhad University of Science and
    Information Technology
Peshawar, Pakistan
and
Key Laboratory of Remote Sensing
    for Agri-Hazards of Ministry of
    Agriculture, College of Information and
    Electrical Engineering
China Agricultural University
Beijing, People's Republic of China

Dragan D. Kukolj
University of Novi Sad
Novi Sad, Republic of Serbia

Sergey Makov
Don State Technical University
Rostov Region, Russia

Dejan Milic
Faculty of Electrical Engineering
University of Nis
Nis, Republic of Serbia

Zoran Milicevic
Department of Telecommunication
    and IT, GS of the Serbian
    Armed Forces
Belgrade, Serbia

Dragorad A. Milovanovic
University of Belgrade
Belgrade, Republic of Serbia

Varun Mishra
Amity School of Engineering and
    Technology (ASET),
    Department of Computer Science
    and Engineering
Amity University of Madhya Pradesh
Gwalior, Madhya Pradesh, India

Vladimir Mladenovic
Faculty of Technical Sciences
    in Cacak
University of Kragujevac
Kragujevac, Republic of Serbia

Stefan Panic
Faculty of Sciences
University of Pristina
Kosovska Mitrovica, Republic
    of Serbia

Yadunath Pathak
Department of Information
    Technology
Indian Institute of Information
    Technology (IIIT)
Bhopal, Madhya Pradesh, India

Marijana Petrovic
Faculty of Transport and Traffic
    Engineering
University of Belgrade
Belgrade, Republic of Serbia

Aditya Santokhee
Middlesex University Mauritius
Uniciti, Flic-en-Flac, Mauritius

V. Ramnarain-Seetohul
ICT Department, Faculty of Information,
    Communication and Digital
    Technologies
University of Mauritius
Réduit, Republic of Mauritius

Neha Sharma
Cyber Physical Systems
CSIR-Central Electronics Engineering
    Research Institute (CSIR-CEERI)
Pilani, Rajasthan, India

Caslav Stefanovic
Faculty of Sciences
University of Pristina
Kosovska Mitrovica, Republic of Serbia

Mihajlo Stefanovic
Faculty of Electrical Engineering
University of Nis
Nis, Republic of Serbia

Aleksandar Sugaris
ICT College
Belgrade, Republic of Serbia

Vladimir Terzija
Power Systems Engineering, Electrical
  Energy and Power Systems Group,
  Department of Electrical and Electronic
  Engineering
University of Manchester
Manchester, United Kingdom

Gongpu Wang
School of Computer and
  Information Technology,
  Institute of Information
  Communication and Networks
Beijing Jiaotong University
Beijing, China

Chaochao Yao
School of Computer and
  Information Technology,
  Institute of Information
  Communication and Networks
Beijing Jiaotong University
Beijing, China

Jia You
School of Computer and
  Information Technology,
  Institute of Information
  Communication and Networks
Beijing Jiaotong University
Beijing, China

Tanja Zivojinovic
Faculty of Transport and Traffic
  Engineering
University of Belgrade
Belgrade, Republic of Serbia

Section I

Critical Enabling Technology

1

Toward Green Communication in 5G: New Concepts and Research Challenges

Zoran S. Bojkovic and Dragorad A. Milovanovic
University of Belgrade

Tulsi Pawan Fowdur
University of Mauritius

CONTENTS

1.1 Introduction

Fifth-generation (5G) mobile networks delineate from previous generations by providing ultrareliable and low-latency communications, enabling machine-type communications, as well as including enhanced mobile broadband. These three main objectives form the basis of the use cases and applications domain of 5G. Also, there is a clear motivation for addressing energy efficiency (EE) in 5G networks: we want to keep the energy consumption of the mobile network on the current level, or even lower it, while data traffic keeps increasing together with a number of base stations (BSs). However, today's 5G NR typically operating at only 10% power efficiency and consume 3× as much power as the LTE BSs. To improve the EE, new technologies, cell deployment strategies and resource allocation schemes are being developed in 5G. Given the predicted significant increase in the power consumption of the information and communication sector in the next decade, green communication has become one of the main characteristics of 5G systems.

From statistical data, information and communication technology (ICT) industry is responsible for 2% of the total CO_2 emission and 2%–10% of global energy consumption. A striking observation is that radio access networks (RANs) contribute directly to over 60% of the energy consumption of the ICT industry [1]. With significant increase in smart devices and application developments such as high-resolution video streaming [2], tactile Internet, remote monitoring, road safety, real-time control, connected cars and robots [3], data traffic will continue to witness an exponential growth. In this regard, it is expected that the spectral and energy efficiencies of 5G networks will have to increase by a factor of 10, so as to ensure a corresponding increase by a factor of 10 in the battery life of connected devices [4,5].

The cost is a critical factor that will influence the transition from conventional to green power. However, the purchasing, installation, deployment and maintenance costs are expected to fall in the near future, taking into account that the costs of green devices are experiencing a progressive decrease. Renewable energy is considered clean because of no environmental pollution. Reducing CO_2 emission is one of the primary goals in order to deliver a cleaner environment. The maintenance cost of green sources is intractable (solar panels and wind turbines in rural areas compared with water-based turbines). A mobile BS consumes a specific amount of power during its operation. A green power source must be capable of constantly generating the required power, to allow a mobile station to depend on it for its power supply [6–11]. However, there is a major challenge in ensuring the reliability and scalability of green energy because of its dependency on the weather which is very dynamic and difficult to predict [12].

To reduce the costs of high energy usage of wireless systems for both mobile users and service providers, there have been numerous studies and innovations around the world toward energy-efficient (EE) and renewable wireless communications and networking. A range of main wireless technologies are part of 5G networks, leveraging available resources and improving EE. In reality, a 5G network scenario is an integration of the various technologies such as device-to-device (D2D) communication, spectrum sharing, millimeter wave (mmWave) communication, Internet of things (IoT) and ultradense networks. These technologies will help green networking in 5G wireless communications and extend mobile device battery life [13]. The available renewable sources simplify the process of energy transfer to the network sites and the gap between transmission lines, while at the same time reducing maintenance costs. Contrary to traditional energy services, green energy services (sun, wind, heat) are capacity based while highly localized and weather specific, which makes it a major challenge for developing and controlling wireless communication options. As for green technology, it means offering a system with operational challenges, which can be described as follows:

- the input deployment cost is very high;
- green networks require very high expenses such as buying, installing and periodic maintenance;
- green networks are a way of reducing toxic emissions from combustion of fossil fuels;
- green power generation is complex, with low power generation rates compared with simple fuel generators for electricity generation.

Energy-harvesting (EH) technology is another aspect and a promising approach to support green communication considering network lifetime. Evolving 5G networks, the capacity limitations of the cellular networks have been overcome. In EH networks, nodes are in a position to replenish energy from a mobile charge with the idea to overcome variations of removable energy. Namely, nodes simultaneously connect and collect resources. Several key characteristics with regard to EH for 5G green networks have been researched, including the smartphone charger to replenish the energy for nodes in the developing IoT. In order to ensure flexible energy scheduling as well as facilitating energy utilization optimization, software-defined energy-harvesting networking (SD-EHN) for 5G green communication can be used with the benefits such as flexible energy scheduling, efficient energy utilization and sustainable development.

Global green evolution will have profound impact on the research development and applications of wireless communications. There are many research challenges in integrating green energy technologies for developing energy efficiency networking algorithms and protocols as well as service quality provisioning. Of course, all these issues require interdisciplinary efforts from various networks, power systems and devices.

This chapter is organized as follows. In the first part, EE as the main requirement of the 5G technology is introduced. How to boost EE with service quality (QoS) constraint is a major issue in the second part. As one of the solutions for 5G wireless networks, nonorthogonal multiple access (NOMA) superposition coding in power domain is pointed out in the third part. Significant interest has been placed in the fourth part about a new concept of an energy system called green energy Internet (EI).

1.2 5G Mobile Communication

5G communication systems aim at continual improvement in quality of experience (QoE), high data rate for mobile users and a low energy consumption [14]. In achieving this goal, 5G mobile communication systems play one of the most significant roles, especially when speaking about industrial applications. Of course, it is understood that long-term and self-sustainable operations are included in this process [15]. On the other hand, issues with improvements that have to be achieved in the future communications are dealing with increasing research attention. In this case, it includes scalability, power, end-to-end delay, demand heterogeneity and security.

5G technologies become key enablers in the case of initiatives such as smart city, smart grids (SGs) and IoT. The exponential increase in user volume and device numbers has generated interest in low-power, EE communication, energy consumption and so on. The assignment of resources, optimal network planning and renewable power are identified to control energy consumption. The increased energy consumption and carbon footprint of cellular networks have led to various suggestions by telecom providers, government and researchers for green 5G mobile network solutions. In addition to their environmental value, the objective of EE is also linked to reduced operating costs for mobile network operators and greater customer satisfaction through an increase in battery life.

Reducing energy consumption is one of the major issues in the global warming issue. Decreasing mobile communication network energy consumption is of great importance because it takes a major share of the overall energy use of ICT. Different types of electrical elements, including renewables, energy storage, electric vehicles (EVs), smart appliances and so on, will be connected to the electricity grid in the near future. In mobile communications, the BS is the main source of energy use, while the BS energy consumption is determined by traffic charge, which varies depending on the location. The fundamental principle to reduce a BSs' energy use is to turn the BS components off when they are no longer required.

Cellular systems or BSs cooperate with both supply and requirements in the joint energy and communication cooperation, with the objective of maximizing energy cost reduction. Besides cellular networks, heterogeneous communication networks (small cells) reduce overall energy cost owing to cooperation. It should be noted that these networks are different in service, spectrum as well as harvesting. Many green approaches have been proposed in order to reduce the energy consumption in mobile and fixed networks at all layers as well as in all network parts such access, aggregation and core [16,17].

1.2.1 Scenarios and Requirements

The main requirements of the 5G technology include data rates, latency/quality of service (QoS), flexible functionality and EE [18]. One of the most significant requirements is to provide EE resource managements. To meet the focus area and solution relevant and applicable, 5G green partners will help to ensure that all future mobile networks are as green as possible by the corresponding standard bodies such as the third-generation partnership project (3GPP) (3GPP TR38.913, *Study on Scenarios and requirements for Next Generation access technologies*, June 2017). Network EE is the ability of the radio interface technology to minimize the energy consumption of the access network in relation with the supplied traffic capacity. EE of devices is the ability to minimize the power the device modem consumes in relation to traffic features.

For the 5G primary eMBB service performance metrics, EE in the UE means the UE's ability to maintain a far better mobile broadband data rate while reducing the energy consumption of UE modems. The EE of the network enables RAN energy consumption to be minimized while delivering much better traffic capacity. The network performance metric can be evaluated in low load, median load and heavy load. In current ITU-R report (ITU-R M.2410, *Minimum requirements related to technical performance for IMT-2020 radio interface(s)*, Nov. 2017), low load (or zero load) is the first focusing scenario for EE evaluation. In this case, it would be important to guarantee low energy consumption and to estimate low energy consumption through sleep ratio and sleep time. The sleep ratio is the fraction of unoccupied time resources (for the network) or sleeping time (for the device) over a period of time that corresponds to the control signaling cycle (for the network) or the discontinuous transmission cycle (for the device) when no user data is transferred. The length of sleep is the continuous time period without transmission (for networks and equipment) and reception (for equipment). A frame structure that permits sparse signals and broadcast transmission for synchronization can be anticipated as helpful in improving EE for a low-load network. The target is a design with

- the ability to efficiently deliver data,
- the capacity to provide enough granular network disrupted transmission when data are not transmitted and network service is preserved and
- flexibility to enable operators to adjust BS sleep times to load, services and area.

1.2.2 Green Networking

In order to create greener wireless communication systems, the design of 5G wireless systems should take into account of the minimization of energy consumption.

Generally speaking, a mobile network is a system that uses radio BSs/access points (APs) to interface mobile terminals (MTs) with the core network of the Internet. Each of these elements has a specific rate. For instance, radio resource management, user mobility management and Internet access are under the responsibility of the BSs/APs. The core network provides data services, whereas MTs have processing and display possibilities for providing voice services, video streaming and data applications. The goal is to reduce energy consumption through green wireless communications and networking [14]. Since these architectures are often designed to withstand peak loads and deteriorated conditions, they are not used regularly under normal conditions; hence, significant energy savings are made possible. Recently, there have indeed been important efforts to reduce excessive energy expenditure, which is often called the greening of networking technologies. Green networking aims at reducing greenhouse gas emissions from an environmental viewpoint. The first objective is to encourage renewable energy use in ICT as far as possible. Another approach is to develop low-energy modules, which can offer equal efficiency. Green networking may be more likely to be seen as a way to reduce the energy required for a given task while retaining the same level of performance, from the engineering point of view.

It is necessary to identify where the maximum benefits could be obtained before trying to reduce energy consumption or to consider how such a decrease can be accomplished. Nevertheless, saving energy also means reducing network performance or redundancy. In view of this trade-off between network performance and energy savings, it is a real challenge to define efficient energy management approaches to reduce network energy usage.

1.2.3 Green Strategies

Resource consolidation, selective connections, virtualization and proportional computing [13] are the four main paradigms that a network infrastructure can use for the achievement of the green goals. These are described as follows:

- In order to reduce consumption due to devices underutilized at a given time, the goal of resource consolidation is regrouping all the strategies. The targeted level of performance is assured, but sometimes resources that are adapted to the current traffic demand rather than to the peak demand are used.

- The selective connectivity of the devices requires distributed mechanisms that allow the individual components to remain idle for as long as possible in a transparent manner with respect to the rest of the network. Selective connection allows instead the turning off unused resources at the edge of the network.

- As for virtualization, it regroups a set of mechanisms allowing more than one service to operate on the same hardware, thus improving hardware utilization. This is significant because of a lowered energy consumption. Visualization brings the application to multiple resources such as network links, storage devices, software resources and so on. Sharing servers in data centers (DCs) is one of the main examples of virtualization applications.

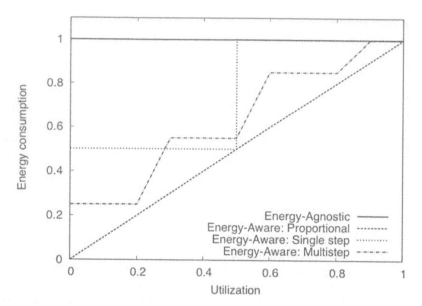

FIGURE 1.1
Energy consumption as a function of utilization.

- Finally, proportional computing may be applied not only to network protocol and individual devices and components but also to a system as a whole. As an example, Figure 1.1 represents different energy consumption as a function of the utilization. These are different energy consumption or cost profiles that a system can show according to its level of use or demand. The usage and energy calculations of the axes varying from 0 to 1 have been normalized.

It is important to point out that different profiles have different opportunities. Firstly, energy-agnostic devices with constant energy consumption, independent of use, are the worst case since they are either active and consume maximum energy or off and inactive. On the other hand, fully energy-aware devices exhibit energy consumption figures proportional to their level of utilization. There are an infinitive number of possible intermediate profiles between these two extreme positions, such as single-step and multistep cases. Energy consumption adjusts roughly to the load. Two modes of operation are available for single-step devices while there are several thresholds [19] for multistage devices.

Green technology has become mandatory for wireless networks in order to meet the challenges arising from the high demands of the wireless domain and energy consumption. Green radio has become an important phenomenon in academia as well as industry as a research orientation in evolving future wireless architectures and technologies toward high EE [20]. It targets solution based on top-down architecture and joint design across all system levels and protocols.

The rate of EE increases without any loss in communication capacity and is important to specific networks with many heterogeneous BSs and connected devices. There have been extensive research efforts to develop not just energy but also spectrum efficiency techniques to address this challenge, including efforts and progress in industry as well as academia, in the field of improving EE and 5G network sustainability. From this point of view, three goals have been emerged:

- to exploit unused and unlicensed spectrum;
- to decrease the distance between the transmitter and receiver distance and increase the frequency reuse; and
- to boost spectral efficiency (SE) by deploying a large number of entities.

These technologies increase the system throughput. For example, with wireless communication systems, EE measured in *bits per Joule* has become the most widely adopted green design metric for wireless communication systems [15].

1.2.4 HetNet Systems

As a possible solution, a small cell network (SCN)–based heterogeneous network (HetNet) can also be included in a design consideration for achieving the 5G challenge [14]. The deployment of HetNet is accomplished by the following characteristics: unsatisfactory EE, severe interface, no-computing enhanced coordination centers, inflexibility and unscalability. A HetNet with its mixed wireless infrastructure is characterized by a combination of a small number of high-power macrocells and several low-power small cells such as micro, pico and femto. This fact brings the network closer to the end user offering high signal-to-interference-plus-noise ratio. On the other hand, link robustness and QoS become improved. High reuse of frequency leads to reducing the bandwidth scarcity. In fact, one of the biggest challenges in 5G is to achieve energy savings of up to 90% [15]. The combined power consumption of access and backhaul networks lead to the overall power use of 5G HetNet systems. More and more noncorrelated and highly charged active SCNs can increase network access energy consumption [21]. Maximum bandwidth becomes available when keeping all the small cells active all the day. This results in oversupply of bandwidth and thus in increased operational costs when the demands are not at peak. In this way, in a 5G HetNet, different traffic loads at different places are used to send calls in a sleep mode. The result is power savings and reduction in operation costs.

The network architecture of a 5G HetNet is shown in Figure 1.2 [22]. An access network is composed of microcell and several SCNs. A backhaul network is formed with the BSs connected to the core network using wires, wireless or mixed architecture of known technologies. Densification with small cells has open the question of the bandwidth scarcity problem of 5G systems. Next, the problem often arises is with higher consumption from both the impact of the access and backhaul networking process.

Green heterogeneous cloud radio access. Cloud RANs (C-RANs), central cell phone architectures with cloud computing (CC) and virtualization techniques have been put in place to coordinate and manage resources across cells and RANs in order to address challenges focused on HetNets. As a result, HetNets and C-RAN combinations known as heterogeneous cloud radio access networks (H-CRANs) are becoming a potential alternative to both spectral and EE transmission. One of H-CRAN's main tasks is to build environmentally friendly and economical communication systems. In these EE transmission scenarios, several innovative strategies can be applied, such as joint processing/allocation, energy balance of traffic offloading, self-organization and network

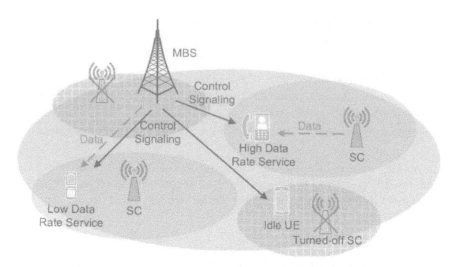

FIGURE 1.2
SC (small cell)-based 5G HetNet architecture. HetNet, heterogeneous network; MBS, macro-base station; UE, user equipment.

deployment. The EE of the network typically enhances at the expense of other metrics such as spectral efficiency (SE), fairness and delay, all of which are equally important for the customer's QoS. It is of interest to investigate the performance trade-offs in H-CRAN (such as EE-SE, EE-Frames and EE-Delay) for establishing rules when greening H-CRANs. C-RAN is expected to be a 5G network architecture by incorporating CC into the RAN [23]. C-RAN is also expected to significantly cut energy consumption to a green RAN because of its revolutionary change from the baseband processing feature to the centralized cloud-based unit pool. In addition, several new functions and RAN implementations with the cloud-based architecture can be introduced that redefine the RAN as a scalable RAN.

Green femtocell. Femtocell technology is a common infrastructure used in HetNet environments. Several applications can use femtocells to improve network coverage and capacity. Macrocell BSs consume a huge amount of energy, while most of the energy is used for driving high-power radiofrequency (RF) signal. Macrocells have a large area coverage while most areas are empty and have no users who require radiated signals [24]. It is of importance to note that femtocells deliver power where there is a need. Thus, because it is closer to the user, they need less RF power for high bandwidth. A femtocell also offers EE over and above common features such as more coverage, better QoS capabilities and greater battery life. Another important factor in the process of femtocells integration with current microcellular network is the so-called traffic modeling. The idea is to provide analytical models in order to describe the most important parameters for a given traffic type in order to demonstrate the advantages of a real femtocell network. Various applications are making use of femtocell. The aim is to boost the network's coverage and capacity. Indoor environments are used for enhanced coverage and for higher data rates in mobile applications within home or office environments [25]. In public transportation vehicles, femtocells can be used to improve coverage and capacity as well as to provide better Internet experience for users being in move. The conception with femtocell inside a vehicle is often called mobile femtocell [26].

1.3 Energy-Efficient Multimedia Communication

Different resource allocation management systems have become a major feature of mobile multimedia communication systems [27–32], including power allocation [33,34], the bandwidth allocation [35–38], the allocation of subchannels [39] and so on, due to the demand for better EE in mobile multimedia communication systems. Multiple-input multiple-output (MIMO) systems will create separate parallel channels for the transmission of data streams, enhancing spectrum efficiency and network capacity, without raising the demand for bandwidth [40]. The multipath effect is reduced by the use of orthogonal frequency-division multiplexing (OFDM) technology, which converts frequency-selective channels into flat channels. MIMO-OFDM technologies are commonly used jointly for mobile multimedia communication systems. Nonetheless, in mobile multimedia communication systems, how to boost EE with QoS constraint is a major issue.

On the other hand, as various intelligent devices are made popular, augmented reality (AR) and virtual reality (VR) can potentially become more demanding than traditional multimedia services. VR is able to imitate the real world by creating an ongoing virtual world that can provide immersive user experience based on high definition (HD) video services/applications. In contrast, AR is able to improve the perception of reality by integrating more knowledge of the real world. As the interactivity of AR/VR applications in real time generates massive amount of information flows, the architectural designs of future networks will face new challenges [33] to accommodate AR/VR online applications. Specifically, AR/VR applications require innovations in cloud network architectures for 5G wireless networks [34] with the goal of significantly improving network performance, transmission delays and wireless capacity.

1.3.1 Energy-Efficient Architecture for 5G Network

Mobile networks are distinguished by their architecture and implementation based on demand for high traffic volumes and are best active irrespective of low usage over various times every day. Traffic management mechanisms are enablers for EE for operations, particularly with regard to multi–radio access technology (RAT) environments and the deployment of 5G systems. For increased battery life, the discontinuous transmission (DT) system has long been used on MTs with the intention of transmitting only when a need occurs and otherwise placing the transmitter in a low power state. This technique is called cell DT on a network side and is based on the hardware disabling feature to allow low power levels. There are two DT versions: fast-cell DT and long-cell DT. Fast cell DT is available in various versions and operates at the subframe level. For example, cell micro-DT is possible in long-term evolution LTE Release 8. This means that the radio is placed in DT (microsleep) between cell-specific reference symbols when no user data are being transferred. Long-cell DT operates at a slower rate and refers to a low activity mode for the cell. It can be considered to be a cell sleep and could be based on a deeper sleeping state that is lower in energy consumption than the lower power conditions in the fast-cell DT variants. One long-cell DT technique is to operate in relatively dense networks where a good coverage, e.g., dense urban, suburban and urban, exists. Long-cell DT can therefore be seen as an enabler/tool not only for network management but also for multi-RAT management and macrocell management. If there is no traffic requirement, long-cell DT can be enabled, or if cells are at low load in low traffic times, then traffic steering strategies can be easily used.

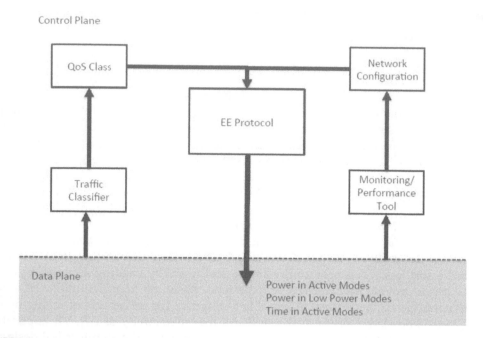

FIGURE 1.3
One solution for the energy efficiency (EE) architecture in a 5G platform. QoS, quality of service.

For implementing EE systems in 5G, it is recommended to consider the insertion of EE in the control plane. As for the control plane, it will make possible both information processing and input information about the data plane. In that way, it will be enabled to take into account the status of the current network for the machine learning analysis. Also, any cognitive-based functionality is involved. Energy efficiency architecture (EEA) is shown in Figure 1.3. In this solution, control plane accommodates the EEA in a 5G. The inputs to the EEA from the control plane are provided. Control plane processes data from the data plane. There are the following different blocks: traffic classifier, QoS class, monitoring/performance tool, network configuration and EE protocol.

First of all, it is necessary for the traffic to be classified before a corresponding QoS is attributed to it. After classifying, the traffic is posing through EE protocol. The EE protocol has got the information about the data rate, latency and QoS requirement of the application. Of course, the network configuration is monitored during the process. The EE protocol commands the data plane with a network protocol. In that way, regulatory messages can enhance energy polices performance with power in low power modes and time in active modes. Information concerning the network traffic and service layer agreement will be useful for the point of view decision-making of EE protocols.

1.3.2 Energy Efficiency Metrics

The mathematical value for EE is the ratio of utilities to energy consumption. In this case, the utility includes every important communication parameter, e.g., flow, outage probability and a number of BSs irrespective of QoS and QoE. For example, in case the power is considered on equipment level, there is a relation between output/input power and energy consumption. It is important to note that in SCNs, researchers pay great attention on the network and BS levels. Also, the ratio of spatial spectrum efficiency and spatial energy consumption is taken into account when considering the area communication performance [41]. With green energy technology and low power consumption, the energy will be cheaper and greener. The hardware applies to both the level of the BS and equipment. The computation is always seen as the energy necessary to assess the efficiency in energy collection. In case when the harvested energy is not sufficient, the BS must use elastic energy from the grid. The weighted sum of energy from different sources represents a metric. Energy-saving techniques, such as efficient collection and selection of channel strategies, are considered more commonly at equipment level than at BS level. The most widely used metrics are EE and power consumption. Two main EE metrics are bit/joule and W/km^2. The most commonly used efficiency metric is bit/joule. It is important to note how its use for the performance evaluation of the wireless network has been extended naturally. Denoting the bit/joule efficiency of the network by Ψ, it will be

$$\Psi = \frac{C_{net}}{P_{net}} \tag{1.1}$$

where C_{net} represents the aggregate network capacity in bits/s, whereas P_{net} is the total power consumption of the network in (W). The second acceptance EE metric is the area power consumption in the form of

$$\Omega = \frac{C_{net}}{A} \tag{1.2}$$

It can be seen that it relates the local power consumption of the network P_{net} to the size of the covered area A. The optimal EE is obtained when the bit/joule metric is maximized or the power per unit area is minimized [42].

1.3.3 Energy-Efficient Optimization with Quality of Service/Quality of Experience Constraints

To achieve green communications, 5G wireless networks are expected to increase the network EE with the guaranteed QoS in advance for time-sensitive multimedia wireless traffic. The key to increase network energy or power efficiency is in optimizing spectrum efficiency per unit power. In order to reach optimal spectrum capacity for each unit power, all different types of power constraints must be addressed for 5G networks. The power/EE of the network is affected by two power limitations: the mean power limit (i.e., the average power limit over a period) and the maximum power constraint. In the case of green 5G wireless networks, both the average power constraint and the peak power constraint are taken into consideration with the final goal to attain the maximum network energy/power efficiency [31].

Green 5G wireless networks have to maximize the energy and power output of the network and to ensure QoS. Several studies have been conducted in traditional mobile multimedia communication systems. With regard to the corresponding QoS demand for various throughput rates in MIMO communications systems, there has been a proposed effective antenna allocation system and access control scheme [40]. From the viewpoint of mobile users in orthogonal frequency-division multiaccess (OFDMA) wireless networks, a downlink QoS evaluation system has been proposed [32]. A statistical QoS restriction model has been developed to analyze the queue properties of data transmissions to guarantee

QoS within wireless networks [33]. The EE was evaluated for fading channels under QoS restrictions in which effective capacity was considered to be a measure of the maximum throughput under certain statistical QoS constraints [34]. Based on the effective capacity of the block fading channel model, for wireless mobile networks, a QoS-regulated power and rate adaptation scheme was proposed [35]. In addition, some QoS-driven power and rate adaptation systems for diversity and multiplexing systems were proposed by integrating information theory with effective capacity [36]. Results from simulations showed that multichannel systems can attain high-performance and strict QoS simultaneously. The major trade-offs between EE and QoS calculations in the various wireless communications scenarios were analyzed to maximize EE [37].

EE in MIMO wireless communication systems has become one of the main research challenges in the past decade. A model of EE for cellular networks was developed, taking into account of spatial load distributions and energy consumption [38]. A study was carried out on the energy–bandwidth efficiency trade-off in MIMO wireless multihop networks and the impact of various antenna numbers on energy–bandwidth efficiency trade-off [39]. By considering various types of energy consumption models, an exact closed-form model was proposed for the trade-off between EE and spectrum quality over the MIMO Rayleigh fading channel [40,41]. A relay cooperation system to investigate the spectral and EE trade-off in multicellular MIMO networks was proposed [42]. The EE spectral efficiency compromise was examined with decode and forward-type protocols on the uplink connection of a multiuser cellular MIMO system [42]. In a relay-assisted multicell MIMO cellular network, a comparison between the signal forwarding and the interference relaying paradigms was investigated to determine the trade-off between spectral and EE [43]. The trade-off between the operating power and the embodied power contained in the manufacturing process of infrastructure equipment from a life cycle perspective is explored [44].

1.4 NOMA Superposition Coding in Power Domain

As one of the solutions for 5G wireless networks, NOMA has attracted interest because of the significant enhancing spectral efficiency. In the open literature, NOMA is characterized in power domain and code domain [45]. NOMA employs superposition code for the transmitter and subsequent cancelation of interference at the receiver side, in the power domain. This allows several users to simultaneously transmit data on the same subcarrier channel. The order of decoding is dependent on the channel characteristics of the wireless connection of each pair of transceivers. In other words, data transferred to the recipient with the highest wireless connection are decoded without the interface. One of NOMA's main challenges is how well aligned it can be with the other techniques of the 5G standards. Other advantages of NOMA can be categorically presented as follows: high bandwidth efficiency, fairness, ultrahigh connectivity and flexibility.

NOMA and HetNet. The most important thing about dense HetNet as one of the 5G technologies is the capability to move the low-power BSs closer to the served users. The idea is to form small cells under the oversailing macrocells. In order to cope with the interference from the cochannel layers, NOMA solution for HetNet is proposed for minimizing the interuser interference [46]. Coming back to the MIMO system, massive MIMO can be seen as a crucial technology in 5G networks that increase the signal-to-noise ratio and the bandwidth efficiency [47]. With the installation of hundreds or thousands of antennas on the macro-BSs, the fundamental idea of massive MIMO-supported HetNets is to give a degree of spatial freedom while using individual antennas on the density-located small-cell BSs. To show the efficiency of bandwidth of small cells, a massive MIMO-based architecture supported by NOMA was proposed in Ref. [47]. Combination of NOMA and massive MIMO-based hybrid HetNets is shown in Figure 1.4. In order to serve N users simultaneously, the macrocells incorporated a massive MIMO scheme. Small cells employed user pairing-based NOMA transmissions. During MIMO–NOMA integration, the potential advantages of MIMO–NOMA design become evident [48]. That means HetNet is with low power consumption and specific spectrum, while NOMA understands high bandwidth efficiency, fairness/throughput trade-off and so on.

NOMA in mmWave communication. mmWave communications are a viable technology in 5G networks because of their large bandwidths in high-frequency spectrum [49]. On the other hand, it becomes clear that the propagation path loss of mmWave channel and low penetration lead to redesign multiple access (MA) techniques, first of all to support massive connectivity in dense networks. There are enough reasons for supporting NOMA as MA technique coexisting together with mmWave networks. In mmWave communications, the highly directional beams, for example, may lead to correlated channels which may decrease the efficiency of traditional orthogonal multiaccess systems (OMA) but at the same time are NOMA capable. Furthermore, sharp beams in mmWave networks eliminate the interbeam interference between users, which becomes conducive to support NOMA in each beam. Finally, the use of NOMA in mmWave can increase the efficiency of bandwidth and support large-scale connectivity. In a number of research activities and contributions, it is demonstrated that the NOMA–mmWave system outperforms conventional OMA mmWave systems [50].

NOMA and cognitive radio networks. Cognitive radio (CR) has the characteristic of allowing unlicensed secondary users (SUs) to access the licensed range of primary users (PUs) at a certain time. There are three categories of CR techniques, i.e., interwave, overlay and underlay [51]. An interwave CR is a model for preventing interference, whereby before entering channels, SUs monitor the provisional slices of the space frequency domain of PUs [44,52]. The overlay paradigm is used as an interference mitigating technique. CR guarantees that a cognitive user is able to transfer with a noncognitive PU concurrently.

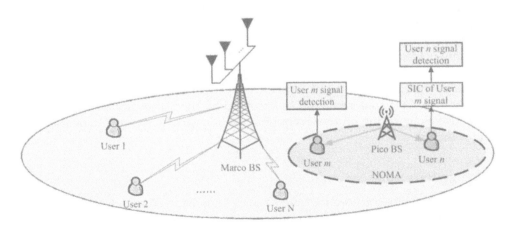

FIGURE 1.4
Combination of NOMA and massive MIMO-based hybrid NetNets. BS, base station; MIMO, multiple-input multiple-output; NOMA, nonorthogonal multiple access.

SUs can forward the information of PU to the PU receivers. The underlay CR functions like a model of intelligent interference management. SUs are allowed to access the spectrum allocated to PUs as along as the constraint on the interference power of the PUs is respected [44,53]. Interference management represents one of the main challenges in CR and NOMA. The bandwidth efficiency has to be improved, too. Studies on CR-NOMA have been conducted only in the context of the underlay CR paradigm.

NOMA and D2D communication. The main characteristics of invoking and integrating D2D communication are as follows:

- low-power support of services in order to improve EE,
- reusing the frequency of the overselling cellular networks for increasing bandwidth efficiency and
- possibility of promoting the deployment of new types of P2P services [44,54].

D2D and NOMA both increase the bandwidth efficiency using the process of managing the interference among users. Starting from this point of view, joint interference management strategies are desirable for fully extracting the potential advantages not only of D2D but also of NOMA.

1.5 Green Energy Internet

Significant interest has been placed on a new concept of an energy system called green EI [54,55]. The EI, which borrows attractive features from the Internet, promises highly efficient interconnection of distributed energy resources and the promotion on a large scale of peer-to-peer (P2P) transactions. EI routers, also known as power flow routers or grid routers that execute the physical control of energy and allow energy routing [56], are the key elements of the EI.

There are some problems which arise in P2P energy transactions. Firstly, the consumers and sellers may be found in varying locations and distribution grids. Secondly, buyer consumption and seller output may not occur simultaneously, which necessitates the use of buffers to allow energy storage. The buyers and sellers may also have different demands on elasticity, such as DC and AC, three phases or single phase, varying voltage levels, etc. This is the reason why in the green, software-defined (SD) energy controller must be able to calculate the optimum configuration of the energy routers to guarantee the power quality of every user. Data content should be observed by the SD controller. This is of importance because of satisfying different requirements for various types of data. Response on demand is included.

1.5.1 Energy-Harvesting Network

EH, in particular with low-power wireless nodes, is considered as one of the most important techniques. At the same time, it provides a solution for the energy needs of wireless devices. In comparison with electronic equipment that consume high amounts of electric power generated from known solar or wind energy, EH is a concept that defines the technology for acquiring energy from low-power electronic devices (milliwatt or microwatt levels). In an EH communication system, wireless devices act as energy sources. The energy collection system consists of energy collection circuits, energy storage units and energy management devices, along with the power source to be used in the system. The final objective is to improve as much as possible the performance of the EH system to ensure sufficient energy supply and an efficient, durable and functional wireless systems. The combined case appears when using varying energy sources in the EH system. A stable energy supply and more harvested energy can be obtained using EH systems. For various application services, energy consumption patterns determine the energy consumption profiles of the various wireless devices. Clean green communications are enabled, leading to high-density topologies as well as high-spatial reuse [57]. Sustainable green 5G wireless networks are made possible through EH technologies which use the energy of renewables or radio frequency signals. Comparison of EH technologies is presented from different points of view such as energy source, compatibility, efficiency, distance and 5G applications (Table 1.1).

In order to complete the process, integration of renewable energy, mobile charge and bidirectional process is inevitable. This means that several renewable energy sources can be incorporated into an EH network.

One of the main roles for SD-EHN is to abstract the network function and at the same time to manage the network in a centralized manner. Taking into account control of both the energy and data flows, we obtain energy scheduling and energy utilization as a huge benefit for 5G green communications, enabling sustainable development of the network. The benefits from SD-EHN procedures can be summarized through the following characteristics: flexible energy scheduling, efficient energy utilization and sustainable development.

Energy cooperation is scheduled to optimize energy transfer in the energy plane. More nodes are enabled to join and get benefits in that way. Energy utilization is improved because the SD data and energy controllers are in a position to obtain global information concerning not only workloads but energy states too. Finally, with the energy scheduled on demand and also utilized with higher efficiency, green communications contribute in improving network quality in the sense of throughput and saving sustainability and prolonging network lifetime. EHN has two queues: a queue for the data buffer and a battery energy queue. With regard to sensory data, the nodes generate them, and then multihop routing is used to transfer the data to a sink node. Sensory data transmission and reception contribute to energy consumption. On the other hand, harvesting renewable energy such as solar, wind and geothermal contribute to the energy queue.

In the SD-EHN, the main idea is that three planes should be separated: the control plane, the data plane and the energy plane (Figure 1.5) [58]. In that way, an EE architecture is

TABLE 1.1

Comparison of Energy-Harvesting Technologies

Technology	Energy Source Controllability	Efficiency	Distance	5G Applications
Renewable energy harvesting	Noncontrollable	Low	Long	Hybrid base stations
Radio frequency energy harvesting	Controllable	Low	Long	IoT sensor networks
Inductive coupling	Controllable	Very high	Very short	Cell phone charging
Magnetic resonant coupling	Controllable	High	Short	Cell phone charging
Laser beaming	Controllable	High	Long	

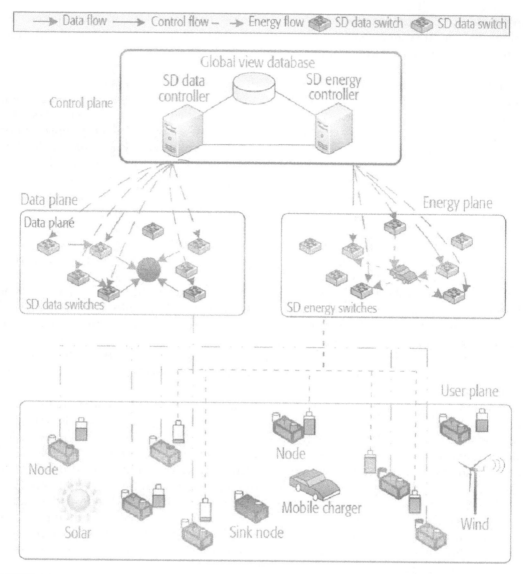

FIGURE 1.5
SD-EHN architecture with planes separation. SD-EHN, software-defined energy-harvesting networking.

obtained for supporting flexible energy scheduling and facilitating energy optimization in 5G networks [59].

In the control plane are mounted a global view database, an SD data controller and a SD data energy controller. The SD data controller and energy SD controllers can control the data and energy flows by first collecting and analyzing global information in real time. The two SD networking controllers interact and collaborate with each other in an effort to provide central control. The nodes are connected to the SD data switches which relay data in the data plane. The routing optimization of data packets in the network is carried out using the nodes. The energy flow is bidirectional between the nodes and the mobile charger. The nodes decide whether they remain inactive or function in energy downstream or energy upstream mode. All physical networks entities are included in the user plane. In compliance with the topology nodes' location and properties, thermal characteristics, renewable energy forms and the charge rate of the moving capacitor, SD-EHN guides the flows along the appropriate pathways from or to entities. The mobile charger has the function of using a high capacity battery to transport the energy.

1.5.2 Software-Defined Green Energy Internet

In SD green EI, there are three planes separation: control, data and energy planes [60]. SD green EI architecture is presented in Figure 1.6. The function of control plane is to present an interface for software control of green. In the control plane, there are SD data controllers (for data flow control) and energy controllers (for energy flow control). In order to realize the coordination between the data and energy planes, the two SD controllers communicate each other. The function of SD data controller in the control plane is to provide recon-figuration of the network elements. In that way, energy-related data management will be optimized according to the network topology and traffic. The SD data controller does not control the energy routers which are essentially end devices. The SD energy controllers control the energy routers since they are grid devices. To offer best effort services and to focus on the reliability of services, EI is used [61]. Centralized energy control is recommended for use to maintain the reliability of the whole EI. The objective is to guarantee that each user can manage its energy consumption and production. It is recommended to use the centrally controlled energy system with a full view of power systems for achieving efficient energy routing and scheduling.

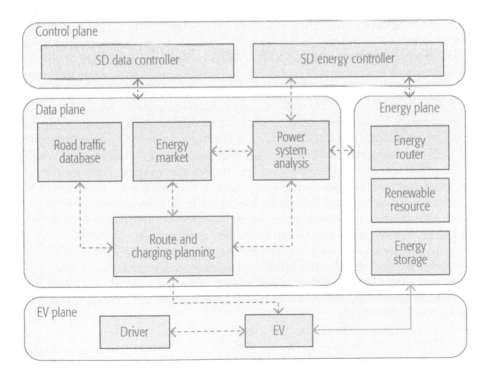

FIGURE 1.6
SD green energy architecture. EV, electric vehicle; SD, software-defined.

When considering software-defined energy Internet (SDEI), it must be taken into account that data will be generated and transmitted in the data plane [62]. The route and charge management entity analyzes the information of EVs, electricity systems, energy markets and road traffic simultaneously. In the data plane, the SD data control system optimizes the utilization of computation and network resources. The goal is to enhance the user experience and quality of service of the system. Traffic flow information is stored on the edges of the local network. This decreases the data traffic load while accessing database. It should be emphasized that the 5G mobile networks invoke different wireless technologies such as D2D, long-term evolution-advanced (LTE-A) and so on. The SD data controller handles heterogeneous wireless settings, while at the same time optimizing computation and spectrum resources allocation in order to meet mobile EVs data traffic requirements [62].

1.5.3 Green Internet of Things

The IoT was recognized as one of the big innovations over the past decade. It makes it possible to connect individuals and things anywhere, anytime, with anybody or anything using any link and service. At the same time, a sensor and device platform is seamlessly connected to a smart environment to deliver advanced and intelligent human services [63,64]. On the other hand, green IoT represents one of the EE procedures, including hardware or software which is adopted by IoT. The aim is to concentrate green IoT's life cycle on green design, manufacturing, utilization and green disposal/recycling, with little or no environmental impact. In order to achieve the necessary green IoT aspects, there are some green enabling technologies. This results in the generation of enormous amounts of data in data centers, which need to be stored, processed and presented energy efficiently.

Green radio frequency identification. A small electronic device consisting of a small chip and an antenna to automatically identify and track the tags attached to objects is called radio frequency identification (RFID). RFID contains a number of RFID tags and a small proportion of tag readers. The function of RFID tags is to keep information about the things they are connected to. RFID tag readers activate the information flow by sending a test signal accompanied by replies from neighboring RFID tags. The RFID system transmission range is just a few meters, while various constraints are used for transmission from low frequencies at 124–135 KHz to ultrahigh frequencies at 860–960 MHz. Two types of RFID tags are available: active tags and passive tags. Active tags contain a battery which enhances signal transmission and boosts the range of transmission. Passive tags, however, have no batteries on board and use the principle of induction to collect energy from the reader signal [64]. RFID has become of considerable importance in an application that promotes a more green system which would reduce vehicle emissions, monitor health, conserve energy use in buildings and improve waste disposal. As most of the examples have an economic benefit, further deployment of RFID systems is recommended. Nevertheless, the RFID technology invokes some problems such as recycling the objects to which they are attached. However, there are a lot of initiatives to address this inconvenience to offer green solutions for RFID systems. The simplest solution is to decrease the size of RFID tags.

Green wireless sensor networks. The combination of sensors and wireless communications leads to wireless sensor networks (WSNs), one of the technologies that helps in developing IoT for different applications such as object tracking, environment monitoring, fire detection, etc. The commercial use of WSNs is increasing daily. The main characteristic of WSN is that sensors are distributed autonomously for monitoring the physical environmental conditions, such as temperature, sound, vibration, pressure and motion. WSN is composed of the BS and the sensor nodes. Low speed, limited power and storage capacities characterize the sensor nodes. Also, sensor nodes are going to register parameters such as temperature, humidity and acceleration from the surroundings. They generally provide the sensory data to the BS in an *ad hoc* way after cooperation with each other. Various methods should be considered to obtain green WSN [65]:

- For the required activity, the sensor uses energy and the sensor nodes work only if necessary, to save energy consumption.
- Wireless charging with energy collection mechanisms is characterized by sun power generation, kinetic energy, vibration, temperature, etc.
- Use techniques of EE optimization (control of transmitter power, optimization of modulation, cooperative communication, directional antennas, CR).
- Use data reduction mechanisms including aggregation, adaptive sampling, compression and coding. Also data and context algorithms to reduce the data size and thus reduce the storage capacity have to be taken into account.
- Use EE routing schemes to decrease the mobility power consumption (cluster architectures, energy as a routing metric, multipath routing, relay node placement, node mobility).

Green cloud computing. Green CC is a model of computing that provides easy access to a common pool of configurable resources, including networks, servers, storage, applications and services on an on-demand basis. Mobile CC can integrate CC into a mobile environment and offload many data processing and storage tasks from mobile devices to the cloud. The data centralization for every sensor and object is carried out by the CC system. They can also communicate and interact with a ubiquitous network criterion. The cloud also enables big data analysis to be integrated to achieve an insight that can be used for determining the patterns of human dynamics. In addition, the human evolutionary model provides channels and feedback mechanisms that can encourage behavioral change. CC offers various resources based on users' requirements: high-performance computing resources and storage. Further resources have to be employed with increasing applications switched to the cloud. It means that more power is consumed resulting in CO_2 emissions. As for green CC, there are some potential solutions as follows [66]:

- Hardware solutions are designed to produce EE devices. Software solutions, in contrast, must attempt to provide effective software designs that consume less energy with the minimum use of resources.
- Virtual machine (VM) power-saving strategies related to transformation, relocation, positioning and distribution
- Different systems for the allocation of resources in an EE way and associated scheduling processes for tasks
- Energy-saving policies models and methods for assessment
- Green CC mechanisms based on networks, communications, etc., supported by cloud.

Green machine-to-machine communications. Machine-to-machine (M2M) communications is a standard that allows both wireless and wired devices to be connected and to exchange data of similar types of devices [66]. Massive M2M nodes gather the monitored data in M2M environment. The wireless/cable network transmits data to the BS in a network domain. Several M2M applications are supported by the BS. In the M2M domain, a lot of energy is consumed by green M2M together with massive machine in M2M communications. To increase EE, there are some methods such as follows [66]:

- tuning the transmit power to the minimum required level,
- employing algorithmic and distributed techniques to develop routing protocols,
- turning other nodes to low power mode, so that just a few attached nodes remain active while maintaining the collection of data, so that the network usability is preserved and
- employing EH and the advantages of CR.

Green data center. Generally speaking, green DC represents a repository for storing, managing, as well as disseminating data and information created by users, things and systems [67,68]. A rise in demand for online services has led to architectural design, congestion registration and VM integration to routing challenges in DC networks [69–71]. Operators of DCs evaluate the sustainability of DCs. As the metric, the effectiveness of carbon and power usage are used. Here, carbon usage effectiveness is defined as the ratio of the total CO_2 emission caused by the total DC energy consumption to information technology equipment energy consumption. For example, when carbon usage effectiveness has a value of zero, it implies that carbon usage has an association with the operations of the DC [72]. With the growing generation of tremendous data on the path to the intelligent planet, EE for DCs is becoming more and more demanding. For the selection of the DCs, the influence of different parameters such as server/availability of content, the communication distance between front end and DC, the cost of the electricity and availability of renewable energy is taken into account. On the way to green, renewable power generators must be installed in the location of the DC. Different distribution strategies at each front end can be adopted in a manner to consider different objectives such as optimizing green energy use, reducing electricity costs or optimizing benefits obtained by operating DC networks. The quality of service at a DC is guaranteed by enforcing a user constraint on the queuing delay at a DC [73]. To improve EE in green DC, some techniques can be used as follows [66]:

- Use renewable/green energy sources (wind, water, solar energy, heat pumps, etc.).
- Use effective and adaptable technological innovations for power management.
- Increase EE hardware development.
- Develop EE DC architectures for power conservation.
- Energy-aware routing algorithms need to be designed to merge traffic flows to a subset of the network and to switch off idle devices.
- Construct accurate DC power models.
- Get communication and computing technology support.

1.5.4 Green Industrial Internet of Things

IoT has become an important and promising component when considering the future transformation in contemporary industrial systems. One of the goals for introducing industrial

Internet of things (IIoT) was to decrease the consumption of resources and CO_2 emissions in digital systems. IIoT becomes more complicated, leading to several important issues that have to be addressed. For example, IIoT systems with a density of sensing, processing and communications devices consume significant amounts of energy, which leads to an increased carbon footprint. The IIoT systems, in contrast, are made up of battery-supported low-power devices. This fact constrains the continuous operation of IIoT system. In the IIoT domain, data collection relies on the sensor nodes and smart devices. Therefore, the efficient optimization of sensing processing and communication for IoT devices reduces energy consumption effectively [74]. Today, WSNs are the core of IIoT networks and are the primary source of energy consumption [75]. Collection of data in the IIoT network depends on sensor nodes and intelligent devices. In this way by the optimization of IoT device sensing, processing and communication, we arrive to the process of reducing energy consumption. The nodes can be categorized as sensory nodes, gateways and control nodes to optimize energy savings. The corresponding IIoT sensing entities domain architecture with the layers is presented in Figure 1.7. These are the sense layer, gateway layer and control layer. Here, nodes involved in the IIoT architecture are placed hierarchically. Nodes serving for collecting the desired data from the area of interest are from part of the sensor layer. Nodes sense the target environment sending information to gateway nodes. The gateway layer is composed of collection of nodes featuring high processing power for running a complex routing protocol. Gateway nodes act as a transmission route to the cloud server for additional analysis and processing of data collected by their sensory nodes.

1.5.5 Green Smart Grid

The SG is a power grid consisting of smart nodes that can autonomously operate, communicate and interact to efficiently supply their consumers with electricity. It has omnipresent

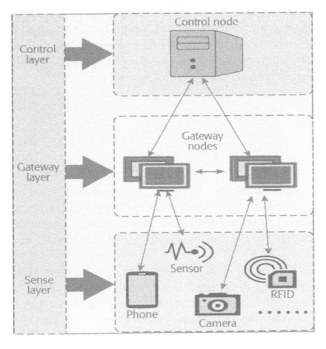

FIGURE 1.7
IIoT sensing entities domain with three layers. IIoT, industrial Internet of things; RFID, radio frequency identification.

interconnections to power devices to allow two-way data and electricity flows in order to balance in real time both the supply and demand. By the incorporation of advanced sensing and communications in everyday network operations, SG has modernized how electricity is generated, transported, distributed and consumed. Electricity is a fundamental utility for the running of society and for the ICT industry. For the purpose of improving efficiency, reliability, security and reducing emissions, SG combines sophisticated sensing, communications and control functions in the power grid operation. Smart monitoring, control, communication and technology allow generators and distributors to have improved communication and operations, as well as dependability and security of electricity supplies. In brief, SGs are a complex cyberphysical system by their nature, impacting the way energy as generated, transported and used.

Smart grid and grid communications (SGGCs) are designed to promote new insights that promote computer and communication systems design. Although SGs are required to meet the challenges faced by energy demand and supply, energy and service management, SGGC is bringing data communications and utility networks together to build an efficient energy supply network with an emphasis on supporting new products, services and technology.

Through software and hardware technologies, the grid regulates the decentralized units. Different IoT-enabled smart meters, sensor devices, electricity generating and distribution machinery and storage power reservoirs are used in the hardware technology. Semantic web applications and semantic storage facilities such as big data platforms [76] are part of the software technology. In a standard grid environment, the smart meter is expected to generate information records with sufficient durability and use of optimized energy consumption within the prescribed life span. The growing requirement for intelligent meters (home users and plug-in hybrid EVs) will produce a massive energy consumption when interacting with components in the 5G. The design of an EE scheme which can decrease CO_2 emissions and cost-effectively manage the energy in the 5G is important. The consumption of energy by intelligent meters can be reduced, and green wireless communication in the 5G can be achieved. The standard of urban life is enhanced by supplying people with knowledge of interest more conveniently and easily. In line with individual needs, for instance, several green interconnected systems provide people globally with the valuable services, such as transport, healthcare, utilities and so on.

1.6 Concluding Remarks

There are many research challenges in integrity green energy technologies with wireless networking technologies. Of course, all the issues require interdisciplinary efforts from various networks, power systems, devices and clients. Protecting the environment and energy conservation are becoming worldwide requirements and common practice every day.

Wireless networks of the 5G would differ from prior generations, as they include enhanced mobile broadband, offer ultrahigh reliability and low latency communications as well as enable massive machine-type communications. Those priorities come from the 5G usage scenarios and application areas. Additionally, 5G should provide high maximum data rates, spectrum efficiency, traffic capacity, link density, EE in the network and substantially decreased latency. To integrate advanced technology, a scalable and flexible communication framework is needed.

Different resource allocation management systems have become a major feature of mobile multimedia communication systems, including power allocation, due to the demand for better EE in mobile multimedia communications systems. On the other hand, as various intelligent devices are made popular, AR and VR can potentially become more demanding than traditional multimedia services. In order to reach optimal power allocation, all different types of power constraints must be addressed, including QoS/QoE constraint.

Green technology is now an urgently needed wireless networking approach to meet the challenges posed by the high demands of wireless traffic and energy consumption. Green communication is a major trend in both the academic and industrial networks, as a research path for the development of future mobile architectures and innovations to achieve high EE.

References

1. Y.Li, et al., "Green heterogeneous cloud radio access networks: Potential techniques, performance trade-offs and challenges", *IEEE Communication Magazine*, vol.55, no.11, pp.35–39, Nov. 2017.
2. Cisco Visual networking index: *Global mobile data traffic forecast update, 2016–2021*, White paper.
3. P.Agyapong, et al., "Design considerations for a 5G network architecture", *IEEE Communications Magazine*, vol.52, no.11, pp.65–75, Nov. 2014.
4. A.Fehske, et al., "The global foot print of mobile communications: The ecological and economic perspective", *IEEE Communication Magazine*, vol.49, no.8, pp.55–62, Aug. 2011.
5. ITU-T Focus Group on IMT-2020, Requirements of IMT-2020 from network perspective, Dec. 2016.
6. X.Shen. "Green wireless communication networks [Editor's note]", *IEEE Network*, vol.27, no.2, pp.2–3, 2013.
7. S.Rehan, D.Grace. "Combined green resource and topology management for beyond next generation mobile broadband systems", *International Conference on Computing, Networking and Communications (ICNC)*, 2013.
8. Q.Wu, G.Y.Li, W.Chen, D.W.Kwan Ng, R.Schober. "An overview of sustainable green 5G networks", *IEEE Wireless Communications*, vol.24, no.4, pp.72–80, 2017.
9. D.Liu, L.Wang, Y.Chen, M.Elkashlan, K.-K.Wong, R.Schober, L.Hanzo. "User association in 5G networks: A survey and an outlook", *IEEE Communications Surveys & Tutorials*, vol.18, no.2, pp.1018–1044, 2016.
10. A.K.Kiani, N.Ansari. "On the fundamental energy trade-offs of geographical load balancing", *IEEE Communications*, vol.55, no.5, pp.170–175, May 2017.
11. Y.Ramamoorthi, A.Kumar. "Resource allocation for energy efficient next generation cellular networks", *CSI Transactions on ICT*, vol.5, pp.179–187, 2017.
12. Y.Al-Dunainawi, R.S.Alhumaima, H.S.Al-Rawershidi, "Green network costs of 5G and beyond: Expectations vs reality", *IEEE Access*, vol.6, no.8, Nov. 2018.
13. P.Gandota, R.K.Iha, "A survey on green communication and security challenges in 5G wireless communication networks", *Journal on Network and Computer Applications*, vol.96, pp.39–61, 2017.
14. J.Andrews, et al., "What will 5G be?" *IEEE JSAC*, vol.12, no.6, pp.1065–1082, June 2014.
15. M.R.Palatteli, et al., "Internet of Things in the 5G era: Enablers, architecture and business models", *IEEE JSAC*, vol.34, no.3, pp.510–527, 2016.
16. W.Verecken, et al., "Power consumption in telecommunication networks: Overview and reduction strategies", *IEEE Communication Magazine*, vol.49, no.6, pp.62–69, June 2011.
17. L.Chiaraviglio, et al., "Is green networking beneficial in terms of device lifetime?" *IEEE Communication Magazine*, vol.53, no.5, pp.232–240, May 2015.

18. F.Meshkati, H.V.Poor, S.C.Schwartz, "Energy-efficient resource allocation in wireless networks", *IEEE Signal Processing Magazine*, vol.24, no.3, pp.58–68, May 2007.
19. A.P.Bainzino, et al., "A survey of green networking research", *IEEE Communications Surveys & Tutorials*, vol.14, no.1, 2012.
20. C.Han, et al., "Green radio: Radio techniques to enable energy efficient wireless networks", *IEEE Communication Magazine*, vol.49, no.6, pp.46–54, June 2011.
21. Md M.Mowla, et al., "A green communication model for 5G systems", *IEEE Transactions on Green Communications and Networking*, vol.1, no.3, pp.1–16, Sept. 2017.
22. S.Zhang, et al., "How many small cells can be turned off via vertical offloading under a separation architecture?" *IEEE Transactions on Wireless Communications*, vol.14, no.4, pp.5440–5453, Oct. 2015.
23. Y.Tang, et al., "Full exploiting cloud computing to achieve a green and flexible C-RAN", *IEEE Communications Magazine*, vol.55, no.11, pp.40–46, Nov. 2017.
24. R.Baines, *Femtocells-reducing power consumption mobile networks*, 2016.
25. F.Cao, Z.Fan, "The tradeoffs between energy efficiency and system performance of femtocell deployment", in *Proc. Int. Symposium Wireless Communication Systems*, pp.315–319, Sept. 2010.
26. F.Helder, et al., "Spectral energy efficiency tradeoff of cellular system performance of femtocell deployment", *IEEE Transactions on Vehicular Technology*, vol.55, pp.3389–3400, 2016.
27. X.Ge, W.Zhang, *5G Green Mobile Communication Networks*, Chapter 4 *Energy Efficiency of 5G Multimedia Communications*, Singapore:Springer, 2019.
28. W.Cheng, X.Zhang, H.Zhang, "Statistical-QoS driven energy-efficiency optimization over Green 5G mobile wireless networks", *IEEE Journal on Selected Areas in Communications*, vol.34, no.12, pp.3092–3107, 2016.
29. J.Tang, X.Zhang, "QoS-driven adaptive power and rate allocation for multichannel communications in mobile wireless networks", *IEEE Int. Symposium on Information Theory*, 2006. pp.2516–2520.
30. X.Ge, X.Huang, Y.Wang, M.Chen, Q.Li, T.Han, C.-X.Wang, "Energy-efficiency optimization for MIMO-OFDM mobile multimedia communication systems with QoS constraints", *IEEE Transactions on Vehicular Technology*, vol.63, no.5, pp.2127–2138, 2014.
31. A.Chehri, G.Jeon, "Optimal matching between energy saving and traffic load for mobile multimedia communication", *Concurrency and Computation: Practice and Experience*, pp.1–9, 2018.
32. A.Mukherjee, "Energy efficiency and delay in 5G ultra-reliable low-latency communications system architectures", *IEEE Network*, vol.32, no.2, pp.55–61, Mar.–Apr. 2018.
33. C.-X.Wang, et al., "Cellular architecture and key technologies for 5G wireless communication networks", *IEEE Communications Magazine*, vol.52, no.2, pp.122–130, 2014.
34. S.Raghavendra, B.Daneshrad, "Performance analysis of energy efficient power allocation for MIMO-MRC systems", *IEEE Transactions on Communications*, vol.60, no.8, pp.2048–2053, Aug. 2012.
35. J.Liu, Y.T.Hou, Y.Shi, D.S. Hanif, "Cross-layer optimization for MIMO-based wireless ad hoc networks: Routing, power allocation, and bandwidth allocation", *IEEE Journal on Selected Areas in Communications*, vol.26, no.6, pp.913–926, 2008.
36. J.Tang, D.K.C. So, E.Alsusa, K.A.Hamdi, A.Shojaeifard, "Energy efficiency in multi-cell MIMO broadcast channels with interference alignment", *2014 IEEE Global Communications Conference*, 2014. pp.4036–4041.
37. L.Xiang, X.Ge, C.-X.Wang, F.Li, F.Reichert, "Energy efficiency evaluation of cellular networks based on spatial distributions of traffic load and power consumption", *IEEE Transactions on Wireless Communications*, vol.12, no.3, pp.961–973, 2013.
38. C.Chen, W.Stark, S.Chen, "Energy-bandwidth efficiency tradeoff in MIMO multihop wireless networks", *IEEE Journal on Selected Areas in Communications*, vol.29, no.8, pp.1537–1546, 2011.
39. F.Heliot, M.A.Imran, R. Tafazolli, "On the energy efficiency spectral efficiency trade-off over the MIMO Rayleigh fading channel", *IEEE Transactions on Communications*, vol.60, no.5, pp.1345–1356, 2012.
40. I.Ku, C.Wang, J.S. Thompson, "Spectral-energy efficiency tradeoff in relay-aided cellular networks", *IEEE Transactions on Wireless Communications*, vol.12, no.10, pp.4970–4982, 2013.

41. X.Hong, Y.Jie, C.Wang, J.Shi, X.Ge, "Energy-spectral efficiency trade-off in virtual MIMO cellular systems", *IEEE Journal on Selected Areas in Communications*, vol.31, no.10, pp.2128–2140, 2013.

42. I.Ku, C.Wang, J.S.Thompson, "Spectral, energy and economic efficiency of relay aided cellular networks", *IET Communications*, vol.7, no.14, pp.1476–1487, 2013.

43. I.Humar, X.Ge, X.Lin, M.Jo, M.Chen, "Rethinking energy efficiency models of cellular networks with embodied energy", *IEEE Network Magazine*, vol.25, no.2, pp.40–49, 2011.

44. Y.Lin, et al., "Nonorthogonal multiple access for 5G and beyond", *Proceedings of the IEEE*, vol.105, no.12, pp.2347–2381, Dec. 2017.

45. Y.Xu, et al., "Cooperative non-orthogonal multiple access in heterogeneous networks", in *Proc. IEEE Globecom*, pp.1–6, Dec. 2015.

46. Y.Lin, et al., "Non-orthogonal multiple access in massive-MIMO aided heterogeneous networks", in *Proc. IEE Globecom*, pp.1–6, Dec. 2016.

47. Z.Ding, F.Adachi, H.V.Poor, "The application of MIMO in non-orthogonal multiple access", *IEEE Transactions on Wireless Communication*, vol.15, no.1, pp.537–552, Jan. 2016.

48. Z.Pi, F.Khan, "An introduction to millimeter-Wave mobile broadband systems", *IEEE Communications Magazine*, vol.49, no.6, pp.101–107, June 2011.

49. J.Cui, et al., "Optimal user scheduling and power allocation for millimeter wave NOMA systems", *IEEE Transactions on Wireless Communication*, vol.17, no.33, pp.1502–1517, Mar. 2018.

50. A.Goldsmith, et al., "Breaking spectrum gridlock with cognitive radios: An information theoretic perspective", *Proceedings of the IEEE*, vol.97, no.5, pp.894–914, Apr. 2009.

51. Z.Qin, et al., "Wideband spectrum sensing on real-time signals at sub-Nyquist sampling rates in single and cooperative multiple nodes", *Transactions on Signal Processing*, vol.64, no.12, pp.3106–3117, Dec. 2016.

52. Q.Zhao, B.M.Sadler, "A survey of dynamic assisted device-to-device communications", *IEEE Communications Magazine*, vol.50, no.3, pp.170–177, Mar. 2012.

53. R.Abe, H.Toke, D.mcQuilkin, "Digital grid: Communicative electrical grids of the future", *IEEE Transactions on Smart Grid*, vol.2, no.2, pp.399–410, 2011.

54. A.Q.Huang, et al., "The future renewable electric energy delivery and management system" the energy Internet", *Proceedings of the IEEE*, vol.39, no.1, pp.143–148, 2011.

55. J.Lin, et al. "Architectural design and load flow study of power flow routers", in *Proc. Int. Conference Smart grid Communications*, pp.31–36, 2011.

56. D.Gunduz, et al., "Designing intelligent energy harvesting communication systems", *IEEE Communication Magazine*, vol.52, no.1, pp.210–216, Jan. 2014.

57. X.Huang, et al., "Software defined energy harvesting networking for 5G green communication", *IEEE Wireless Communicaitons*, vol.24, no.4, pp.38–45, Aug. 2017.

58. D.Altinel, G.K.Kurt, "Modeling of hybrid energy harvesting communication systems", *IEEE Transactions on Green Communications and Networking*, vol.3, no.2, pp.523–534, June 2019.

59. C.Sernados, et al., "An architecture for software defined wireless networking", *IEEE Wireless Communications*, vol.21, no.3, pp.32–61, 2014.

60. W.Zhong, et al., "Software defined networking for flexible and green energy Internet", *IEEE Communications Magazine*, vol.54, no.12, pp.68–74, Dec. 2016.

61. L.Tsoukalas, R.Gao, "From smart grid to an energy Internet: Assumption, architectures and requirements", in *Proc. IEEE Int. Conference Electric Utility Deregulation and Reconstructing and Power Technologies*, pp.94–98, 2008.

62. H.Jiag, et al., "Energy Big Data: A survey", *IEEE Access*, vol.4, 2016.

63. X.Cao, L.Liu, Y.Cheng, X.Shen, "Towards energy-efficient wireless networking in the Big Data era: A survey", *IEEE Communications Surveys & Tutorials*, vol.20, no.1, pp.303–332, 2018.

64. X.Huang, et al., "Software defined networking with pseudonym systems for secure vehicular clouds", *IEEE Access*, vol.4, 2016

65. M.M.Albream, et al., "Green Internet of Things (IoT): An overview", in *Proc. IEEE Int. Conference on Smart Instrumentations, Measurements and Applications*, Nov. 2017. pp.1–6.

66. C.Zhu, et al., "Green Internet of Things for smart world", *IEEE Access*, vol.3, 2015. pp.2151–2162.

67. F.Al-Turjman, E.Ever, H.Zahmatkesh, "Green Femtocells in the IoT Era: Traffic modeling and challenges: An overview", *IEEE Network*, vol.31, no.6, pp.48–55, 2017.
68. E.Exposito, "Enabling technologies for green Internet of Things", *IEEE Systems Journal*, vol.11, no.2, June 2017. pp.983–994.
69. F.Al-Turjman, E.Ever, H.Zahmatkesh, "Small cells in the forthcoming 5G/IoT: Traffic modelling and deployment overview", *IEEE Communications Surveys & Tutorials*, vol.21, no.1, pp.28–65, 2019.
70. T.Wang, Y.Xia, J.Muppala, M.Hamdi, S.Foufou, "A general framework for performance guaranteed green data center networking", *IEEE Global Communications Conference*, 2014. pp. 2510-2515
71. Y.Zhang, N.Ansari, "On architecture design, congestion notification, TCP Incast and power consumption in data centers", *IEEE Communication Surveys and Tutorials*, vol.15, no.1, pp.39–64, Jan. 2013.
72. C.Baladi (editor), *Carbon usage effectiveness (CUE): A green grid data center sustainability metric.* The Green grid white paper, 2010.
73. Z.Lin, et al., "Greening geographical load balancing?" *IEEE ACM Transactions on Networks*, vol.23, no.2, pp.657–671, 2015.
74. H.Chen, Y.Chen, J.Wu, "Power saving for machine-to-machine communications in cellular networks", in *Proc. IEEE Globecom*, pp.389–393, 2011.
75. K.Wang, et al., "Green Industrial Internet of Things architecture: An energy-efficient perspective", *IEEE Communication Magazine – Communication Standard Supplement*, vol.54, no.12, pp.48–54, Dec. 2016.
76. J.F.Sidaliqui, et al., "Optimization lifespan and energy consumption by smart meters in green-cloud-based smart grids", *IEEE Access*, vol.5, Oct. 2017.

Additional Reading

H.Bogucka, A.Conti, "Degrees of freedom for energy savings in practical adaptive wireless systems", *IEEE Communications Magazine*, vol.49, no.6, pp.38–45, 2011.
J.Ding, D.Deng, T.Wu, H.Chen, "Quality-aware bandwidth allocation for scalable on-demand streaming in wireless networks", *IEEE Journal on Selected Areas in Communications*, vol.28, no.3, pp.366–376, 2010.
M.C.Gursoy, D.Qiao, S.Velipasalar, "Analysis of energy efficiency in fading channels under QoS constraints", *IEEE Transactions on Wireless Communications*, vol.8, no.8, pp.4252–4263, 2009.
D.Helonde, V.Wadhai, V.S.Deshpande, H.S.Ohal, "Performance analysis of hybrid channel allocation scheme for mobile cellular network", *in Proceedings of IEEE ICRTIT 2011*, pp.245–250, June 2011.
M.K.Karray, "Analytical evaluation of QoS in the downlink of OFDMA wireless cellular networks serving streaming and elastic traffic", *IEEE Transactions on Communications*, vol.9, no.5, pp.1799–1807, 2010.
D.Niyato, E.Hossain, K.Dong, "Joint admission control and antenna assignment for multiclass QoS in spatial multiplexing MIMO wireless networks", *IEEE Communications Magazine*, vol.8, no.9, pp.4855–4865, 2010.
D.Sabella, et al., "Energy management in mobile networks towards 5G", in M.Z.Shakir et al. (eds), Energy Management in Wireless Cellular and Ad-Hoc Networks, Studies in Systems, Decision and Control 50, Springer 2016.
T.Sibel, K.W.Sung, J.Zander, "On metrics and models for energy-efficient design of wireless access networks", *IEEE Wireless Communications Letters*, vol.3, no.6, pp.649–6522, Dec. 2014.
X.Su, S.Chan, J.H. Manton, "Bandwidth allocation in wireless *ad hoc* networks: Challenges and prospects", *IEEE Communications Magazine*, vol.48, no.1, pp.80–85, 2010.

J.Tang, X.Zhang, "Quality-of-service driven power and rate adaptation for multichannel communications over wireless links", *IEEE Transactions on Wireless Communications*, vol.6, no.12, pp.4349–4360, 2007.

J.Tang, X.Zhang, "Quality-of-service driven power and rate adaptation over wireless links", *IEEE Transactions on Wireless Communications*, vol.6, no.8, pp.3058–3068, 2007.

D.Wu, R.Negi, "Effective capacity: A wireless link model for support of quality of service", *IEEE Transactions on Wireless Communications*, vol.2, no.4, pp.630–643, 2003.

Y.Zhang, et al., "Energy efficient of small networks: Metrics, methods and market", *IEEE Access*, vol.5, 2017.

2

A Framework for Statistical Channel Modeling
in 5G Wireless Communication Systems

Caslav Stefanovic, Danijel Djosic, and Stefan Panic
University of Pristina

Dejan Milic and Mihajlo Stefanovic
University of Nis

CONTENTS

2.1 Introduction

The wireless communications are among the crucial technologies for our civilization in performing everyday tasks. The increasing number of connected devices includes not only smartphones but also connected machines, sensors, vehicles, trains, drones and so on. Next generation of communication systems already require new technologies in order to enable wireless connections at any transmission rates, for any kind of wireless nodes and in any type of scenarios. In order to provide such communications, one of the main tasks is to model and develop reliable mathematical framework for the variety of wireless channels that can be envisioned for 5G and beyond 5G communications (B5G) in terms of reliability, capacity, mobility, latency and other relevant performance measures (Wang et al., 2018; He et al., 2019; Rappaport et al., 2019; Gustafson et al., 2019; Douik et al., 2016; Wang et al., 2016). Moreover, 5G channel models need to incorporate coexistence of different technologies for 5G and B5G communications such as radio frequency (RF), mmWave, cmWave, THz, visible light (VL), optical and free space optical (FSO) communications. Thus, there are several research approaches dependent not only on technology but also on particular scenario. The literature available methods are mainly (1) non–geometry-based stochastic (NGBS), (2) geometry-based stochastic (GBS) and (3) deterministic models.

5G communications are often exposed to rapidly time-varying channels due to increased mobility and density of moving objects (Wu and Fan, 2016), especially for the types of communications named as device-to-device (D2D), mobile-to-mobile (M2M), vehicle-to-everything (V2X), high-speed railway (HSR) and so on. In general, the antennas on the moving objects are approximately at the same height that additionally perplexes communications between moving objects. The channel models that address fading and that are in accordance with the experiments can be modeled as the product of two or more random processes (RPs) (Bithas, Kanatas et al., 2018; Talha and Pätzold, 2011; Hajri et al., 2018; Al-Ahmadi, 2014; Bhargav et al., 2018), which makes performance analysis demanding, especially when higher-order statistical measures are considered. It is important to note that in addition to first-order performance analysis (outage probability, bit error rate, channel capacity and so on), higher-order statistics (level crossing rate [LCR] and average fade duration [AFD]) can deepen understanding and provide new insights about dynamic 5G propagation environments. In particular, the LCR indeed addresses time-variant fading channels, characterized by Doppler spread. It describes the time rate of change of the output signal. On the other hand, AFD is characterized as the average time for which the output signal is below a specified threshold and is dependent on the speed of the moving object.

Among the last one to be proposed for vehicle-to-vehicle (V2V) communications is NGBS model with double generalized gamma (DGG) RP, modeled as the product of two general α–μ RPs (Bithas et al., 2018). The Bithas (2018) provides LCR expression of double generalized α–μ RP, obtained as onefold integral expression. In Stefanovic (2017a), the closed-form approximate LCR expression of the product of two α–μ RPs has been efficiently calculated. Moreover, the general DGG RP can be reduced to double Rayleigh, double Nakagami-m and double Weibull RPs by appropriate parameter settings. Moreover, α–μ, double α–μ and multiple α–μ RPs have already been proposed to model next-generation propagation environments for direct or relaying communications in RF, FSO, mixed RF-FSO and unmanned aerial vehicle (UAV) communication systems (Leonardo and Yacoub, 2015; Badarneh and Almehmadi, 2016; Kashani et al., 2015; Amer and Al-Dharrab, 2019; Dautov et al., 2018). Dos Anjos et al. (2019) show that general distributions are more likely to fit well with experimental data obtained from measurements for 5G millimeter-wave channel modeling in indoor environment, where comparison of exact analytical and experimental results is also provided for α–μ RP.

Salous et al. (2019) present the channel modeling of body area networks (BANs) that includes the impact of multipath and shadowing (large-scale fading), where shadowing is modeled with log-normal RP. The gamma (G) RP, due to its mathematical tractability and its ability to fit accurately with experimental data, is often used to account for shadowing instead of log-normal RP (Kostić, 2005). In Shankar (2004) and Yilmaz and Alouini (2010), generalized composite fading channels are modeled with Nakagami-m and gamma RPs. Badarneh (2016) proposes α–μ/α–μ RP over composite fading channels, whereas α–μ/gamma RPs address multipath/shadowing scenario in Al-Hmood and Al-Raweshidy

(2017) and Goswami and Ashok (2019). UAV-to-ground communications is considered over composite fading in Bithas, Nikolaidis and Kanatas (2019), modeled as the product of double Nakagami-m and inverse G RPs. Those papers mainly provide closed-form mathematical expressions for the first- and the second-order statistics, expressed through Meijer's G and Fox's H functions.

The Laplace approximation–based method (LABM) for derivation of fast-computing closed-form performance metrics has been exploited in Hajri et al. (2018), Zlatanov et al. (2008) and Hadzi-Velkov et al. (2009), where the products of two Hoyt, two Nagkagami-m and multiple Rayleigh RPs for derivation of second-order performance metrics have been considered, respectively. The relevant higher-order statistics of two and more DGG RPs by applying LABM have been considered in Stefanovic (2017b), Stefanovic, Pratesi and Santucci (2018) and Stefanovic, Pratesi and Santucci (2019). In particular, Stefanovic, Pratesi and Santucci (2019) investigate higher-order statistics of mixed RF/FSO/RF V2V cascaded relay link modeled as the product of Nakagami-m RPs (for the case of RF links) and DG RP (for the case of FSO links) by LABM and exponential LABM for derivation of closed-form measures by evaluating up to threefold integrals. Accordingly, the LABM and exponential LABM can be applied as precise approximate tools for solving complex many-fold integrals by providing closed-form statistical expressions, thus decreasing computational time. The proof that LABM provides accurate results under subasymptotic conditions is given in Butler and Wood (2002), while accurate and efficient applications of exponential LABM are given in Wang (2010).

The relevant impairment in 5G communications is cochannel interference, coexisting due to ultra dense heterogeneity of the 5G networks (Kim et al., 2017; Meng et al., 2019). Performance analysis of V2V communications over fading channels in the presence of interference is analyzed in Bithas, Efthymoglou and Kanatas (2018). The interference-limited environment (when the impact of noise is negligible) can be modeled as the ratio of RPs (Paris, 2013; Da Silva et al., 2019). Thus, performance analysis of first- and higher-order statistics of M2M communications based on LABM in interference-limited environment is further discussed in Milosevic et al. (2018a), Milosevic et al. (2018b) and Stefanovic et al. (2019).

In this chapter, we provide derivation of unified mathematical framework for evaluating novel exact expressions as well as novel closed-form approximate expressions for (1) probability density function (PDF), (2) cumulative distribution function (CDF), (3) outage probability, (4) average LCR and (5) AFD of the products and the ratios of DGG and G RPs by applying NGBS models. The considered performances are directly related to the impact of multipath, shadowing and interference in rapidly time-variant fading channels on envisioned 5G and B5G systems. We rely on LABM and exponential LABM for derivation of closed-form analytical expressions for the first- and the second-order statistical measures. The obtained analytical and approximate results are compared numerically and evaluated for the different system model parameters.

2.2 Double Scattering Fading Model

The fading can be modeled as a DGG RP, denoted as z_{DGG} (the dependence on time t is excluded if not stated otherwise). According to Yacoub (2007) and Panic et al. (2013), z_{DGG} can be expressed as the product of two generalized G RPs:

$$z_{DGG} = z_{G1}z_{G2} = \underbrace{(x_{N1}x_{N2})^{\frac{2}{\alpha}}}_{\text{multipath}} \tag{2.1}$$

where α is nonlinearity parameter. The $x_{N,i}$, i=1,2 fading envelopes are independent but not necessarily identically distributed (i.n.i.d) Nakagami-m (N) RPs (Simon and Alouini, 2000), whose fading severity parameters and average powers are denoted as m_i and Ω_i, respectively:

$$p_{x_{N,i}}\left(x_{N,i}\right) = \frac{2\left(m_i/\Omega_i\right)^{m_i}}{\Gamma(m_i)} x_{N,i}^{2m_i-1} e^{-\frac{m_i x_{N,i}^2}{\Omega_i}}, i = 1,2; \tag{2.2}$$

2.2.1 Performance Evaluation

PDF of z_{DGG} can be obtained according to Panic et al. (2013) and Simon and Alouini (2000) and using Gradshteyn and Ryzhik (2000):

$$p_{z_{DGG}}\left(z_{DGG}\right) = \int_0^\infty \left|\frac{dx_{N1}}{dz_{DGG}}\right| p_{x_{N1}} \left(\frac{z_{DGG}^{\frac{a}{2}}}{x_{N2}}\right) p_{x_{N2}}\left(x_{N2}\right) dx_{N2} = \frac{2\alpha}{\Gamma(m_1)\Gamma(m_2)}$$

$$\cdot \left(\frac{m_1}{\Omega_1}\right)^{m_1} \left(\frac{m_2}{\Omega_2}\right)^{m_2} z_{DGG}^{\alpha m_1 - 1} \left(\frac{m_1\Omega_2}{m_2\Omega_1} z_{DGG}^\alpha\right)^{\frac{m_2-m_1}{2}} K_{m_2-m_1}\left(2\sqrt{\frac{m_1 m_2}{\Omega_1 \Omega_2}} z_{DGG}^\alpha\right) \tag{2.3}$$

where $K_n(\cdot)$ is the modified Bessel function of nth order and second kind.

CDF of z_{DGG} can be derived according to Panic et al. (2013) and Simon and Alouini (2000) as:

$$F_{Z_{DGG}}\left(z_{DGG}\right) = \int_0^{z_{DGG}} p_{Z_{DGG}}(h) dh \tag{2.4}$$

The closed-form $F_{Z_{DGG}}\left(z_{DGG}\right)$ can be then obtained using Gradshteyn and Ryzhik (2000), for the case where m_1 is positive integer, as:

$$F_{Z_{DGG}}\left(z_{DGG}\right) = \frac{1}{\Gamma(m_1)\Gamma(m_2)}(m_1-1)!\Gamma(m_2) - \frac{2}{\Gamma(m_1)\Gamma(m_2)}\left(\frac{m_2}{\Omega_2}\right)^{m_2}$$

$$\cdot(m_1-1)!\sum_{k=0}^{m_1-1}\frac{\left(\frac{m_1 z_{DGG}^\alpha}{\Omega_1}\right)^k}{k!}\left(\frac{m_1\Omega_2}{m_2\Omega_1}z_{DGG}^\alpha\right)^{\frac{m_2-k}{2}}K_{m_2-k}\left(2\sqrt{\frac{m_1 m_2}{\Omega_1 \Omega_2}}z_{DGG}^\alpha\right). \tag{2.5}$$

The system outage probability for a given threshold $z_{th,DGG}$ can be easily obtained and expressed as $F_{Z_{DGG}}(z_{th,DGG})$.

2.2.2 Second-Order Statistics Measures

The LCR for a given threshold $z_{th,DGG}$ can be derived through integration of the joint distribution of RPs, z_{DGG} and its first derivative $\dot{z}_{DGG}$, as:

$$N_{Z_{DGG}}(z_{th}) = \int_0^{\infty} \dot{z}_{DGG} p_{Z_{DGG}\dot{Z}_{DGG}}(z_{th,DGG}\dot{z}_{DGG}) d\dot{z}_{DGG} \tag{2.6}$$

The detailed derivation of $N_{Z_{DGG}}(z_{th,DGG})$ is provided in the appendix. Here we provide exact analytical expression for $N_{Z_{DGG}}(z_{th,DGG})$ obtained as:

$$N_{Z_{DGG}}(z_{th,DGG}) = \frac{4\left(\dfrac{m_1}{\Omega_1}\right)^{m_1}\left(\dfrac{m_2}{\Omega_2}\right)^{m_2}\sigma_{\dot{X}_{N1}}}{\sqrt{2\pi}\Gamma(m_1)\Gamma(m_2)} z_{th,DGG}^{\alpha(m_1-1/2)}$$

$$\cdot \underbrace{\int_0^{\infty} \sqrt{1+\frac{\sigma_{\dot{X}_{N2}}^2 z_{th,DGG}^{\alpha}}{\sigma_{\dot{X}_{N1}}^2 x_{N2}^4}}\, e^{-\frac{m_1 z_{th,DGG}^{\alpha}}{\Omega_1 x_{N2}^2}-\frac{m_2}{\Omega_2}x_{N2}^2+2(m_2-m_1)\ln(x_{N2})}\, dx_{N2}}_{j_1}. \tag{2.7}$$

while closed form approximate expression for $N_{Z_{DGG}}(z_{th,DGG})$ is derived by LABM and is given as:

$$N_{Z_{DGG}}(z_{th,DGG}) \approx \frac{4\left(\dfrac{m_1}{\Omega_1}\right)^{m_1}\left(\dfrac{m_2}{\Omega_2}\right)^{m_2} z_{th,DGG}^{\alpha(m_1-1/2)}\sigma_{X_{N1}}}{\sqrt{2\pi}\Gamma(m_1)\Gamma(m_2)}\left(\frac{2\pi}{\gamma}\right)^{1/2}$$

$$\cdot \frac{\left(1+\dfrac{\sigma_{\dot{X}_{N2}}^2 z_{th,DGG}^{\alpha}}{\sigma_{\dot{X}_{N1}}^2 x_{N20}^4}\right)^{1/2} e^{-\gamma\left(\frac{m_1 z_{th,DGG}^{\alpha}}{\Omega_1 x_{N20}^2}+\frac{m_2}{\Omega_2}x_{N20}^2-2(m_2-m_1)\ln(x_{N20})\right)}}{\left(\dfrac{6m_1 z_{th,DGG}^{\alpha}}{\Omega_1 x_{N20}^4}+2\dfrac{m_2}{\Omega_2}+\dfrac{2(m_2-m_1)}{x_{N20}^2}\right)^{1/2}}. \tag{2.8}$$

where $\gamma=1$, and $x_{N20} = \left(\dfrac{\Omega_1\Omega_2(m_2-m_1)+\sqrt{\Omega_1\Omega_2(m_2-m_1)^2-4(\Omega_1\Omega_2 m_1 m_2 z_{th,DGG}^{\alpha})}}{2\Omega_1 m_2}\right)^{\frac{1}{2}}$.

It is important to note that approximate closed-form $N_{Z_{DGG}}\left(z_{th,DGG}\right)$ enables faster computing time than exact $N_{Z_{DGG}}\left(z_{th,DGG}\right)$.

In the case of RF scenario, the variances of $\dot{X}_{N1}$ and $\dot{X}_{N2}$ can be expressed as $\sigma^2_{\dot{X}_{N,i}} = \pi^2 \left(f_{m,i}\right)^2 \Omega_i / m_i$, $i=1,2$ where the maximum Doppler frequencies of transmitting moving object and receiving moving object are assumed to be the same, $f_m = f_{m,i} = \sqrt{\left(f_{mTR}\right)^2 + \left(f_{mRR}\right)^2}$, $i=1,2$, as expressed in Hadzi-Velkov et al. (2009), where f_{mTR} and f_{mRR} are maximum Doppler frequencies for direct RF mobile communications of the transmitting and receiving moving objects, respectively. Beside RF V2V application (Bithas, et al., 2018), the DGG can be also applied to account for turbulence induced fading in FSO fixed communications (Kashani et al., 2015; Amer and Al-Dharrab, 2019), where variances can be modeled as zero mean Gaussian RP $\sigma^2_{\dot{X}} = \sigma^2_{\dot{X}_{N1}} = \sigma^2_{\dot{X}_{N2}} = v_0^2 \pi^2 \sigma^2_{DG} \langle X \rangle$, as obtained in Jurado-Navas et al. (2017), where mean value $\langle X \rangle = 1$ is assumed for DGG RP. v_0 is quasi frequency and can be further expressed as $v_0 = 1/\left(\pi \tau_0 \sqrt{2}\right)$ (Jurado-Navas et al., 2017). Furthermore, $\tau_0 = \sqrt{\lambda L / ut}$ is turbulence correlation time, where λ is optical window, L is optical distance and ut is average wind speed. Since, in this chapter, we focused on developing general mathematical framework for performance evaluation, the presented results for the second-order statistics are normalized by $\sigma^2_{\dot{X}_{N,i}}$

Figure 2.1 provides behavior of the $N_{Z_{DGG}}\left(z_{th,DGG}\right)$ for a given threshold $z_{th,DGG}$ normalized by $\sigma_{\dot{X}_{N1}}$ in relation to various values of DGG system model parameters. It is obvious that increase in α and m_i enables decrease in normalized $N_{Z_{DGG}}\left(z_{th,DGG}\right)$, which in turn provides more stable system performances (for the case of higher $z_{th,DGG}$ values) in terms of $N_{Z_{DGG}}\left(z_{th,DGG}\right)$.

The average fade duration $\text{AFD}z_{DGG}$ is obtained as

$$\text{AFD}z_{DGG}\left(z_{th,DGG}\right) = \frac{F_{z_{DGG}}\left(z_{th,DGG}\right)}{N_{z_{DGG}}\left(z_{th,DGG}\right)} \tag{2.9}$$

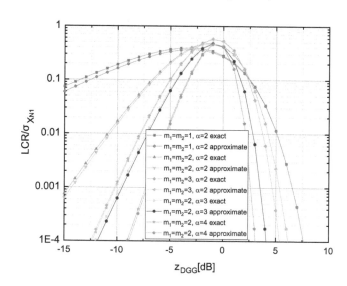

FIGURE 2.1

Normalized LCR of $z_{th,DGG}$ for exact and approximate results regarding different fading severity and nonlinearity parameters for $\Omega i=1$. LCR, level crossing rate.

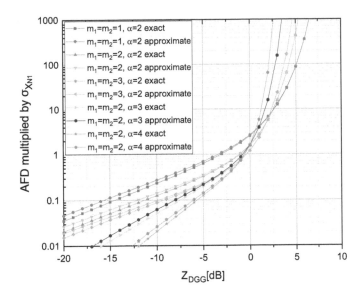

FIGURE 2.2

AFD of $z_{th,DGG}$ for exact and approximate results regarding different fading severity and nonlinearity parameters for $\Omega i = 1$. AFD, average fade duration.

The AFD$z_{DGG}(z_{th,DGG})$ is an important higher-order system performance measure evaluated as closed-form as well as exact analytical expression describes fast time-variant fading channels characterized by Doppler spread modeled as DGG RP.

AFD multiplied by $\sigma^2_{\dot{X}_{N1}}$, presented in Figure 2.2, shows that the decrease in α and m_i leads to increase in AFD for lower $z_{th,DGG}$ dB values and decrease in AFD$z_{DGG}(z_{th,DGG})$ for higher $z_{th,DGG}$ dB values. Thus, the system performance improvement in lower dB $z_{th,DGG}$ output regime can be achieved for a higher number of clusters (m_i) as well as for higher values of nonlinearity parameter α. Moreover, α has slightly higher impact than m_i for considered $z_{th,DGG}$ on $N_{Z_{DGG}}(z_{th,DGG})$ and AFD$z_{DGG}(z_{th,DGG})$ over DGG RP.

2.3 Composite Fading Model

In this part of the chapter, composite fading channel (channel where multipath and shadowing are present at the same time) is considered. The composite fading RP is denoted as $z_{DGG/G}$ and modeled as the product of DGG and G RPs where G RP is expressed as squared Nakagami-m (N) RP (Simon and Alouini, 2000):

$$z_{DGG/G} = z_{DGG}z_G = \underbrace{(x_{N1}x_{N2})^{\frac{2}{a}}}_{\text{multipath}} \underbrace{x^2_{N3}}_{\text{shadowing}} \tag{2.10}$$

with the following PDFs, respectively (Simon and Alouini, 2000):

$$p_{x_{N,i}}(x_{N,i}) = \frac{2(m_i/\Omega_i)^{m_i}}{\Gamma(m_i)} x^{2m_i-1}_{N,i} e^{-\frac{m_i x^2_{N,i}}{\Omega_i}} , \ i = 1,3; \tag{2.11}$$

where without loss of generality, multipath severities and mean powers are denoted as, respectively, $m_1 = m_{\text{mul}1}, m_2 = m_{\text{mul}2}, \Omega_1 = \Omega_{\text{mul}1}$ and $\Omega_2 = \Omega_{\text{mul}2}$ while shadowing severity and mean power are denoted as, respectively, $m_3 = m_{\text{sh}}, \Omega_3 = \Omega_{\text{sh}}$.

2.3.1 Performance Evaluation

PDF of $z_{\text{DGG/G}}$ can be expressed through joint and conditional probabilities as:

$$
\begin{aligned}
p_{Z_{\text{DGG/G}}}\left(z_{\text{DGG/G}}\right) &= \int_0^\infty dx_{N1} \int_0^\infty \left| \frac{dx_{N3}}{dz_{\text{DGG/G}}} \right| p_{X_{N1}}(x_{N1}) p_{X_{N2}}(x_{N2}) p_{X_{N3}}\left(\frac{z_{\text{DGG}}^{1/2}}{x_{N1}^{1/\alpha} x_{N2}^{1/\alpha}} \right) dx_{N2} \\
&= \frac{4\left(m_{\text{mul}1}/\Omega_{\text{mul}1}\right)^{m_{\text{mul}1}} \left(m_{\text{mul}2}/\Omega_{\text{mul}2}\right)^{m_{\text{mul}2}} \left(m_{\text{sh}}/\Omega_{\text{sh}}\right)^{m_{\text{sh}}} z_{\text{DGG/G}}^{m_{\text{sh}}-1}}{\Gamma\left(m_{\text{mul}1}\right)\Gamma\left(m_{\text{mul}2}\right)\Gamma\left(m_{\text{sh}}\right)} J_2
\end{aligned}
\tag{2.12}
$$

where

$$
J_2 = \int_0^\infty dx_{N1} \int_0^\infty e^{-\frac{m_{\text{mul}1}}{\Omega_{\text{mul}1}}x_{N1}^2 - \frac{m_{\text{mul}2}}{\Omega_{\text{mul}2}}x_{N2}^2 - \frac{m_{\text{sh}}}{\Omega_{\text{sh}}}\frac{z_{\text{DGG/G}}}{x_{N1}^{2/\alpha} x_{N2}^{2/\alpha}} + \left(2m_{\text{mul}1}-\frac{2m_{\text{sh}}}{\alpha}-1\right)\ln x_{N1} + \left(2m_{\text{mul}2}-\frac{2m_{\text{sh}}}{\alpha}-1\right)\ln x_{N2}} dx_{N2}.
$$

The PDF of $z_{\text{DGG/G}}$ evaluated for exact and approximate analytical expressions is presented in Figure 2.3. It can be seen that approximations fit well with exact analytical results for the observed system values, which is more evident for higher values of $z_{\text{DGG/G}}$.

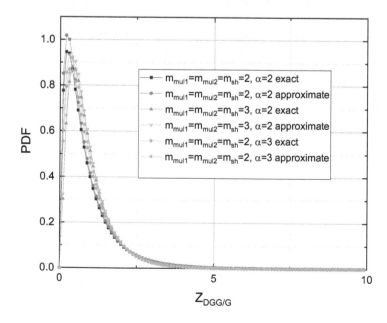

FIGURE 2.3
PDF of $z_{\text{DGG/G}}$ for exact and approximate results regarding different fading severity and nonlinearity parameters for $\Omega i = 1$. PDF, probability density function.

CDF of $z_{DGG/G}$ is expressed using Simon and Alouini (2000), Panic et al. (2013) and Gradshteyn and Ryzhik (2000), respectively, for the case where m_{sh} is positive integer as

$$F_{Z_{DGG/G}}\left(z_{DGG/G}\right) = \int_0^{z_{DGG/G}} p_{Z_{DGG/G}}(h)\,dh = \frac{4\left(\dfrac{m_{mul1}}{\Omega_{mul1}}\right)^{m_{mul1}}\left(\dfrac{m_{mul2}}{\Omega_{mul2}}\right)^{m_{mul2}}(m_{sh}-1)!}{\Gamma(m_{mul1})\Gamma(m_{mul2})\Gamma(m_{sh})}$$

$$\left(\frac{\Gamma(m_{mul1})\Gamma(m_{mul2})}{4\left(\dfrac{m_{mul1}}{\Omega_{mul1}}\right)^{m_{mul1}}\left(\dfrac{m_{mul2}}{\Omega_{mul2}}\right)^{m_{mul2}}} - \sum_{k=0}^{m_{sh}-1}\frac{\left(\dfrac{m_{sh}}{\Omega_{sh}}z_{DGG/G}\right)^k}{k!}J_3\right)$$

(2.13)

where J_3 is expressed as a twofold integral of exponential function:

$$J_3 = \int_0^\infty dx_{N1}\int_0^\infty e^{-\frac{m_{mul1}}{\Omega_{mul1}}x_{N1}^2 - \frac{m_{mul2}}{\Omega_{mul2}}x_{N2}^2 - \frac{m_{sh}}{\Omega_{sh}}\frac{z_{DGG/G}}{x_{N1}^{2/\alpha}x_{N2}^{2/\alpha}} + \left(2m_{mul1}-\frac{2k}{\alpha}-1\right)\ln x_{N1} + \left(2m_{mul2}-\frac{2k}{\alpha}-1\right)\ln x_{N2}}\,dx_{N2}.$$

Performance evaluation in terms of outage probability is then determined as $F_{Z_{DGG/G}}\left(z_{th,DGG/G}\right)$, where $z_{th,DGG/G}$ is predetermined threshold.

2.3.2 Second-Order Statistics Measures

The LCR of fading signal threshold $z_{th,DGG/G}$ can be calculated as the average value of the first derivative of $z_{DGG/G}$, denoted as $\dot{z}_{DGG/G}$ with the following expression

$$N_{Z_{DGG/G}}\left(z_{th,DGG/G}\right) = \int_0^\infty \dot{z}_{DGG/G}\,p_{Z_{DGG/G}\dot{z}_{DGG/G}}\left(z_{th,DGG/G},\dot{z}_{DGG/G}\right)d\dot{z}_{DGG/G}$$

(2.14)

After derivation, which is presented in the appendix, the $N_{Z_{DGG/G}}\left(z_{th,DGG/G}\right)$ is obtained as:

$$N_{Z_{DGG/G}}\left(z_{th,DGG/G}\right) = \frac{8\left(\dfrac{m_{mul1}}{\Omega_{mul1}}\right)^{m_{mul1}}\left(\dfrac{m_{mul2}}{\Omega_{mul2}}\right)^{m_{mul2}}\left(\dfrac{m_{sh}}{\Omega_{sh}}\right)^{m_{sh}}z_{th,DGG/G}^{m_{sh}}\sigma_{\dot{x}_{N1}}}{a\sqrt{2\pi}\Gamma(m_{mul1})\Gamma(m_{mul2})\Gamma(m_{sh})}J_4.$$

(2.15)

where,

$$J_4 = \int_0^\infty dx_{N1}\int_0^\infty\left(1 + \frac{x_{N1}^2}{x_{N2}^2}\frac{\sigma_{\dot{X}_{N2}}^2}{\sigma_{\dot{X}_{N1}}^2} + \frac{x_{N1}^{2/\alpha+2}x_{N2}^{2/\alpha}}{z_{DGG/G}}\frac{\sigma_{\dot{X}_{N3}}^2}{\sigma_{\dot{X}_{N1}}^2}\right)^{1/2}$$

(2.16)

$$\cdot x_{N1}^{2m_{mul1}-\frac{2k}{\alpha}-1}x_{N2}^{2m_{mul2}-\frac{2k}{\alpha}-1}e^{-\frac{m_{mul1}}{\Omega_{mul1}}x_{N1}^2 - \frac{m_{mul2}}{\Omega_{mul2}}x_{N2}^2 - \frac{m_{sh}}{\Omega_{sh}}\frac{z_{DGG/G}}{x_{N1}^{2/\alpha}x_{N2}^{2/\alpha}}}\,dx_{N2}.$$

The closed-form approximate $N_{Z_{DGG/G}}\left(z_{th,DGG/G}\right)$ and its derivation are also given in appendix. Similarly, the variances of $\dot{X}_{N1}, \dot{X}_{N2}$ and $\dot{X}_{N3}$ for RF system can be expressed as

$\sigma^2_{\dot{X}_{N,i}} = \pi^2 \left(f_{m,i}\right)^2 \Omega_i / m_i$, $i = 1, 2, 3$ where the maximal Doppler frequencies can be expressed as, respectively, $f_m = f_{m,i} = \sqrt{\left(f_{mTR}\right)^2 + \left(f_{mRR}\right)^2}$, $i = 1, 2, 3$.

In addition to apply DGG/G model for RF scenario, it can be also applied to account for mixed RF-FSO dual hop relying system over turbulence-induced fading for FSO (modeled as DGG RP) and non–turbulence-induced fading for RF (modeled as squared Nakagami RP) where variances can be modeled as zero mean Gaussian RP $\sigma^2_{\dot{X}} = \sigma^2_{\dot{X}_{N1}} = \sigma^2_{\dot{X}_{N2}} = v_0^2 \pi^2 \sigma^2_{DG} \langle X \rangle$, as obtained in Jurado-Navas et al. (2017) while variance of $\dot{X}_{N3}$ can be expressed as $\sigma^2_{\dot{X}_{N3}} = \pi^2 \left(f_{m,3}\right)^2 \Omega_3 / m_3$. Similar scenario has been considered in Stefanovic, Pratesi and Santucci (2019). In this chapter, presented numerical results for higher-order statistics are normalized by $\sigma^2_{\dot{X}_{N,i}}$.

Figure 2.4 presents LCR versus $z_{th,DGG/G}$ normalized by $\sigma^2_{\dot{X}_{N,i}}$ for various $m_{mul,i}, m_{sh}$ and α. As expected, increasing in $m_{mul,i}, m_{sh}$ and α enables LCR decreasing which is more evident for lower $z_{th,DGG/G}$ dB values. The AFD is directly obtained from

$$\text{AFD}z_{DGG/G}\left(z_{th,DGG/G}\right) = \frac{F_{Z_{DGG/G}}\left(z_{th,DGG/G}\right)}{N_{Z_{DGG/G}}\left(z_{th,DGG/G}\right)} \tag{2.17}$$

AFD of $z_{th,DGG/G}$ multiplied by $\sigma^2_{\dot{X}_{N,i}}$ for various $m_{mul,i}, m_{sh}$ and α is provided in Figure 2.5. As expected, increasing in $m_{mul,i}, m_{sh}$ and α enables $\text{AFD}z_{DGG/G}\left(z_{th,DGG/G}\right)$ decrease for lower $z_{th,DGG/G}$ dB values and $\text{AFD}z_{DGG/G}\left(z_{th,DGG/G}\right)$ increase for higher $z_{th,DGG/G}$ dB values. It can be concluded that exact and approximate curves for LCR as well as for $\text{AFD}z_{DGG/G}\left(z_{th,DGG/G}\right)$ fit well in higher $z_{th,DGG/G}$ dB output regime. Moreover, fitting improvement is noticeable by increasing the values of composite fading severities as well as composite fading nonlinearity. Moreover, the impact of composite fading severities is higher than nonlinearity parameter on higher-order statistics.

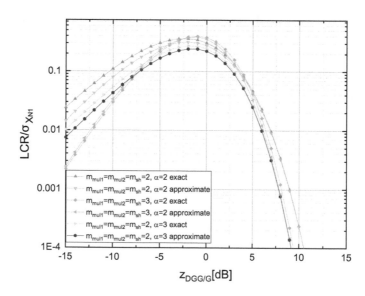

FIGURE 2.4
Normalized LCR of $z_{th,DGG/G} \cdot z_{th,DGG/G}$ for exact and approximate results regarding different fading severity and nonlinearity parameters for $\Omega i=1$. LCR, level crossing rate.

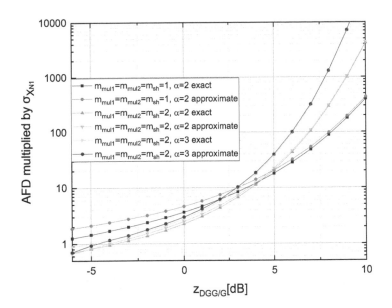

FIGURE 2.5
AFD of $z_{th,DGG/G}$ for exact and approximate results regarding various fading severity and nonlinearity parameters for $\Omega_i=1$. AFD, average fade duration.

2.4 Composite Fading Model in Interference-Limited Environment

We model fading signal envelope as the ratio of two composite DGG/G independent but not necessarily identical (i.n.i.d) RPs in order to address the interference-limited environment. The ratio of two composite fading RP is denoted as $z_{R(DGG/G)}$ and modeled as ratio of the products of DGG and G RPs for desired signal $z_{(DGG/G)d}$ and interfering signal $z_{(DGG/G)I}$ expressed through Nakagami-m (N) RPs (Simon and Alouini, 2000):

$$z_{R(DGG/G)} = \frac{z_{(DGG/G)d}}{z_{(DGG/G)I}} = \frac{(x_{N1}x_{N2})^{\frac{2}{a}} x_{N3}^2}{(x_{N4}x_{N5})^{\frac{2}{a}} x_{N6}^2}. \tag{2.18}$$

with PDFs, respectively (Simon and Alouini, 2000):

$$p_{X_{N,i}}(x_{N,i}) = \frac{2(m_i/\Omega_i)^{m_i}}{\Gamma(m_i)} x_{N,i}^{2m_i-1} e^{-\frac{m_i x_{N,i}^2}{\Omega_i}}, \ i=1,6; \tag{2.19}$$

where without loss of generality, multipath severities and mean powers of desired signal are denoted, respectively, $m_1 = m_{mul1}^d$, $m_2 = m_{mul2}^d$, $\Omega_1 = \Omega_{mul1}^d$, and $\Omega_2 = \Omega_{mul2}^d$ while shadowing severity and mean power of desired signal are denoted, respectively, $m_3 = m_{sh}^d$, $\Omega_3 = \Omega_{sh}^d$.

Similarly, the parameters of interfering composite fading signal are denoted, respectively: $m_4 = m_{mul1}^I, m_5 = m_{mul2}^I, m_6 = m_{sh}^I, \Omega_4 = \Omega_{mul1}^I, \Omega_5 = \Omega_{mul2}^I$ and $\Omega_6 = \Omega_{sh}^I$.

2.4.1 Performance Measures

PDF of $z_{R(DGG/G)}$ can be transformed and expressed through joint and conditional probabilities:

$$p_{Z_{R(DGG/G)}}\left(z_{R(DGG/G)}\right) = \int_0^\infty dx_{N1} \int_0^\infty dx_{N2} \int_0^\infty dx_{N4} \int_0^\infty dx_{N5}$$

$$\cdot \int_0^\infty \frac{\frac{1}{2} z_{R(DGG/G)}^{-1/2} x_{N4}^{1/\alpha} x_{N5}^{1/\alpha} x_{N6}}{x_{N1}^{1/\alpha} x_{N2}^{1/\alpha}} p_{X_{N1}}(x_{N1}) p_{X_{N2}}(x_{N2}) p_{X_{N4}}(x_{N4})$$

$$\cdot p_{X_{N5}}(x_{N5}) p_{X_{N6}}(x_{N6}) p_{X_{N3}}\left(\frac{\sqrt{z_{R(DGG/G)}} x_{N4}^{1/\alpha} x_{N5}^{1/\alpha} x_{N6}}{x_{N1}^{1/\alpha} x_{N2}^{1/\alpha}}\right) dx_{N6} \qquad (2.20)$$

$$= \frac{32\left(m_{mul1}^d/\Omega_{mul1}^d\right)^{m_{mul1}^d} \left(m_{mul2}^d/\Omega_{mul2}^d\right)^{m_{mul2}^d} \left(m_{sh}^d/\Omega_{sh}^d\right)^{m_{sh}^d} \left(m_{mul1}^I/\Omega_{mul1}^I\right)^{m_{mul1}^I}}{\Gamma\left(m_{mul1}^d\right)\Gamma\left(m_{mul2}^d\right)\Gamma\left(m_{sh}^d\right)\Gamma\left(m_{mul1}^I\right)\Gamma\left(m_{mul2}^I\right)\Gamma\left(m_{sh}^I\right)}$$

$$\cdot \left(m_{mul2}^I/\Omega_{mul2}^I\right)^{m_{mul2}^I} \left(m_{sh}^I/\Omega_{sh}^I\right)^{m_{sh}^I} z_{R(DGG/G)}^{m_{sh}^d-1} J_5$$

where,

$$J_5 = \int_0^\infty dx_{N1} \int_0^\infty dx_{N2} \int_0^\infty dx_{N4} \int_0^\infty dx_{N5}$$

$$\cdot \int_0^\infty e^{\frac{m_{mul1}^d x_{N1}^2}{\Omega_{mul1}^d} - \frac{m_{mul2}^d x_{N2}^2}{\Omega_{mul2}^d} - \frac{m_{sh}^d z_{R(DGG/G)} x_{N4}^{2/\alpha} x_{N5}^{2/\alpha} x_{N6}^{2/\alpha}}{\Omega_{sh}^d x_{N1}^{2/\alpha} x_{N2}^{2/\alpha}} - \frac{m_{mul1}^I x_{N4}^2}{\Omega_{mul1}^I} - \frac{m_{mul2}^I x_{N5}^2}{\Omega_{mul2}^I} - \frac{m_{sh}^I x_{N6}^2}{\Omega_{sh}^I}}$$

$$\cdot e^{\left(2m_{mul1}^d - \frac{2}{a} m_{sh}^d - 1\right)\ln(x_{N1}) + \left(2m_{mul2}^d - \frac{2}{a} m_{sh}^d - 1\right)\ln(x_{N2}) + \left(2m_{mul1}^I + \frac{2}{a} m_{sh}^d - 1\right)\ln(x_{N4})}$$

$$\cdot e^{\left(2m_{mul2}^I + \frac{2}{a} m_{sh}^d - 1\right)\ln(x_{N5}) + \left(2m_{sh}^I + 2m_{sh}^d - 1\right)\ln(x_{N6})} dx_{N6}$$

Figure 2.6 presents PDF of the ratio of the product of DGG and G RPs evaluated from exact and approximate analytical expressions.

CDF of $z_{R(DGG/G)}$ is derived using Simon and Alouini (2000):

$$F_{Z_{R(DGG/G)}}\left(z_{R(DGG/G)}\right) = \int_0^{z_{R(DGG/G)}} p_{Z_{R(DGG/G)}}(h) dh \qquad (2.21)$$

In accordance with Gradshteyn and Ryzhik (2000), integral form of $F_{Z_{R(DGG/G)}}\left(z_{R(DGG/G)}\right)$ is derived as:

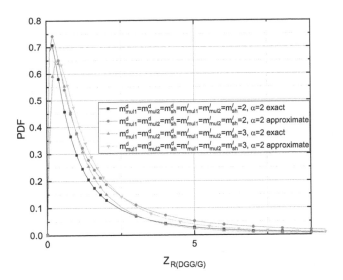

FIGURE 2.6
PDF of the ratio of the product of double generalized gamma and gamma RPs evaluated from exact and approximate analytical expressions for $\Omega i=1$. PDF, probability density function; RP, random process.

$$F_{Z_{R(DGG/G)}}\left(z_{R(DGG/G)}\right) = \frac{32\left(\frac{m_{mul1}^d}{\Omega_{mul1}^d}\right)^{m_{mul1}^d}\left(\frac{m_{mul2}^d}{\Omega_{mul2}^d}\right)^{m_{mul2}^d}\left(m_{mul1}^I/\Omega_{mul1}^I\right)^{m_{mul1}^I}}{\Gamma\left(m_{mul1}^d\right)\Gamma\left(m_{mul2}^d\right)\Gamma\left(m_{sh}^d\right)\Gamma\left(m_{mul1}^I\right)\Gamma\left(m_{mul2}^I\right)\Gamma\left(m_{sh}^I\right)}$$

$$\cdot\left(m_{mul2}^I/\Omega_{mul2}^I\right)^{m_{mul2}^I}\left(m_{mul2}^I/\Omega_{mul2}^I\right)^{m_{mul2}^I}\left(m_{sh}^I/\Omega_{sh}^I\right)^{m_{sh}^I}\left(\frac{\Gamma\left(m_{mul1}^d\right)\Gamma\left(m_{mul2}^d\right)\Gamma\left(m_{mul1}^I\right)}{32\left(\frac{m_{mul1}^d}{\Omega_{mul1}^d}\right)^{m_{mul1}^d}\left(\frac{m_{mul2}^d}{\Omega_{mul2}^d}\right)^{m_{mul2}^d}}\right.$$

(2.22)

$$\cdot\frac{\Gamma\left(m_{mul2}^I\right)\Gamma\left(m_{sh}^I\right)}{\left(\frac{m_{mul1}^I}{\Omega_{mul1}^I}\right)^{m_{mul1}^I}\left(\frac{m_{mul2}^I}{\Omega_{mul2}^I}\right)^{m_{mul2}^I}\left(\frac{m_{sh}^I}{\Omega_{sh}^I}\right)^{m_{sh}^I}} - \sum_{k=0}^{m_{sh}^d-1}\frac{\left(\frac{m_{sh}^d}{\Omega_{sh}^d}z_{R(DGG/G)}\right)^k}{k!}J_6\right)\left(m_{sh}^d-1\right)!$$

where,

$$J_6 = \int_0^\infty dx_{N1}\int_0^\infty dx_{N2}\int_0^\infty dx_{N4}\int_0^\infty dx_{N5}$$

$$\cdot\int_0^\infty e^{-\frac{m_{mul1}^d x_{N1}^2}{\Omega_{mul1}^d}-\frac{m_{mul2}^d x_{N2}^2}{\Omega_{mul2}^d}-\frac{m_{sh}^d z_{R(DGG/G)}x_{N4}^{2/\alpha}x_{N5}^{2/\alpha}x_{N6}^2}{\Omega_{sh}^d x_{N1}^{2/\alpha}x_{N2}^{2/\alpha}}-\frac{m_{mul1}^I x_{N4}^2}{\Omega_{mul1}^I}-\frac{m_{mul2}^I x_{N5}^2}{\Omega_{mul2}^I}-\frac{m_{sh}^I x_{N6}^2}{\Omega_{sh}^I}}$$

$$\cdot e^{\left(2m_{mul1}^d-\frac{2}{a}k-1\right)\ln(x_{N1})+\left(2m_{mul2}^d-\frac{2}{a}k-1\right)\ln(x_{N2})+\left(2m_{mul1}^I+\frac{2}{a}k-1\right)\ln(x_{N4})}$$

$$\cdot e^{\left(2m_{mul2}^I+\frac{2}{a}k-1\right)\ln(x_{N5})+\left(2m_{sh}^I+2k-1\right)\ln(x_{N6})}dx_{N6}$$

where m_{sh}^d is positive integer. The closed-form approximation whose derivation is provided in the appendix is also obtained by the LABM through direct application of Equation (2.22). The outage probability over composite fading in the presence of interference modeled as the ratio of two DGG/G RPs can be expressed as $F_{Z_{R(DGG/G)}}\left(z_{th,\,R(DGG/G)}\right)$ for a given threshold $z_{th,\,R(DGG/G)}$.

2.4.2 Second-Order Statistics Measures

The LCR of the $z_{th,\,R(DGG/G)}$ can be evaluated as:

$$N_{Z_{R(DGG/G)}}\left(z_{th,R(DGG/G)}\right) = \int_0^\infty \dot{z}_{R(DGG/G)} p_{Z_{R(DGG/G)}\dot{Z}_{R(DGG/G)}}\left(z_{th,R(DGG/G)}\dot{z}_{R(DGG/G)}\right) d\dot{z}_{R(DGG/G)} \quad (2.23)$$

The $N_{Z_{R(DGG/G)}}\left(z_{th,R(DGG/G)}\right)$ derivation is presented in the appendix. Here, we present the fivefold integral expression for $N_{Z_{R(DGG/G)}}\left(z_{th,R(DGG/G)}\right)$:

$$N_{Z_{R(DGG/G)}}\left(z_{th,R(DGG/G)}\right) = \frac{64\sigma_{\dot{x}N1}\left(m_{mul1}^d/\Omega_{mul1}^d\right)^{m_{mul1}^d}\left(m_{mul2}^d/\Omega_{mul2}^d\right)^{m_{mul2}^d}}{a\sqrt{2\pi}\ \Gamma\left(m_{mul1}^d\right)\Gamma\left(m_{mul2}^d\right)\Gamma\left(m_{sh}^d\right)\Gamma\left(m_{mul1}^l\right)}$$
$$\cdot\frac{\left(m_{sh}^d/\Omega_{sh}^d\right)^{m_{sh}^d}}{\Gamma\left(m_{mul2}^l\right)\Gamma\left(m_{sh}^l\right)}\left(\frac{m_{mul1}^l}{\Omega_{mul1}^l}\right)^{m_{mul1}^l}\left(\frac{m_{mul2}^l}{\Omega_{mul2}^l}\right)^{m_{mul2}^l}\left(\frac{m_{sh}^l}{\Omega_{sh}^l}\right)^{m_{sh}^l} z_{th,R(DGG/G)}^{m_{sh}^d} J_7 \quad (2.24)$$

where,

$$J_7 = \int_0^\infty dx_{N1} \int_0^\infty dx_{N2} \int_0^\infty dx_{N4} \int_0^\infty dx_{N5} \int_0^\infty \left(1+\frac{x_{N1}^2}{x_{N2}^2}\frac{\sigma_{X_{N2}}^2}{\sigma_{X_{N1}}^2}+\alpha^2\right.$$
$$\cdot\frac{x_{N1}^{2(1/\alpha+1)}x_{N2}^{2/\alpha}}{z_{R(DGG/G)}x_{N4}^{2/\alpha}x_{N5}^{2/\alpha}x_{N6}^2}\frac{\sigma_{X_{N3}}^2}{\sigma_{X_{N1}}^2}+\frac{x_{N1}^2}{x_{N4}^2}\frac{\sigma_{X_{Nm4}}^2}{\sigma_{X_{N1}}^2}+\frac{x_{N1}^2}{x_{N5}^2}\frac{\sigma_{X_{N5}}^2}{\sigma_{X_{N1}}^2}+\alpha^2\left.\frac{x_{N1}^2}{x_{N6}^2}\frac{\sigma_{X_{N6}}^2}{\sigma_{X_{N1}}^2}\right)^{\frac{1}{2}}$$
$$\cdot e^{\frac{m_{mul1}^d x_{N1}^2}{\Omega_{mul1}^d}-\frac{m_{mul2}^d x_{N2}^2}{\Omega_{mul2}^d}-\frac{m_{sh}^d z_{R(DGG/G)}x_{N4}^{2/\alpha}x_{N5}^{2/\alpha}x_{N6}^2}{\Omega_{sh}^d x_{N1}^{2/\alpha}x_{N2}^{2/\alpha}}-\frac{m_{mul1}^l x_{N4}^2}{\Omega_{mul1}^l}-\frac{m_{mul2}^l x_{N5}^2}{\Omega_{mul2}^l}-\frac{m_{sh}^l x_{N6}^2}{\Omega_{sh}^l}}$$
$$\cdot e^{\left(2m_{mul1}^d-\frac{2}{a}m_{sh}^d-1\right)\ln(x_{N1})+\left(2m_{mul2}^d-\frac{2}{a}m_{sh}^d-1\right)\ln(x_{N2})+\left(2m_{mul1}^l+\frac{2}{a}m_{sh}^d-1\right)\ln(x_{N4})}$$
$$\cdot e^{\left(2m_{mul2}^l+\frac{2}{a}m_{sh}^d-1\right)\ln(x_{N5})+\left(2m_{sh}^l+2m_{sh}^d-1\right)\ln(x_{N6})} dx_{N6} \quad (2.25)$$

In order to calculate closed-form LCR of $z_{R(DGG/G)}$, we applied LABM to evaluate fivefold integral expression whose derivation is also given in the appendix.

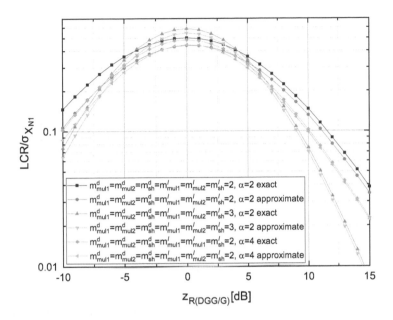

FIGURE 2.7

Normalized LCR of $z_{th,R(DGG/G)}$ for exact and approximate results regarding different fading severity and non-linearity parameters for $\Omega i=1$. LCR, level crossing rate.

Figure 2.7 shows $N_{Z_{R(DGG/G)}}\left(z_{th,R(DGG/G)}\right)$ versus threshold $z_{th,R(DGG/G)}$ for various sets of system model parameters. It is noticeable that increase in desired as well as in interference severity parameters causes $N_{Z_{R(DGG/G)}}\left(z_{th,R(DGG/G)}\right)$ decreases. Moreover, further decrease in $N_{Z_{R(DGG/G)}}\left(z_{th,R(DGG/G)}\right)N_{Z_{R(DGG/G)}}\left(z_{th,R(DGG/G)}\right)$ is provided for higher values of nonlinearity parameter, as expected.

AFD is then simply evaluated as

$$\text{AFD}_{Z_{R(DGG/G)}}\left(z_{th,R(DGG/G)}\right) = \frac{F_{Z_{R(DGG/G)}}\left(z_{th,R(DGG/G)}\right)}{N_{Z_{R(DGG/G)}}\left(z_{th,R(DGG/G)}\right)} \tag{2.26}$$

The $\text{AFD}_{Z_{R(DGG/G)}}\left(z_{th,R(DGG/G)}\right)$ for the limited interference scenario of the ratio of two DGG/G RPs is obtained as exact as well as fast computing approximate expression whose derivation is given in the appendix.

Figure 2.8 presents $\text{AFD}_{Z_{R(DGG/G)}}\left(z_{th,R(DGG/G)}\right)$ behavior in terms of $m^d_{mul1}, m^d_{mul2},$ $m^d_{sh}, m^I_{mul1}, m^I_{mul2}, m^I_{sh}, m^I_{mul1},$ and α. It can be concluded that approximations fit well with exact integral form results. Moreover, the system performance improvement in terms of $\text{AFD}_{Z_{R(DGG/G)}}\left(z_{th,R(DGG/G)}\right)$ for lower $z_{th,R(DGG/G)}$ dB values can be achieved for higher values of severity and nonlinearity parameters.

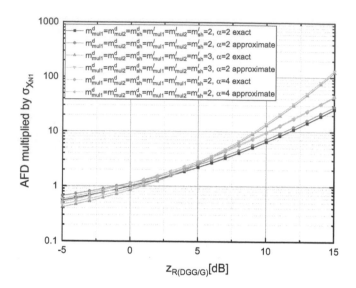

FIGURE 2.8

AFD of $z_{th,R(DGG/G)}$ for exact and approximate results regarding various severity and nonlinearity parameters for $\Omega i=1$. AFD, average fade duration.

2.5 Conclusions

A unified framework for channel modeling and statistical performance evaluation based on Laplace approximation method for envisioned high-mobility 5G and beyond 5G communications has been developed. We model fading signals as DGG as well as the product of DGG and G (denoted as DGG/G) RPs in order to incorporate double scattered and composite fading into the proposed model for high-mobility 5G systems. Moreover, ratio of two DGG/G is considered in order to examine interference-limited scenario over composite fading. The first- and higher-order exact and approximate statistics are efficiently derived and presented for various sets of fading severities and nonlinearity parameters. The numerical results show that accurate fit between approximate and exact analytical results for considered metrics has been achieved especially for higher output dB signal values. Finally, one can conclude that in order to model complex propagation environments, more random variables are needed, which in turn leads to complex formulas for performance evaluation. In general, by comparing considered scenarios, modeling with more random variables leads to LCR increase by reducing the system parameters impact. On the other hand, more random variables cause AFD increase for lower threshold values, whereas AFD decreases for higher threshold values, by reducing the impact of the system parameters.

Appendix

Here, we provide detailed derivation of all provided analytical results.

A.1 Derivation of Level Crossing Rate for Double Generalized Gamma Random Process

The $N_{Z_{DGG}}(z_{th,DGG})$ provided can be easily obtained through averaging of conditional and joint PDFs:

$$p_{Z_{DGG}\dot{Z}_{DGG}}(z_{DGG}\dot{z}_{DGG}) = \int_0^\infty p_{Z_{DGG}\dot{Z}_{DGG}X_{N2}}(z_{DGG}\dot{z}_{DGG}x_{N2})dx_{N2}$$

$$= \int_0^\infty p_{\dot{Z}_{DGG}|Z_{DGG}X_{N2}}(\dot{z}_{DGG}|z_{DGG}x_{N2})p_{Z_{DGG}|X_{N2}}(z_{DGG}|x_{N2})p_{X_{N2}}(x_{N2})dx_{N2}$$

$$(2.27)$$

After a simple transformation of $p_{Z_{DGG}|X_{N2}}(z_{DGG}|x_{N2}) = \left|\dfrac{dx_{N1}}{dz_{DGG}}\right|p_{X_{N1}}\left(\dfrac{z_{DGG}^{\alpha/2}}{x_{N2}}\right)$, we obtain:

$$p_{Z_{DGG}|X_{N2}}(z_{DGG}|x_{N2}) = \frac{\alpha z_{DGG}^{\alpha/2-1}}{2x_{N2}}p_{X_{N1}}\left(\frac{z_{DGG}^{\alpha/2}}{x_{N2}}\right) \qquad (2.28)$$

Furthermore, for $p_{\dot{Z}_{DGG}|Z_{DGG}X_{N2}}(\dot{z}_{DGG}|z_{DGG}x_{N2})$, we observe that, for given values of $x_{N2},\dot{z}_{DGG}$ and z_{DGG} the expression is a zero mean Gaussian RP, whose variance can be expressed as:

$$\sigma_{\dot{z}_{DGG}}^2 = \frac{4x_{N2}^2\sigma_{\dot{X}_{N1}}^2}{\alpha^2 z_{DGG}^{\alpha-2}}\left(1+\frac{z_{DGG}^\alpha}{x_{N2}^4}\frac{\sigma_{\dot{X}_{N2}}^2}{\sigma_{\dot{X}_{N1}}^2}\right) \qquad (2.29)$$

Since,

$$\int_0^\infty \dot{z}_{DGG}p_{\dot{Z}_{DGG}|Z_{DGG}X_{N2}}(\dot{z}_{DGG}|z_{DGG}x_{N2})d\dot{z}_{DGG} = \frac{\sigma_{\dot{z}_{DGG}}}{\sqrt{2\pi}},$$

the $N_{Z_{DGG}}$ for a given threshold, $z_{th,DGG}$ after adequate substitutions is expressed in Equation (2.7). The integral J_1 can be solved by applying the LABM for onefold integrals (Zlatanov et al., 2008):

$$\int_0^\infty g(x_{N2})e^{-\gamma f(x_{N2})}dx_{N2} \approx \sqrt{\frac{2\pi}{\gamma}}\frac{g(x_{N20})}{\sqrt{\dfrac{d^2 f(x_{N20})}{dx_{N20}^2}}}e^{-\gamma f(x_{N20})} \qquad (2.30)$$

where the arguments in previous equation are, respectively, $\gamma=1, g(x_{N2}) = \sqrt{1+\dfrac{\sigma_{\dot{X}_{N2}}^2 z_{th,DGG}^\alpha}{\sigma_{\dot{X}_{N1}}^2 x_{N2}^4}}$

$$f(x_{N2}) = \frac{m_1 z_{th,DGG}^\alpha}{\Omega_1 x_{N2}^2}+\frac{m_2}{\Omega_2}x_{N2}^2 - 2(m_2-m_1)\ln(x_{N2}) \qquad (2.31)$$

while x_{N20} is obtained from $\dfrac{df(x_{N20})}{dx_{N20}} = 0$ and is provided in Equation (2.8).

A.2 Derivation of Probability Density Function, Cumulative Density Function and Level Crossing Rate of Double Generalized Gamma/Gamma Random Process

In order to calculate closed-form PDF, we transform J_2 in Equation (2.12) as a twofold integral of exponential function and then apply exponential LABM (Hadzi-Velkov and Karagiannidis, 2008):

$$\int_0^\infty dx_{N1} \int_0^\infty f_1(x_{N1}, x_{N2}) e^{\lambda f_2(x_{N1}, x_{N2})} dx_{N2} \approx \sqrt{\frac{2\pi}{\gamma}} \frac{f_1(x_{N10}, x_{N20})}{\sqrt{\det H}} e^{-\gamma f_2(x_{N10}, x_{N20})} \qquad (2.32)$$

where, H is Hessian matrix given as:

$$H = \begin{vmatrix} \dfrac{\partial^2 f_2(x_{N10}, x_{N20})}{\partial x_{N10}^2} & \dfrac{\partial^2 f_2(x_{N10}, x_{N20})}{\partial x_{N10} \partial x_{N20}} \\[4mm] \dfrac{\partial^2 f_2(x_{N10}, x_{N20})}{\partial x_{N20} \partial x_{N10}} & \dfrac{\partial^2 f_2(x_{N10}, x_{N20})}{\partial x_{N20}^2} \end{vmatrix} \qquad (2.33)$$

and where x_{N10} and x_{N20} are positive and real values obtained from the following set of expressions:

$$\frac{\partial f_2(x_{N10}, x_{N20})}{\partial x_{Ni0}} = 0, \; i = 1,2 \qquad (2.34)$$

The exponential LABM for constant multivariate function f_1 and nonconstant multivariate function f_2 has been efficiently considered in Wang (2010). Thus, the arguments of exponential LABM in J_2 are taken as $\gamma = 1$, $f_1(x_{N1}, x_{N2}) = 1$,

$$\begin{aligned} f_2(x_{N1}, x_{N2}) &= \frac{m_{\text{mul1}}}{\Omega_{\text{mul1}}} x_{N1}^2 + \frac{m_{\text{mul2}}}{\Omega_{\text{mul2}}} x_{N2}^2 + \frac{m_{\text{sh}}}{\Omega_{\text{sh}}} \frac{z_{\text{DGG/G}}}{x_{N1}^{2/\alpha} x_{N2}^{2/\alpha}} \\ &- \left(2m_{\text{mul1}} - \frac{2m_{\text{sh}}}{a} - 1\right) \ln x_{N1} - \left(2m_{\text{mul2}} - \frac{2m_{\text{sh}}}{a} - 1\right) \ln x_{N2} \end{aligned} \qquad (2.35)$$

The J_3 in Equation (2.13) for CDF exact analytical expression is further evaluated by exponential LABM, which is already provided in Equation (2.32). The arguments in Equation (2.13) for the particular case are $\lambda = 1$, $f_1(x_{N1}, x_{N2}) = 1$,

$$\begin{aligned} f_2(x_{N1}, x_{N2}) &= \frac{m_{\text{mul1}}}{\Omega_{\text{mul1}}} x_{N1}^2 + \frac{m_{\text{mul2}}}{\Omega_{\text{mul2}}} x_{N2}^2 + \frac{m_{\text{sh}}}{\Omega_{\text{sh}}} \frac{z_{\text{DGG/G}}}{x_{N1}^{2/\alpha} x_{N2}^{2/\alpha}} \\ &- \left(2m_{\text{mul1}} - \frac{2k}{a} - 1\right) \ln x_{N1} - \left(2m_{\text{mul2}} - \frac{2k}{a} - 1\right) \ln x_{N2} \end{aligned} \qquad (2.36)$$

The LCR of DGG/G RP can be derived as previously through some mathematical manipulations based on joint and conditional PDFs:

$$p_{Z_{DGG/G}\,\dot{Z}_{DGG/G}\,x_{N1}\,x_{N2}}\left(z_{DGG/G}\,\dot{z}_{DGG/G}\,x_{N1}\,x_{N2}\right)=p_{\dot{z}_{DGG/G}|Z_{DGG/G}\,x_{N1}\,x_{N2}}\left(\dot{z}_{DGG/G}\mid z_{DGG/G}\,x_{N1}\,x_{N2}\right)$$

$$p_{Z_{DGG/G}\mid x_{N1}\,x_{N2}}\left(z_{DGG/G}\mid x_{N1}\,x_{N2}\right)p_{x_{N1}}\left(x_{N1}\right)p_{x_{N2}}\left(x_{N2}\right) \tag{2.37}$$

where

$$p_{Z_{DGG/G}\mid x_{N1}\,x_{N2}}\left(z_{DGG/G}\mid x_{N1}\,x_{N2}\right)=\left|\frac{dx_{N3}}{dZ_{DGG/G}}\right|p_{x_{N3}}\left(\frac{z_{DGG}^{1/2}}{x_{N1}^{1/\alpha}x_{N2}^{1/\alpha}}\right) \tag{2.38}$$

The joint PDF $p_{Z_{DGG/G}\dot{Z}_{DGG/G}}\left(z_{DGG/G},\dot{z}_{DGG/G}\right)$ after simple mathematical transformations can be expressed as:

$$p_{Z_{DGG/G}\dot{Z}_{DGG/G}}\left(z_{DGG/G},\dot{z}_{DGG/G}\right)=$$

$$\int_{0}^{\infty}dx_{N1}\int_{0}^{\infty}p_{Z_{DGG/G}\,\dot{Z}_{DGG/G}x_{N1}x_{N2}}\left(z_{DGG/G}\dot{z}_{DGG/G}x_{N1}x_{N2}\right)dx_{N2} \tag{2.39}$$

The variance of $\dot{z}_{DGG/G}$ is a zero mean Gaussian RP and can be expressed through the variances of $\dot{x}_{N1},\dot{x}_{N2}$ and $\dot{x}_{N3}$.

$$\sigma_{\dot{z}_{DGG/G}}^{2}=\frac{4\,\sigma_{\dot{x}_{N1}}^{2}}{\alpha^{2}\,x_{N1}^{2}}\left(1+\frac{x_{N1}^{2}}{x_{N2}^{2}}\frac{\sigma_{\dot{x}_{N2}}^{2}}{\sigma_{\dot{x}_{N1}}^{2}}+\frac{x_{N1}^{2/\alpha+2}x_{N2}^{2/\alpha}}{z_{DGG/G}}\frac{\sigma_{\dot{x}_{N3}}^{2}}{\sigma_{\dot{x}_{N1}}^{2}}\right)^{\frac{1}{2}} \tag{2.40}$$

Finally, after adequate substitutions, and after evaluating the integral,

$$\int_{0}^{\infty}\dot{z}_{DGG/G}p_{\dot{z}_{DGG/G}|Z_{DGG/G}\,x_{N1}\,x_{N2}}\left(\dot{z}_{DGG/G}\mid z_{DGG/G}\,x_{N1}\,x_{N2}\right)d\dot{z}_{DGG/G}=\frac{\sigma_{\dot{z}_{DGG/G}}}{\sqrt{2\pi}}$$

the exact $N_{Z_{DGG/G}}\left(z_{th,DGG/G}\right)$ is given in Equation (2.15).

The LABM for twofold integrals, given in Equation (2.32) is applied again for evaluation of J_4, where the LABM parameters are, respectively, $\gamma=1$,

$$f_1\left(x_{Nm2},x_{Nm3},x_{Nm4}\right)=\left(1+\frac{x_{N1}^{2}}{x_{N2}^{2}}\frac{\sigma_{\dot{x}_{N2}}^{2}}{\sigma_{\dot{x}_{N1}}^{2}}+\frac{x_{N1}^{2/\alpha+2}x_{N1}^{2/\alpha}}{z_{DGG/G}}\frac{\sigma_{\dot{x}_{N3}}^{2}}{\sigma_{\dot{x}_{N1}}^{2}}\right)^{\frac{1}{2}} \tag{2.41}$$

$$f_2\left(x_{N1},x_{N2}\right)=\frac{m_{mul1}}{\Omega_{mul1}}x_{N1}^{2}+\frac{m_{mul2}}{\Omega_{mul2}}x_{N2}^{2}+\frac{m_{sh}}{\Omega_{sh}}\frac{z_{DGG/G}}{x_{N1}^{2/\alpha}x_{N2}^{2/\alpha}}$$

$$-\left(2m_{mul1}-\frac{2m_{sh}}{\alpha}-2\right)\ln x_{N1}-\left(2m_{mul2}-\frac{2m_{sh}}{\alpha}-1\right)\ln x_{N2} \tag{2.42}$$

where $f_2(x_{N1}, x_{N2})$ is taken from Equation (2.16) and expressed as fully exponential function. Furthermore, the Hessian matrix H, x_{N10} and x_{N20} for the considered integral J_4 are calculated by substituting Equation (2.42) in Equations (2.33) and (2.34), respectively.

A.3 Derivation of Probability Density Function, Cumulative Density and Level Crossing Rate of the Ratio of Two Double Generalized Gamma/Gamma Random Processes

In order to calculate closed-form PDF of $z_{R(DGG/G)}$, we applied exponential LABM to evaluate fivefold integral given as J_5:

$$\int_0^\infty dx_{N1} \int_0^\infty dx_{N2} \int_0^\infty dx_{N4} \int_0^\infty dx_{N5} \int_0^\infty f_1(x_{N1}, x_{N2}, x_{N4}, x_{N5}, x_{N6}) e^{-\gamma f_2(x_{N1}, x_{N2}, x_{N4}, x_{N5}, x_{N6})} dx_{N6}$$

$$\approx \left(\frac{2\pi}{\gamma}\right)^{\frac{5}{2}} \frac{f_1(x_{N10}, x_{N20}, x_{N40}, x_{N50}, x_{N60})}{\sqrt{\det H}} e^{-\gamma f_2(x_{N10}, x_{N20}, x_{N40}, x_{N50}, x_{N60})} \tag{2.43}$$

where the arguments for the particular case are, respectively, $\gamma = 1$, $f_1(x_{N1}, x_{N2}, x_{N4}, x_{N5}, x_{N6}) = 1$,

$$f_2(x_{N1}, x_{N2}, x_{N4}, x_{N5}, x_{N6}) = \frac{m_{mul1}^d x_{N1}^2}{\Omega_{mul1}^d} + \frac{m_{mul2}^d x_{N2}^2}{\Omega_{mul2}^d} + \frac{m_{sh}^d z_{R(DGG/G)} x_{N4}^{2/\alpha} x_{N5}^{2/\alpha} x_{N6}^2}{\Omega_{sh}^d x_{N1}^{\frac{2}{\alpha}} x_{N2}^{\frac{2}{\alpha}}}$$

$$+ \frac{m_{mul1}^I x_{N4}^2}{\Omega_{mul1}^I} + \frac{m_{mul2}^I x_{N5}^2}{\Omega_{mul2}^I} + \frac{m_{sh}^I x_{N6}^2}{\Omega_{sh}^I} - \left(2m_{mul1}^d - \frac{2}{\alpha} m_{sh}^d - 1\right) \ln(x_{N1})$$

$$- \left(2m_{mul2}^d - \frac{2}{\alpha} m_{sh}^d - 1\right) \ln(x_{N2}) - \left(2m_{mul1}^I + \frac{2}{\alpha} m_{sh}^d - 1\right) \ln(x_{N4})$$

$$- \left(2m_{mul2}^I + \frac{2}{\alpha} m_{sh}^d - 1\right) \ln(x_{N5}) - \left(2m_{sh}^I + 2m_{sh}^d - 1\right) \ln(x_{N6}) \tag{2.44}$$

Furthermore, $x_{N10}, x_{N20}, x_{N40}, x_{N50}, x_{N60}$ and H can be calculated from the following equations as, respectively:

$$\frac{\partial f_2(x_{N10}, x_{N20}, x_{N40}, x_{N50}, x_{N60})}{\partial x_{Ni0}} = 0, \ i = 1, 2, 4, 5, 6 \tag{2.45}$$

$$H = \begin{vmatrix} \dfrac{\partial^2 f_2(x_{N10}, ..., x_{N60})}{\partial x_{N10}^2} & \dfrac{\partial^2 f_2(x_{N10}, ..., x_{N60})}{\partial x_{N10} \partial x_{N20}} & \cdots & \dfrac{\partial^2 f_2(x_{N10}, ..., x_{N60})}{\partial x_{N10} \partial x_{N60}} \\[2ex] \dfrac{\partial^2 f_2(x_{N10}, ..., x_{N60})}{\partial x_{N20} \partial x_{N10}} & \dfrac{\partial^2 f_2(x_{N10}, ..., x_{N60})}{\partial x_{N20}^2} & \cdots & \dfrac{\partial^2 f_2(x_{N10}, ..., x_{N60})}{\partial x_{N20} \partial x_{N60}} \\[2ex] \vdots & \vdots & \vdots & \vdots \\[2ex] \dfrac{\partial^2 f_2(x_{N10}, ..., x_{N60})}{\partial x_{N60} \partial x_{N10}} & \dfrac{\partial^2 f_2(x_{N10}, ..., x_{N60})}{\partial x_{N60} \partial x_{N20}} & \cdots & \dfrac{\partial^2 f_2(x_{N10}, ..., x_{N60})}{\partial x_{N60}^2} \end{vmatrix} \tag{2.46}$$

The LCR of the ratio of two DGG/G RPs can be derived relaying on mathematical framework based on joint and conditional PDFs of i.n.i.d RPs, $Z_{R(DGG/G)}$, $\dot{Z}_{R(DGG/G)}$, $X_{N1}, X_{N2}, X_{N4}, X_{N5}$ and X_{N6} that can be expressed as:

$$p_{Z_{R(DGG/G)} \dot{Z}_{R(DGG/G)} X_{N1} X_{N2} X_{N4} X_{N5} X_{N6}} \left(z_{R(DGG/G)} \dot{z}_{R(DGG/G)} x_{N1} x_{N2} x_{N4} x_{N5} x_{N6} \right)$$

$$= p_{\dot{Z}_{R(DGG/G)}|Z_{R(DGG/G)} X_{N1} X_{N2} X_{N4} X_{N5} X_{N6}} \left(\dot{z}_{R(DGG/G)} \mid z_{R(DGG/G)} x_{N1} x_{N2} x_{N4} x_{N5} x_{N6} \right)$$

$$\cdot p_{Z_{R(DGG/G)}|X_{N1} X_{N2} X_{N4} X_{N5} X_{N6}} \left(z_{R(DGG/G)} \mid x_{N1} x_{N2} x_{N4} x_{N5} x_{N6} \right)$$

$$\cdot p_{X_{N1}}(x_{N1}) \, p_{X_{N2}}(x_{N2}) p_{X_{N4}}(x_{N4}) p_{X_{N5}}(x_{N5}) p_{X_{N6}}(x_{N6})$$

(2.47)

where after the simple transformation of the conditional probability function we obtained

$$p_{Z_{R(DGG/G)}|X_{N1} X_{N2} X_{N4} X_{N5} X_{N6}} = \left| \frac{dx_{N3}}{dz_{R(DGG/G)}} \right| p_{X_{N3}} \left(\frac{\sqrt{z_{R(DGG/G)}} x_{N4}^{1/\alpha} x_{N5}^{1/\alpha} x_{N6}}{x_{N1}^{1/\alpha} x_{N2}^{1/\alpha}} \right).$$

Since the variance of the first derivative of zero-mean Gaussian RP is zero-mean Gaussian RP as well as the linear transformation of the Gaussian RPs is also zero-mean Gaussian RP, the variance of $\dot{Z}_{R(DGG/G)}$, after simple manipulation, can be expressed through the variances of $\dot{X}_{N,i}$, $i = 1,6$:

$$\sigma_{\dot{Z}_{R(DGG/G)}}^2 = \frac{4 z_{R(DGG/G)}^2}{\alpha^2 x_{N1}^2} \sigma_{\dot{X}_{N1}}^2 \left(1 + \frac{x_{N1}^2}{x_{N2}^2} \frac{\sigma_{\dot{X}_{N2}}^2}{\sigma_{\dot{X}_{N1}}^2} + \alpha^2 \frac{x_{N1}^{2(1/\alpha+1)} x_{N2}^{2/\alpha}}{z_{R(DGG/G)} x_{N4}^{2/\alpha} x_{N5}^{2/\alpha} x_{N6}^2} \right.$$

$$\left. \cdot \frac{\sigma_{\dot{X}_{N3}}^2}{\sigma_{\dot{X}_{N1}}^2} + \frac{x_{N1}^2}{x_{N4}^2} \frac{\sigma_{\dot{X}_{Nm4}}^2}{\sigma_{\dot{X}_{N1}}^2} + \frac{x_{N1}^2}{x_{N5}^2} \frac{\sigma_{\dot{X}_{N5}}^2}{\sigma_{\dot{X}_{N1}}^2} + a^2 \frac{x_{N1}^2}{x_{N6}^2} \frac{\sigma_{\dot{X}_{N6}}^2}{\sigma_{\dot{X}_{N1}}^2} \right)$$

(2.48)

Since,

$$\int_0^\infty \dot{Z}_{R(DGG/G)} p_{\dot{Z}_{R(DGG/G)}|Z_{R(DGG/G)} X_{N1} X_{N2} X_{N4} X_{N5} X_{N6}} d\dot{Z}_{R(DGG/G)} = \frac{1}{\sqrt{2\pi}} \sigma_{\dot{Z}_{R(DGG/G)}},$$

the exact integral form $N_{Z_{R(DGG/G)}} \left(z_{th,R(DGG/G)} \right)$ is obtained after some algebra and presented in Equation (2.24). Application of LABM to evaluate fivefold integral expression from equations is already provided in (2.43) where the arguments the particular case are, respectively, $\gamma = 1$,

$$f_1 \left(x_{N1}, x_{N2}, x_{N4}, x_{N5}, x_{N6} \right) = \left(1 + \frac{x_{N1}^2}{x_{N2}^2} \frac{\sigma_{\dot{X}_{N2}}^2}{\sigma_{\dot{X}_{N1}}^2} + \alpha^2 \frac{x_{N1}^{2(1/\alpha+1)} x_{N2}^{2/\alpha} x_{N4}^{2/\alpha} x_{N5}^{2/\alpha} x_{N6}^2}{z_{P(DGG/G)}} \right.$$

(2.49)

$$\left. \cdot \frac{\sigma_{\dot{X}_{N3}}^2}{\sigma_{\dot{X}_{N1}}^2} + \frac{x_{N1}^2}{x_{N4}^2} \frac{\sigma_{\dot{X}_{N4}}^2}{\sigma_{\dot{X}_{N1}}^2} + \frac{x_{N1}^2}{x_{N5}^2} \frac{\sigma_{\dot{X}_{N5}}^2}{\sigma_{\dot{X}_{N1}}^2} + \alpha^2 \frac{x_{N1}^2}{x_{N5}^2} \frac{\sigma_{\dot{X}_{N6}}^2}{\sigma_{\dot{X}_{N1}}^2} \right)^{\frac{1}{2}}$$

$$f_2\left(x_{N1},x_{N2},x_{N4},x_{N5},x_{N6}\right) = \frac{m_{\text{mul1}}^d x_{N1}^2}{\Omega_{\text{mul1}}^d} + \frac{m_{\text{mul2}}^d x_{N2}^2}{\Omega_{\text{mul2}}^d} + \frac{m_{\text{sh}}^d z_{P(\text{DGG/G})}}{\Omega_{\text{sh}}^d x_{N1}^{2/\alpha} x_{N2}^{2/\alpha} x_{N4}^{2/\alpha} x_{N5}^{2/\alpha} x_{N6}^2}$$

$$+\frac{m_{\text{mul1}}^l x_{N4}^2}{\Omega_{\text{mul1}}^l} + \frac{m_{\text{mul2}}^l x_{N5}^2}{\Omega_{\text{mul2}}^l} + \frac{m_{\text{sh}}^l x_{N6}^2}{\Omega_{\text{sh}}^l} - \left(2m_{\text{mul1}}^d - \frac{2}{a}m_{\text{sh}}^d - 1\right)\ln\left(x_{N1}\right)$$

$$-\left(2m_{\text{mul2}}^d - \frac{2}{a}m_{\text{sh}}^d - 1\right)\ln\left(x_{N2}\right) - \left(2m_{\text{mul1}}^l - \frac{2}{a}m_{\text{sh}}^d - 1\right)\ln\left(x_{N4}\right)$$

$$-\left(2m_{\text{mul2}}^l - \frac{2}{a}m_{\text{sh}}^d - 1\right)\ln\left(x_{N5}\right) - \left(2m_{\text{sh}}^l - 2m_{\text{sh}}^d - 1\right)\ln\left(x_{N6}\right)$$

(2.50)

References

Al-Ahmadi, S. (2014). The gamma-gamma signal fading model: A survey, *IEEE Antennas and Propagation Magazine*, vol. 56, no. 5, 245–260.

Al-Hmood, H., Al-Raweshidy, H. (2017). Unified modeling of composite kappa-mu/gamma, eta-mu/gamma, and alpha-mu/gamma fading channels using a mixture gamma distribution with applications to energy detection, *IEEE Antennas and Wireless Propagation Letters*, vol. 16, 104–108.

Amer, M.A., Al-Dharrab, S. (2019). Performance of two-way relaying over α-μ fading channels in hybrid RF/FSO wireless networks, arXiv preprint arXiv:1911.05959.

Badarneh, O.S. (2016) The α-μ/α-μ composite multipath-shadowing distribution and its connection with the extended generalized K distribution, *AEU-International Journal of Electronics and Communications*, vol. 70, no. 9, 1211–1218.

Badarneh,O.S., Almehmadi, F.S. (2016). Performance of multihop wireless networks in α-μ fading channels perturbed by an additive generalized Gaussian noise, *IEEE Communications Letters*, vol. 20, no. 5, 986–989.

Bithas, P.S., Efthymoglou, G.P., Kanatas, A.G. (2018). V2V cooperative relaying communications under interference and outdated csi, *IEEE Transactions on Vehicular Technology*, vol. 67, no. 4, 3466–3480.

Bithas, P.S., Kanatas, A.G., da Costa, D.B., Upadhyay, P.K., Dias, U.S. (2018). On the double-generalized gamma statistics and their application to the performance analysis of V2V communications, *IEEE Transactions on Communications*, vol. 66, no. 1, 448–460.

Bithas, P.S., Nikolaidis, V., Kanatas, A.G. (2019). A new shadowed double-scattering model with application to uav-to-ground communications. In *Proceedings of the International Conference Wireless Communications and Networking Conference (WCNC)*, 1–6.

Bhargav, N., da Silva, C.R.N., Chun, Y.J., Leonardo, E.J., Cotton, S.L., Yacoub, M.D. (2018). On the product of two k-μ random variables and its application to double and composite fading channels, *IEEE Transactions on Wireless Communications*, vol. 17, no. 4, 2457–2470.

Butler, R.W., Wood, A.T. (2002). Laplace approximations for hypergeometric functions with matrix argument, *The Annals of Statistics*, vol. 30, no. 4, 1155–1177.

Da Silva, C.R.N., Bhargav, N., Leonardo, E.J., Yacoub, M.D. (2019), Ratio of two envelopes taken from α-μ, κ-μ, and η-μ variates and some practical applications, *IEEE Access*, vol. 7, 54449–54463.

Dautov, K., Arzykulov, S., Nauryzbayev, G., Kizilirmak, R.C. (2018). On the performance of UAV-enabled multihopV2V FSO systems over generalized α-μ channels, In *Proceedings of the International Conference on Computing and Network Communications (CoCoNet)*, 69–73.

Dos Anjos, A.A., Marins, T.R.R., Da Silva, C.R.N., Peñarrocha, V.M.R., Rubio, L., Reig, J., De Souza, R.A.A., Yacoub, M.D. (2019). Higher order statistics in a mmWave propagation environment, *IEEE Access*, vol. 7, 103876–103892.

Douik, A., Dahrouj, H., Al-Naffouri, T.Y., Alouini, M.S. (2016). Hybrid radio/free-space optical design for next generation backhaul systems, *IEEE Transactions on Communications*, vol. 64, no. 6, 2563–2577.

Goswami, A., Ashok, K. (2019). Performance analysis of wireless sensor networks ober α-μ/Gamma, η-μ/Gamma and κ-μ/Gamma composite channels. In *Proceedings of the International Conference on Computing, Communication and Networking Technologies (ICCCNT)*, 1–5.

Gradshteyn, I.S., Ryzhik, I.M. (2000). *Table of Integrals, Series, and Products,*6thed.New York: Academic.

Gustafson, C., Mahler, K., Bolin, D., Tufvesson, F. (2019). The COST IRACON geometry-based stochastic channel model for vehicle-to-vehicle communication in intersections, *IEEE Transactions on Vehicular Technology*, DOI:10.1109/TVT.2020.2964277.

He, R., Yang, M., Ai, B., Zhong, Z. (2019). Device-to-device channels, Wiley *5G Ref: The Essential 5G Reference Online*, 1–18.

Hadzi-Velkov, Z., Zlatanov, N., Karagiannidis, G.K. (2009). On the second order statistics of the multihop Rayleigh fading channel, *IEEE Transactions on Communications*, vol. 57, no. 6, 1815–1823.

Hajri, N., Youssef, N., Kawabata, T., Pätzold, M., Dahech, W. (2018). Statistical properties of double Hoyt fading with applications to the performance analysis of wireless communication systems, *IEEE Access*, vol.6, 19597–19609.

Jurado-Navas, A., GarridoBalsells, J.M., Castillo-Vazquez, M., Puerta-Notario, A., Monroy, I.T., Olmos, J.J.V. (2017). Fade statistics of M-turbulent optical links, *EURASIP Journal on Wireless Communications and Networking*, 2017, 112, DOI:10.1186/s13638–0170898–z.

Kashani, M.A., Uysal, M., Kavehrad, M. (2015). On the performance of MIMO FSO communications over double generalized gamma fading channels. In *Proceedings of the International Conference on Communications (ICC)*, 5144–5149.

Kim, S., Visotsky, E., Moorut, P., Bechta, K., Ghosh, A., Dietrich, C. (2017). Coexistence of 5G with the incumbents in the 28 and 70GHz bands, *IEEE Journal on Selected Areas in Communications*, vol. 35, no. 6, 1254–1268.

Kostić, I.M. (2005). Analytical approach to performance analysis for channel subject to shadowing and fading, *IEE Proceedings-Communications*, vol. 152, no. 6, 821–827.

Leonardo, E.J., Yacoub, M.D. (2015). Product of α-μ variates, *IEEE Communications Letters*, vol. 4, no. 6, 637–640.

Meng, X., Zhong, L., Zhou, D., Yang, D. (2019). Co-channel coexistence analysis between 5G IoT system and fixed-satellite service at 40GHz, *Wireless Communications and Mobile Computing*, vol. 2019, Article ID: 9790219, 1–9, https://doi.org/10.1155/2019/9790219

Milosevic, N., Stefanovic, C., Nikolic, Z., Bandjur, M., Stefanovic, M. (2018a). First-and second-order statistics of interference-limited mobile-to-mobile Weibull fading channel, *Journal of Circuits, Systems and Computers*, vol. 27, no. 11, 1850168.

Milosevic, N., Stefanovic, M., Nikolic, Z., Spalevic, P., Stefanovic, C. (2018b). Performance analysis of interference-limited mobile-to-mobile κ–μ fading channel, *Wireless Personal Communications*, vol. 101, no.3, 1685–1701.

Panic, S., Stefanovic, M., Anastasov, J., Spalevic, P., (2013). *Fading and Interference Mitigation in Wireless Communications*. New York: CRC Press.

Paris, J.F. (2013). Outage probability in η-μ/η-μ and κ-μ/η-μ interference limited scenarios, *IEEE Transactions on Communications*, vol. 61, no. 1, 335–343.

Rappaport, T.S., Xing, Y., Kanhere, O., Ju, S., Madanayake, A., Mandal, S., Trichopoulos, G.C. (2019). Wireless communications and applications above 100GHz: Opportunities and challenges for 6G and beyond, *IEEE Access*, vol. 7, 78729–78757.

Salous, S., Tufvesson, F., Turbic, K., Correia, L.M., Kürner, T., Dupleich, D., Schneider, C., Czaniera, D., Villacieros, B.M. (2019). IRACON propagation measurements and channel models for 5G and beyond, *ITU Journal: ICT Discoveries*, vol. 2, no. 1.

Shankar, P.M. (2004). Error rates in generalized shadowed fading channels, *Wireless Personal Communications*, vol. 28, no. 3, 233–238.

Simon, M.K., Alouini, M.-S. (2000). *Digital Communication over Fading Channels*, 1sted.New York: Wiley.

Stefanović, C (2017a). *Statističkekarakteristikeprvogidrugogredasignala u bežičnomtelekomunikacionomsistem usaselekcionimkombinovanjem* (Doctoral dissertation, Универзитет у Нишу, Електронскифакултет).

Stefanovic, C. (2017b). LCR of amplify and forward wireless relay systems in general alpha-Mu fading environment, In *Proceedings of the International Conference Telecommunication Forum (TELFOR)*, 1–6.

Stefanovic, C., Pratesi, M., Santucci, F. (2018). Performance evaluation of cooperative communications over fading channels in vehicular networks, In *Proceedings of the International Conference on Atlantic Science Radio Meeting (AT-RASC)*, 1–4.

Stefanovic, C., Pratesi, M., Santucci, F. (2019). Second order statistics of Mixed RF-FSO relay systems and its application to vehicular networks, In *Proceedings of the International Conference on Communications (ICC)*, 1–6.

Stefanovic, D., Stefanovic, C., Djosic, D., Milic, D., Rancic, D., Stefanovic, M. (2019). LCR of the ratio of the product of two squared Nakagami-m random processes and its application to wireless communication systems, In *Proceedings of the International Symposium INFOTEH-JAHORINA (INFOTEH)*,1–4.

Talha, B., Pätzold, M. (2011). Channel models for mobile-to-mobile cooperative communication systems: A state of the art review, *IEEE Vehicular Technology Magazine*, vol. 6, no. 2, 33–43.

Wang, C.X., Bian, J., Sun, J., Zhang, W., Zhang, M. (2018). A survey of 5G channel measurements and models, *IEEE Communications Surveys & Tutorials*, vol. 20, no. 4, 3142–3168.

Wang, C.X., Ghazal, A., Ai, B., Liu, Y., Fan, P. (2016). Channel measurements and models for high-speed train communication systems: A survey, *IEEE Communications Surveys &Tutorials*, vol. 18, no. 2, 974–987.

Wang, J. (2010). Dirichlet processes in nonlinear mixed effects models, *Communications in Statistics—Simulation and Computation*, vol. 39, no. 3, 539–556.

Wu, J., Fan, P. (2016). A survey on high mobility wireless communications: Challenges, opportunities and solutions, *IEEE Access*, vol. 4, 450–476.

Yacoub, M.D. (2007). The α-μ distribution: A physical fading model for the stacy distribution, *IEEE Transactions on Vehicular Technology*, vol. 56, no. 1, 27–34.

Yilmaz, F., Alouini, M.S. (2010). A new simple model for composite fading channels: Second order statistics and channel capacity, In *Proceedings of the International Symposium on Wireless Communication Systems*, 676–680.

Zlatanov, N., Hadzi-Velkov, Z., Karagiannidis, G.K. (2008).Level crossing rate and average fade duration of the double nakagami-m random process and application in MIMO keyhole fading channels, *IEEE Communications Letters*, vol. 12, no. 11, 822–824.

3

Massive MIMO Technologies and Beam-Division Multiple Access for 5G

Mussawir Ahmad Hosany

University of Mauritius

CONTENTS

3.1 Introduction

Wireless networks have recently deployed the multiple-input multiple-output (MIMO) technology to increase data rates and spectral as well as energy efficiencies. A MIMO system requires equipping the base station (BS) with multiple antennas, and a variety of MIMO systems have been implemented recently that give significant advantages as opposed to their single-input single-output (SISO) systems. In particular, MIMO provides diversity benefits and improved communication link reliability. Also, MIMO systems yield multiplexing gains by simultaneously combining various users over similar transmission resource. MIMO systems are presently deployed in 3G, 4G as well as LTE-advanced networks for both the time-division duplex (TDD) and frequency-division duplex (FDD) transmission modes [1].

Massive MIMO is a technology regarded as a fundamental 5G enabler to achieve high data rates with high spectral and energy efficiencies. Marzetta [2] introduced massive MIMO which is also known as very large MIMO systems, where a large number of antennas are implemented at the BSs serving single-antenna users. In the massive MIMO regime, a high number of BS antennas concentrate energy into smaller locations giving low interference levels so that all user terminals (UTs) can produce significant improvements in throughput [3–5].

In the following sections of this chapter, the state-of-the-art research on the principles and mathematical analysis of massive MIMO for 5G will be discussed, and the latest development in beam-division multiple access (BDMA) as a key enabler for 5G will be summarized.

3.2 Massive MIMO in 5G

Massive MIMO is the sub-6GHz physical layer technology for the upcoming 5G to achieve huge channel capacity and spectral efficiencies as well as energy efficiencies [6,7]. 5G will be operational in two layers, namely the macrolayer in the microwave bands to provide control signaling and a microlayer in the mmWave bands to carry user traffic. It has been shown that massive MIMO is very efficient at high-frequency bands and provides high antenna array gains to overwhelm path losses and give spatial multiplexing gains. Some typical massive MIMO antennas for the BS could range from 256 to 1024 for the mmWave bands.

The performance of 5G is largely dependent on the massive MIMO propagation channels. Some recent studies have already been carried out in the measurements of the propagation characteristics such as spatial nonstationarities, elevation characteristics and model simplifications. Transforming the measurement results from various hardware configurations into appropriate 5G models have been undergone in Refs. [8–14].

3.2.1 Mathematical Foundations

The operation of massive MIMO systems is accompanied by a mathematical analysis of the channel capacity. In a single-cell scenario, as the number of antenna elements at the BS grows large, the effect of interference vanishes, but in a multicell one, a unique problem arises called pilot contamination or intercell interference [15–20].

3.2.2 Massive MIMO System Model

A comprehensive mathematical analysis of massive MIMO under the TDD operation as introduced by Marzetta [2] is described in this section. Consider a scenario with the hexagonal cells where a very large number of antennas are located at the BSs and the UTs possessing only one receive antenna employing orthogonal frequency-division multiplexing (OFDM). Also, assume TDD protocol and fast fading overlaid by attenuation and log-norm fading. Assume that each cell has a radius (from center to vertex) of r_c, and consider K users that are located within the cell, with the radius r_h.

Consider an OFDM symbol with interval of T_s with a symbol duration given by T_u and the guard interval given by $T_g = T_s - T_u$. Assume that the channel coefficient between the *m*th BS and the *k*th terminal in the lth cell within the *n*th subcarrier is given by g_{nmjkl} and computed as $g_{nmjkl} = h_{nmjkl} \cdots \beta_{jkl}^{1/2}$ where *n*=1, …, N_{FFT}, *m*=1, …, *M*, *j*=1, …, *L*, *k*=1, …, *K*, l=1, …, *L*, and N_{FFT} is the number of subcarriers, *L* is the number of active cells. The fading coefficients, h_{nmjkl}, are assumed to have a mean of 0 and variance of 1.

The fading coefficient is given as follows: $\beta_{jkl} = \dfrac{Z_{jkl}}{r_{jkl}^{\gamma}}$ where r_{jkl} is the distance between the *k*th terminal in the lth cell and the BS in the *j*th cell, γ is a decay exponent. Z_{jkl} is a *log-normal* random variable and $10\log 10(Z_{jkl})$ is *Gaussian* distributed with a mean of 0 and standard deviation of σ_{shad}. Assume that *T* is the length of the coherence interval and τ is the number of symbols employed for pilots. Many studies have focused on how to multiplex a large number of users on similar time/frequency resources that yield considerable high spectral efficiencies [21–26].

3.2.3 Channel Estimation in Time-Division Duplex Systems

The BS usually needs channel state information (CSI) to detect the transmitted signals from all the *K* users for the uplink transmission. The CSI is computed at the BS by using minimum mean square error (MMSE) algorithm. Afterward, the BS estimates the channels based on the received pilot signals, and it requires a minimum of *K* channel users. Considering the forward link, the BS uses CSI to beamform the signals that need to be transmitted, and because of channel reciprocity, the calculated CSI at the BS in the reverse link can be used to beamform or precode the transmit symbols [27]. The standard TDD operation normally employs three operations during a coherence time, and these are as follows: channel estimation (including the reverse link training and the forward training), reverse link traffic transmission and forward link traffic transmission. The TDD massive MIMO operation is shown in Figure 3.1.

3.2.4 Point-to-Point Multiple-Input Multiple-Output Systems

Shannon's noisy-channel coding theorem states that *for any communication link, there is an attainable rate such that for any transmission rate less than the capacity, there exists a coding*

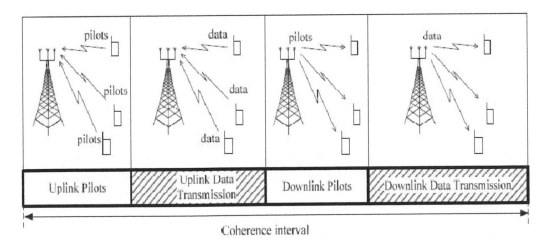

FIGURE 3.1
TDD massive MIMO protocol. MIMO, multiple-input multiple-output; TDD, time-division duplex.

scheme that makes the error rate arbitrarily small [3]. Consider a point-to-point wireless link with a MIMO employing n_t transmit antennas and n_r receive antennas. The narrowband memoryless channel has the mathematical representation given by Eq. (3.1) where x refers to the transmitted n_t signal component vector, y refers to the received n_r signal component vector, H is an $n_r \times n_t$ channel matrix and w is the n_r signal vector of additive noise.

$$y = \sqrt{\rho} \; Hx + w \tag{3.1}$$

The scalar ρ is used to measure the signal-to-noise ratio (SNR) of the wireless link and assume that the total transmit power is in normalized mode as well as the additive noise is independent and identically distributed (i.i.d) with mean of 0 and variance of 1 from a circularly symmetric complex-Gaussian random variables $(CN(0,1))$; then the attainable rate is given by Eq. (3.2) where I_{n_r} is an $n_r \times n_r$ identity matrix and H^H denotes the Hermitian transpose of the channel propagation matrix.

$$C = \log_2 \; \det\left(I_{n_r} + \frac{\rho}{n_t} \; HH^H \right) \tag{3.2}$$

Consider the number of antennas at the transmitter to be large while the number of antennas at the receiver to be constant, and let the channel propagation matrix be asymptotically orthogonal; then Eq. (3.3) will hold, and as a result, the attainable rate reduces to Equation (3.6) based on Equations (3.4) and (3.5).

$$\left(\frac{HH^H}{n_t} \right)_{n_t \gg n_r} \approx I_{n_r} \tag{3.3}$$

$$C_{n_t \gg n_r} \approx n_r \; \log_2 \; (1 + \rho) \tag{3.4}$$

$$\left(\frac{H^H H}{n_r} \right)_{n_r \gg n_t} \approx I_{n_t} \tag{3.5}$$

$$C_{n_r \gg n_t} \approx n_t \; \log_2 \left(1 + \rho \; \frac{n_r}{n_t} \right) \tag{3.6}$$

3.2.5 Multiuser Multiple-Input Multiple-Output Systems

Consider a MIMO system employing multiple users each with only one antenna communicating with a BS having M antennas. Assuming TDD protocol operation, we analyzed the attainable rates for the reverse and forward links. This protocol is employed in our analysis due to its simple transpose nature of the channel propagation matrix of the reverse link. Moreover, under the TDD operation regime, acquisition of the CSI through pilots is possible and enables reliable links [3].

The reverse link channel matrix denoted as H of dimension $M \times K$ is the product of an $M \times K$ matrix G and a $K \times K$ diagonal matrix. G is for small-scale fading, and the diagonal matrix is denoted as $D_\beta^{1/2}$, diagonal elements of which constitute a $K \times 1$ vector, β, of large-scale fading coefficients, $H = G \; D_\beta^{1/2}$. In the large-scale fading model, the path loss and shadow fading are included, and hence the kth column vector of G defines the small-scale fading between the kth user and the M BS antennas, while the kth diagonal element of $D_\beta^{1/2}$ describes the large-scale fading coefficient. For such MIMO systems, as the number of antennas becomes infinite compared with the number of users and under favorable conditions, it can be shown that the column vectors of the propagation matrix are given by Equation (3.7) and are asymptotically orthogonal.

$$\left(\frac{H^H H}{M} \right)_{M \gg K} \approx D_\beta \tag{3.7}$$

Consider K terminals transmitting a $K \times 1$ vector of modulated symbols, q_r, in the uplink and the BS antennas receiving a $M \times 1$ vector, x_r, given by Equation (3.8) where w_r is the $M \times 1$ vector of additive receiver noise modeled as $CN(0,1)$. Each user is forced to have an expected power of unity, and the attainable sum rate for the uplink multiuser MIMO system is computed by Equation (3.9) assuming noncooperative communication [2]. Under the circumstance that there is a large number of BS antennas compared with those of the UTs, the asymptotic sum rate can be found using Equation (3.10).

$$x_r = \sqrt{\rho_r}\, H\, q_r + w_r \tag{3.8}$$

$$C_{\text{sum}} = \log_2 \det\left(I_K + \rho_r\, H^H H\right) \tag{3.9}$$

$$C_{\text{sum}(M \gg K)} = \sum_{k=1}^{K} \log_2 \left(1 + M\rho_r \beta_k\right) \tag{3.10}$$

The downlink or forward link communication uses a BS that transmits a $M \times 1$ vector, s_f, through its M antennas, and the K terminals altogether receive a $K \times 1$ vector, x_f, given as follows. Let D_γ be a matrix whose diagonal elements establish a $K \times 1$ vector γ given by Equation (3.11); then the sum rate capacity can be found from Equation (3.12) assuming a large number of BS antennas compared with the UTs.

$$\sum_{k=1}^{K} \gamma_k = 1, \quad \gamma_k \geq 0, \, \forall\, k \tag{3.11}$$

$$C_{\text{sum-rate}} = \max_{\{\gamma k\}} \sum_{k=1}^{K} \log_2 \left(1 + M\rho_f \gamma_k \beta_k\right) \tag{3.12}$$

3.2.6 Massive Multiple-Input Multiple-Output Linear Precoding

This section reviews the most popular linear precoding techniques such as matched filter (MF) or maximum ratio transmission (MRT), zero forcing (ZF) and MMSE precoders for massive MIMO wireless systems. Consider a signal vector transmitted as $x = \sqrt{\rho}\, Ws$

for the K users during the forward transmission, where $M > K$ and $W \in \mathbb{C}^{M \times K}$ is a linear precoding matrix, $s \in \mathbb{C}^{K \times 1}$ is the message information that needs to be transmitted prior to precoding and ρ is a scalar that defines the average transmit power by the BS. Under the TDD massive MIMO operation, the forward link channel is equivalent to the transpose of the channel matrix H; hence, the collection of signals received at all the K terminals can be written as $y = H^T x + w = \sqrt{\rho} \ H^T W s + w$ where $w \in \mathbb{C}^{K \times 1}$ represents additive noise.

MRT precoder. The MRT or MF precoder is given as $W_{MF} = \sqrt{\alpha} \ H^H$ where α is a normalized scaled power factor. The signal vector received can be written as $y = II^T x + w = \sqrt{\rho\alpha} \ H^T \ H^H s + w$. This precoder maximizes the gain at the desired user, and at the receiver of the reverse link, a maximum ratio combiner (MRC) is employed. The MRT precoder is near optimal under the assumption that the number of antennas at the BS is very large compared with the number of UTs.

ZF precoder. A ZF precoder is mathematically given as $W_{ZF} = \sqrt{\alpha} \ H^H \left(H^T H^H \right)^{-1}$, and the signal vector received in the forward link can be expressed as $y = H^T x + w = \sqrt{\rho\alpha} \ H^T \ H^H \left(H^T H^H \right)^{-1} s + w = \sqrt{\rho\alpha} \ s + w$.

MMSE precoder. Recently an advanced linear precoder called the MMSE or regularized ZF precoder has been designed for MIMO wireless communication systems that can trade off the advantages of MRT and ZF precoders. The MMSE precoding matrix is given by $W_{MMSE} = \sqrt{\alpha} \ H^H \left(H^T H^H + \lambda I_K \right)^{-1}$, and the received signal vector in the downlink channel can be written as $y = H^T x + w = \sqrt{\rho\alpha} \ H^T \ H^H \left(H^T H^H + \lambda I_K \right)^{-1} s + w$.

Figure 3.2 shows the attainable sum rates of MRT, ZF and MMSE precoding techniques assuming a downlink transmit power of 20 dB for 10 users in a single-cell scenario. It can be observed that with increasing values of M that is the number of BS antennas, the precoder rates also increase, and as expected, the MMSE precoder yields higher rate than the MRT as well as the ZF precoders.

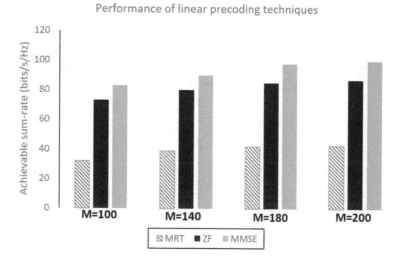

FIGURE 3.2
Aggregate rates of MRT, ZF and MMSE precoding techniques ($K = 10$ and total power = 20 dB) for various number of BS antennas (M). MMSE, minimum mean square error; MRT, maximum ratio transmission; ZF, zero forcing.

3.3 Spectral and Energy Efficiencies of Massive Multiple-Input Multiple-Output Systems

In a multiuser MIMO (MU-MIMO) communication system, the BS normally establishes communication by using the channel orthogonality principle with each user in a separate resource block involving time and frequency. This usually limits the achievable rates of the link [2]. To achieve higher rates, the BS establishes communication in the same time–frequency resource but in the presence of interuser interference, which can be alleviated by applying maximum likelihood multiuser detection methods or dirty paper coding [28–30].

With the MU-MIMO setting and a large number of antennas at the BS, the effect of small scale fading can be averaged out, and the uncorrelated noise as well as the intracell interference disappears completely [28]. In this section, we employ the linear precoders discussed earlier at both the uplink and downlink to find the spectral and energy efficiencies of a massive MIMO communication system. The cases of perfect and imperfect CSI are considered in our analysis.

3.3.1 Uplink Efficiencies under Perfect Channel State Information

Assuming that the BS has perfect CSI that is, the BS knows the channel propagation matrix, then the received signal vector at the BS after employing a linear detector is given by Equation (3.13) where h_k and z_k are the kth column of the matrices H and Z, respectively. Matrix Z is one of the three linear detectors discussed in the previous section. Assume that the additive noise as well as the interference components are random variables with

mean of 0 and variance given by $p_u \sum\limits_{i=1,\ i \neq k}^{K} \left| z_k^H\ h_i \right|^2 + z_k^2$ where the average transmitted power

of each user is denoted as p_u. Assuming an ergodic channel, the achievable ergodic uplink rate of the kth user can be found from Equation (3.14).

$$r_k = \sqrt{p_u}\ z_k^H h_k x_k + \sqrt{p_u} \sum_{i=1,\ i \neq k}^{K} z_k^H\ h_i x_i + n \tag{3.13}$$

$$R_{P,k} = \mathbb{E}\left\{ \log_2 \left[\frac{p_u \left| z_k^H h_k \right|^2}{p_u \sum\limits_{i=1,\ i \neq k}^{K} \left| z_k^H\ h_i \right|^2 + z_k^2} \right] \right\} \tag{3.14}$$

MRC receiver. It can be shown that the achievable rate of the uplink for the kth user employing an MRC receiver is expressed as Equation (3.15) under the circumstance that perfect CSI is available at the BS. As M grows large that is, $M \rightarrow \infty$, the expression of Equation (3.15) reduces to Equation (3.16). The spectral and energy efficiencies can be found from Equations (3.17) and (3.18), respectively.

$$R_{P,k}^{\mathrm{mrc}} = \log_2 \left[1 + \frac{p_u \, (M-1)\beta_k}{p_u \sum_{i=1, \, i\neq k}^{K} \beta_i + 1} \right] \tag{3.15}$$

$$R_{P,k}^{\mathrm{mrc}} \rightarrow \log_2 \left[1 + p_u \, M \, \beta_k \right] \tag{3.16}$$

$$\mathrm{SE}_P^{\mathrm{mrc}} = \sum_{k=1}^{K} R_{P,k}^{\mathrm{mrc}} = \sum_{k=1}^{K} \log_2 \left[1 + \frac{p_u \, (M-1)\beta_k}{p_u \sum_{i=1, \, i\neq k}^{K} \beta_i + 1} \right] \tag{3.17}$$

$$\mathrm{EE}_P^{\mathrm{mrc}} = \frac{1}{p_u} \, \mathrm{SE}_P^{\mathrm{mrc}} = \frac{1}{p_u} \sum_{k=1}^{K} \log_2 \left[1 + \frac{p_u \, (M-1)\beta_k}{p_u \sum_{i=1, \, i\neq k}^{K} \beta_i + 1} \right] \tag{3.18}$$

ZF receiver. The ZF receiver employs an uplink detector matrix given by the transpose of the ZF downlink precoder matrix and is equivalent to $H \left(H^H \, H \right)^{-1}$. It can be shown that the uplink achievable rate is given by equation (3.19), and as M grows large, that is, $M\rightarrow\infty$, Equation (3.19) reduces to Equation (3.20). Furthermore, the spectral and energy efficiencies are computed from Equations (3.21) and (3.22), respectively.

$$R_{P,k}^{\mathrm{zf}} = \log_2 \left[1 + p_u \, (M-K)\beta_k \right] \tag{3.19}$$

$$R_{P,k}^{\mathrm{zf}} \rightarrow \log_2 \left[1 + p_u \, M \, \beta_k \right] \tag{3.20}$$

$$\mathrm{SE}_P^{\mathrm{zf}} = \sum_{k=1}^{K} \log_2 \left[1 + p_u \, (M-K)\beta_k \right] \tag{3.21}$$

$$\mathrm{EE}_P^{\mathrm{zf}} = \frac{1}{p_u} \sum_{k=1}^{K} \log_2 \left[1 + p_u \, (M-K)\beta_k \right] \tag{3.22}$$

MMSE receiver. The MMSE detector matrix employed at the receiver is given by $H \left(H^H H + \frac{1}{p_u} I_K \right)^{-1}$, and it can be shown that the achievable rate is given by Equation (3.23) where $\gamma_k \cong \dfrac{1}{\left[\left(p_u H^H H + I_K \right)^{-1} \right]_{kk}}$. The PDF of γ_k is normally approximated by the well-known gamma distribution [20], where the shape and scale parameters are found using Equation

(3.25) and μ and κ are obtained by solving Equation (3.26). By using Equations (3.23) to (3.26), it can be shown that the uplink achievable rate reduces to Equation (3.27). The spectral and energy efficiencies can be evaluated from Equations (3.28) and (3.29), respectively.

$$R_{P,k}^{\mathrm{mmse}} = \log_2\left(1 + \frac{1}{\mathbb{E}\left\{\frac{1}{\gamma_k}\right\}}\right) \tag{3.23}$$

$$p_{\gamma k}(\gamma) = \frac{\gamma^{\alpha_k - 1}\, e^{-\gamma/\theta_k}}{\Gamma(\alpha_k)\, \theta_k^{\alpha_k}} \tag{3.24}$$

$$\alpha_k = \frac{\left(M - K + 1 + (K-1)\mu\right)^2}{M - K + 1 + (K-1)\kappa}, \theta_k = \frac{M - K + 1 + (K-1)\kappa}{M - K + 1 + (K-1)\mu}\, p_u\, \beta_k \tag{3.25}$$

$$\mu = \frac{1}{K-1} \sum_{i=1,\, i\neq k}^{K} \frac{1}{Mp_u\beta_i\left(1 - \frac{K-1}{M} + \frac{K-1}{M}\mu\right) + 1},$$

$$\kappa\left(1 + \sum_{i=1,\, i\neq k}^{K} \frac{p_u\, \beta_i}{\left(Mp_u\beta_i\left(1 - \frac{K-1}{M} + \frac{K-1}{M}\mu\right) + 1\right)^2}\right) \tag{3.26}$$

$$= \sum_{i=1,\, i\neq k}^{K} \frac{p_u\, \beta_i\, \mu + 1/(K-1)}{\left(Mp_u\beta_i\left(1 - \frac{K-1}{M} + \frac{K-1}{M}\mu\right) + 1\right)^2}$$

$$R_{P,k}^{\mathrm{mmse}} = \log_2\left(1 + (\alpha_k - 1)\theta_k\right) \tag{3.27}$$

$$\mathrm{SE}_P^{\mathrm{mmse}} = \sum_{k=1}^{K} \log_2\left(1 + (\alpha_k - 1)\theta_k\right) \tag{3.28}$$

$$\mathrm{EE}_P^{\mathrm{mmse}} = \frac{1}{p_u} \sum_{k=1}^{K} \log_2\left(1 + (\alpha_k - 1)\theta_k\right) \tag{3.29}$$

3.3.2 Uplink Efficiencies under Imperfect Channel State Information

Consider a massive MIMO system where the BS estimates the propagation matrix as $\hat{H}$. The standard method to obtain $\hat{H}$ is to reserve a section of the coherence interval of the channel for uplink pilot signals and use the linear MMSE estimation technique. If T is the duration of the coherence interval and τ is the number of symbols used for the pilots, then the pilot sequences used by all K users can be represented by a $T \times \tau$ matrix denoted as $\sqrt{p_p}\ \psi$, which satisfies $\psi^H \psi = I_K$ where $p_p = \tau p_u$. Moreover, assume that the error estimation matrix be given as $\epsilon = \hat{H} - H$ and the ith column elements of matrix ϵ be random variables with means of 0 and variances given by $\dfrac{\beta_i}{p_p \beta_i + 1}$. We can further assume that each element of $\hat{h}_k$ be a random variable with zero mean and variance given by $\dfrac{\tau p_u \beta_k^2}{\tau p_u \beta_k + 1}$ [28].

Therefore, the received signal associated with the kth user, using a linear detector at the BS, can be written as Equation (3.30) where $\hat{z}_k$, $\hat{h}_I$ and ϵ_i are the ith column of the detector, channel and error estimation matrices, respectively. In this case, the uplink ergodic achievable rate for the kth user under imperfect CSI using any linear detector is given by Equation (3.31):

$$\hat{r}_k = \sqrt{p_u}\ \hat{z}_k \hat{h}_k x_k + \sqrt{p_u} \sum_{i=1,\ i \neq k}^{K} \hat{z}_k^H h_i x_i - \sqrt{p_u} \sum_{i=1}^{K} \hat{z}_k^H\ \epsilon_i\ x_i + n \tag{3.30}$$

$$R_{IP,k} = \mathbb{E}\left\{ \log_2\left(1 + \frac{p_u \left|\hat{z}_k^H \hat{h}_k\right|^2}{p_u \sum_{i=1,\ i \neq k}^{K} \left|\hat{z}_k^H \hat{h}_i\right|^2 + p_u \left\|\hat{z}_k^H\right\|^2 \sum_{i=1}^{K} \frac{\beta_i}{\tau p_u \beta_i + 1} + \left\|\hat{z}_k\right\|^2}\right) \right\} \tag{3.31}$$

MRC receiver. By using Jensen's inequality and random matrix theory, it can be shown that the uplink achievable rate of the kth user under imperfect CSI is given by Equation (3.32), and as M grows large, we have Equation (3.33). Therefore, the spectral and energy efficiencies for an MRC receiver under imperfect CSI in the uplink can be found from Equations (3.34) and (3.35), respectively.

$$R_{IP,k}^{\mathrm{mrc}} = \log_2\left(1 + \frac{\tau p_u (M-1) \beta_k^2}{(\tau p_u \beta_k + 1) \sum_{i=1,\ i \neq k}^{K} \beta_i + (\tau+1)\beta_k + \dfrac{1}{p_u}}\right) \tag{3.32}$$

$$R_{IP,k}^{\mathrm{mrc}} \rightarrow \log_2\left(1 + \tau p_u^2 M \beta_k^2\right) \tag{3.33}$$

$$SE_{IP}^{\mathrm{mrc}} = \frac{(T-\tau)}{T} \sum_{k=1}^{K} \log_2\left(1 + \frac{\tau p_u (M-1) \beta_k^2}{(\tau p_u \beta_k + 1) \sum_{i=1,\ i \neq k}^{K} \beta_i + (\tau+1)\beta_k + \dfrac{1}{p_u}}\right) \tag{3.34}$$

$$\text{EE}_{\text{IP}}^{\text{mrc}} = \frac{1}{p_u} \text{SE}_{\text{IP}}^{\text{mrc}} = \frac{(T-\tau)}{T.p_u} \sum_{K=1}^{K} log_2 \left(1 + \frac{\tau p_u (M-1) \beta_k^2}{(\tau p_u \beta_k + 1) \sum_{i=1,\ i \neq k}^{K} \beta_i + (\tau+1) \beta_k + \frac{1}{p_u}} \right) \tag{3.35}$$

ZF receiver. With ZF detection at the receiver of the uplink, it can be shown that the rate that can be achieved for the kth user under imperfect CSI is given by Equation (3.36) which reduces to Equation (3.37) as M becomes very large. Also, the spectral and energy efficiencies can be computed from Equations (3.38) and (3.39), respectively.

$$R_{\text{IP},k}^{\text{zf}} = log_2 \left(1 + \frac{\tau p_u^2 (M-K) \beta_k^2}{(\tau p_u \beta_k + 1) \sum_{i=1}^{K} \frac{p_u \beta_i}{\tau p_u \beta_i + 1} + \tau p_u \beta_k + 1} \right) \tag{3.36}$$

$$R_{\text{IP},k}^{\text{zf}} \rightarrow log_2 \left(1 + \tau p_u^2 M \beta_k^2 \right) \tag{3.37}$$

$$\text{SE}_{\text{IP}}^{\text{zf}} = \frac{(T-\tau)}{T} \sum_{K=1}^{K} log_2 \left(1 + \frac{\tau p_u^2 (M-K) \beta_k^2}{(\tau p_u \beta_k + 1) \sum_{i=1}^{K} \frac{p_u \beta_i}{\tau p_u \beta_i + 1} + \tau p_u \beta_k + 1} \right) \tag{3.38}$$

$$= \frac{(T-\tau)}{T.p_u} \sum_{k=1}^{K} log_2 \left(1 + \frac{\tau p_u^2 (M-K) \beta_k^2}{(\tau p_u \beta_k + 1) \sum_{i=1}^{K} \frac{p_u \beta_i}{\tau p_u \beta_i + 1} + \tau p_u \beta_k + 1} \right) \tag{3.39}$$

MMSE receiver. With MMSE processing at the receiver of the uplink, the achievable rate for the kth user under partial CSI is given by Equation (3.40) where the shape and scale parameters are determined as described in Ref. [28, eqs. 49, 50]. Therefore, the spectral and energy efficiencies can be found by using Equations (3.41) and (3.42), respectively.

$$R_{\text{IP},k}^{\text{mmse}} = log_2 \left[1 + (\hat{\alpha}_k - 1) \hat{\theta}_k \right] \tag{3.40}$$

$$\text{SE}_{\text{IP}}^{\text{mmse}} = \frac{(T-\tau)}{T} \sum_{k=1}^{K} log_2 \left[1 + (\hat{\alpha}_k - 1) \hat{\theta}_k \right] \tag{3.41}$$

$$\text{EE}_{\text{IP}}^{\text{mmse}} = \frac{(T-\tau)}{T.p_u} \sum_{k=1}^{K} log_2 \left[1 + (\hat{\alpha}_k - 1) \hat{\theta}_k \right] \tag{3.42}$$

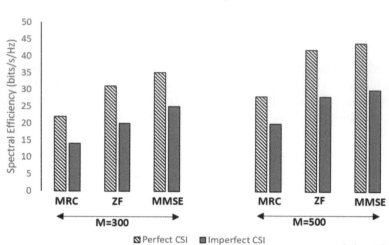

FIGURE 3.3

Uplink spectral efficiency for MRC, ZF and MMSE detection at the receiver, with perfect CSI and with imperfect CSI. CSI, channel state information; MMSE, minimum mean square error; MRC, maximum ratio combiner; ZF, zero forcing.

Figure 3.3 illustrates the variation of spectral efficiencies with increasing number of BS antennas for $K=10$ users and $E_u=20\,$dB. As expected, with $p_u=E_u/M$ for perfect CSI, it can be observed that when M increases, the spectral efficiency reaches a constant value but decreases to 0 for the case of imperfect CSI. However, with $p_u = E_u/\sqrt{M}$, the spectral efficiency increases infinitely with increasing M for the case of perfect CSI. However, it is noted that for imperfect CSI case, the spectral efficiency meets the nonzero limit with increasing M [28].

3.3.3 Downlink Efficiencies under Imperfect Channel State Information

As described in Ref. [27], consider a cellular downlink transmission where a BS with M antennas is serving simultaneously K single-antenna users in similar time–frequency blocks, and assume that $M\gg K$. Assume that the BS employs linear precoding to process the signal prior to transmission in the forward link. Moreover, we assume TDD massive MIMO protocol operation, and hence, we consider that the channels on the reverse and forward links are equal. The partial CSI is obtained from reverse link training based on the technique of MMSE. Considering only one coherence interval used solely for the uplink pilots, then let τ_u be the number of symbols per coherence interval and all users collectively transmit pilot sequences of length τ_u symbols.

Furthermore, let the elements of the channel matrix H be i.i.d. Gaussian distributed with mean of 0 and variance of 1, and the MMSE estimate of H is given by $\hat{H}$. Let the error estimation matrix be given as $\epsilon=\hat{H}-H$ with $\hat{H}$ being i.i.d with mean of 0 and variance $\dfrac{\tau_u p_u}{\tau_u p_u +1}$ and ϵ being i.i.d with zero mean and variance $\dfrac{1}{\tau_u p_u +1}$. The BS employs $\hat{H}$ to linearly precode the symbols, and then it sends the precoded signal vector given as $x = \sqrt{p_d}\, \mathbf{Z}s$ to all the users where $\mathbf{Z}$ is a $M \times K$ linear precoding matrix, s is the $M \times 1$ signal vector and the average transmit power by the BS is denoted as p_d.

In order to coherently detect the transmitted signal, the user should acquire CSI by downlink pilot signals. Let the pilot matrix be $K \times \tau_d$ dimensioned called S_p where τ_d is the number of symbols for downlink training. The pilot matrix is given by $S_p = \sqrt{\tau_d p_d}\, \psi$ with the rows of ψ pairwisely orthogonal, and then the BS precodes the pilot sequence using the precoder matrix $\mathbf{Z}$ giving a $K \times \tau_d$ matrix received by all the K users as $Y_p^T = \sqrt{\tau_d p_d}\, H^T \mathbf{Z}\, \psi + N_p^T$ where N_p is the additive receiver noise whose elements are i.i.d $CN(0,1)$. The $1 \times K$ received pilot vector at user k is given by Equation (3.43) where $a_{ki} = h_k^T\, z_i$, and it can be decomposed as $a_{ki} = \hat{a}_{ki} + \epsilon_{ki}$. It can be shown that the downlink achievable rate with MMSE CSI for the kth user can be found from Equation (3.44).

$$\hat{y}_{p,k}^T = \sqrt{\tau_d p_d}\, a_k^T + \hat{n}_{p,k}^T \tag{3.43}$$

$$R_k = \mathbb{E}\left\{ \log_2\left(1 + \frac{p_d\, |\hat{a}_{kk}|^2}{p_d \sum_{i \neq k}^{K} |\hat{a}_{ki}|^2 + p_d \sum_{i=1}^{K} \mathbb{E}\left\{ |\epsilon_{ki}|^2 \right\} + 1} \right) \right\} \tag{3.44}$$

MF precoder. With an MF precoder, the matrix $\mathbf{Z}$ becomes $\mathbf{Z} = a_{MF}\, \hat{H}^*$ where the power normalization constant is given by $\sqrt{(\tau_u p_u + 1) \big/ (MK\tau_u p_u)}$. The achievable rate can be given by Equation (3.45) where $\hat{a}_{ki} = \frac{\sqrt{\tau_d p_d}}{\tau_d p_d + K}\, \hat{y}_{p,ki} + \frac{K}{\tau_d p_d + K} \sqrt{\frac{(\tau_u p_u M)}{K(\tau_u p_u + 1)}}\, \delta_{ki}, \delta_{ki} = 1$ when $i = k$ and 0 otherwise. The spectral and energy efficiencies of the MF precoder under MMSE CSI for the downlink can be found by using Equations (3.46) and (3.47).

$$R_k^{\text{MF}} = \mathbb{E}\left\{ \log_2\left(1 + \frac{p_d\, |\hat{a}_{kk}|^2}{p_d \sum_{i \neq k}^{K} |\hat{a}_{ki}|^2 + \frac{K p_d}{\tau_d p_d + K} + 1} \right) \right\} \tag{3.45}$$

$$\text{SE}^{\text{MF}} = \frac{T - \tau_u - \tau_d}{T} \sum_{k=1}^{K} R_k^{\text{MF}} \tag{3.46}$$

$$\text{EE}^{\text{MF}} = \frac{T - \tau_u - \tau_d}{T \cdot p_d} \sum_{k=1}^{K} R_k^{\text{MF}} \tag{3.47}$$

ZF precoder. For a ZF precoder, the rate that can be achieved for the kth user is expressed as Eq. (3.48), and the spectral as well as energy efficiencies are given by Equations (3.49) and (3.50).

$$R_k^{\text{ZF}} = \mathbb{E}\left\{ \log_2\left(1 + \frac{p_d\left|\hat{a}_{kk}\right|^2}{p_d\sum_{i\neq k}^{K}\left|\hat{a}_{ki}\right|^2 + \dfrac{Kp_d}{\tau_d p_d + K\left(\tau_u p_u + 1\right)} + 1} \right) \right\} \tag{3.48}$$

where $\hat{a}_{ki} = \dfrac{\sqrt{\tau_d p_d}}{\tau_d p_d + K\left(\tau_u p_u + 1\right)}\,\hat{\mathbf{y}}_{p,ki} + \dfrac{\sqrt{K(M-K)\tau_u p_u\left(\tau_u p_u + 1\right)}}{\tau_d p_d + K\left(\tau_u p_u + 1\right)}\,\delta_{ki}$, $\delta_{ki} = 1$ when $i=k$ and 0 otherwise.

$$\text{SE}^{\text{ZF}} = \frac{T - \tau_u - \tau_d}{T}\sum_{k=1}^{K} R_k^{\text{ZF}} \tag{3.49}$$

$$\text{EE}^{\text{ZF}} = \frac{T - \tau_u - \tau_d}{T.\,p_d}\sum_{k=1}^{K} R_k^{\text{ZF}} \tag{3.50}$$

Simulation results have been carried out by setting the following parameters: $\tau_u = \tau_d = K = 5$, $T = 200$, $p_u = 0$ dB and SNR $= p_d$. Figure 3.4 shows the variation of spectral efficiency versus SNR p_d (dB). Figure 3.4 shows the performance improvement obtained with the aforementioned precoding analysis, and it can be seen that with precoding, higher spectral efficiencies are obtained than without precoding.

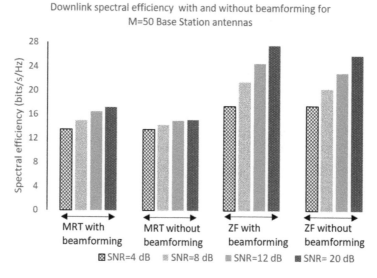

FIGURE 3.4
Downlink spectral efficiency versus SNR (p_d) for with and without beamforming. MRT, maximum ratio transmission; SNR, signal-to-noise ratio; ZF, zero forcing.

3.4 Beam-Division Multiple Access

Multiple access technology employing the orthogonality principle is the technology for all wireless networks. In the second generation of mobile communications, we have time-division multiple access (TDMA) and frequency-division multiple access (FDMA). Third-generation cellular systems employ the code-division multiple access (CDMA) and orthogonal frequency-division multiple access (OFDMA) in the 4G systems. The transmission resource blocks are orthogonally divided in time, frequency or code domains which produces the corresponding orthogonal multiple access (OMA) technology. Under the orthogonality regime, adjacent resource blocks produce negligible interference among themselves, which renders signal detection easy. However, the major limitation of OMA is that it will enable few number of users due to restrictions in the resource blocks, which bounds the spectral efficiencies and the sum rate capacity of current networks.

In FDMA, the frequency resource block is divided and allocated to respective UTs, whereas in TDMA, the resource block in time domain is subdivided and assigned to the UTs. With the introduction of 5G, the user capacity is expected to increase due to an exponential rise in the number of UTs. Unfortunately, because the transmission resource blocks employing both time/frequency domains are limited, additional resource blocks operating in other domains need to be included with the main objective of increasing the sum rate capacity of the system.

3.4.1 Concepts

In line with the one of the objectives of 5G to provide enhanced mobile services to an increased number of users, BDMA is the latest multiple access technology proposed in Ref. [31]. In BDMA, when a BS communicates with the UTs, a beam with orthogonal characteristic is assigned to each UT. The BDMA procedure splits a beam coming from the BS according to the locations of the UTs, hence increasing the capacity of the system significantly. Under the line of sight regime, a BS and the UTs can communicate beams that direct to each other's position without interfering with UTs at cell edge. In the event that UTs are located at different angles with respect to the BS, the latter will transmit the beams at similar angles to the UTs as shown in Figure 3.5. In BDMA, a user does not exclusively use one beam, but users who are located at a similar angle can share a unique beam to communicate with the BS.

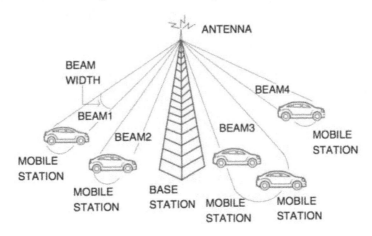

FIGURE 3.5
Beam-division multiple access.

In BDMA, the direction, number and width of the beams can be changed by a BS adaptively according to state of the wireless channel. In practice, the BS and UTs do not have a knowledge of each other's location; hence the UTs detect the BS through their control channel and transmit their locations as well as Doppler speed to the BS. Then the BS computes the path and the size of a beam in the forward link based on the position and speed of each user. Eventually, the BS uses its forward link to send the beam to the UT with the computed direction and width.

The frequency-division duplex (FDD)-BDMA protocol uses the same concept as that of the TDD-BDMA. However, the major dissimilarity is that in the FDD-BDMA, the initial user slot is assigned by splitting a frequency resource, and not by allotting a time resource. Another difference is that for the FDD-BDMA, there is a broadcast carried out by the BS, and in the case of TDD-BDMA, a preamble is used.

3.4.2 Principles

5G networks are under the obligation of supporting huge amount of users with radically diverse classes of traffic. To this end, various nonorthogonal multiple access (NOMA) schemes have been proposed. NOMA introduces a novel access mechanism by carrying out multiplexing within one of the classic time/frequency/code domains. The main principle of NOMA [32] is to use power and/or code domains in multiplexing to house more users in similar transmission resource blocks. There are three major categories of NOMA: power-domain NOMA, code-domain NOMA and multiple-domain NOMA multiplexing. When employing NOMA, the spectrum that is limited can be fully optimized to support more users. Hence, remarkable sum rate capacity improvements can be observed in 5G networks, but at the receiver, additional complexity will be increased.

In Refs. [33,34], the weighted sum rate has been mathematically analyzed to maximize the 5G output data rate subject to the requirements of users. The studies show that NOMA can increase the capacity by partaking resource blocks between users who are near and far to the BS. It is important to note that this result could not be observed in conventional multiple access technologies employing the orthogonality principle.

In Refs. [35,36], NOMA was applied to cellular scenarios comprising of single transmit–receive antenna system or commonly known as SISO scenarios. For example, in Ref. [35], the ergodic sum rate and outage performance were derived in closed form with users equipped with single antenna that were placed randomly. Furthermore, in Ref. [36], it is observed that the SISO-NOMA scenario under imperfect CSI yields enhanced performance over the SISO-OMA case. The studies of Refs. [37,38] focused on combining MIMO with NOMA to achieve high spectral efficiencies.

For a single-user MIMO-NOMA system, it has been shown in Ref. [37] that the ergodic capacity of conventional OMA systems can improve. Moreover, it has been shown in Ref. [38] that with the application of a ZF precoder and detector, it is possible to maximize NOMA gains. The work of Ref. [38] assumes that the transmitter holds perfect CSI, and it has been shown that the analysis minimizes intergroup interference for a MU-MIMO-NOMA system. Also, due to various beams being communicated among near-far users, the studies show that it is possible to cancel the interbeam interference (IBI) completely.

Moreover, at the transmitter of the BS, it is not possible to obtain perfect CSI because of the excessive amount of overhead to provide feedback. This occurs especially in a MIMO system based on the FDD mode with a large amount of antennas and users. Some current studies in BDMA have been able to decrease the overhead of the feedback channel by using an approach, which employs a long-term and a short-term basis. For the long-term basis approach, this deals with optimizing the statistical channel correlation that fluctuates for a long time duration. The long-term feedback basis is subjugated for grouping of users with IBI alleviation, for example, by determining the precoding vector for each user group [39,40].

The astounding work of [35] for a BDMA system shows that by manipulating the channel information statistically, it is possible to schedule users within non-overlapping beams. However, the long term precoding based on feedback hurts from short-term channel deviation thereby acquiring IBI. This has been observed despite reducing the feedback overhead significantly.

Recently, Choi et al. Ding et al. [32] formulated an optimization problem for the combination of NOMA with BDMA; the combined system is termed as BD-NOMA. It is shown that an iterative solution based on the MMSE algorithm can be found by using multiple CQIs (channel quality indicators)-based short-term feedback. Moreover, if the system is modeled using low IBI, it is shown that the proposed BD-NOMA can achieve approximately 20% gain in terms of the aggregate sum rate over its conventional BDMA counterpart. However, if NOMA is combined with BDMA without joint optimization and in the presence of high IBI, it is observed that the BD-NOMA system does not produce significant gain over conventional BDMA. Furthermore, it has been shown that the proposed BD-NOMA scheme with joint power optimization algorithm can achieve approximately 10% aggregate sum rate gain over a conventional BDMA system.

3.5 Conclusion

In this chapter, based on the latest and most comprehensive studies, the state-of-the-art massive MIMO and BDMA for 5G are discussed. These are interesting topics on which many researchers have been making efforts toward achieving the objectives of 5G. The mathematical foundations and spectral as well as energy efficiencies of massive MIMO systems under perfect and imperfect CSI have been described in detail. Recent studies have shown that when the number of BSs in a cellular system grows large and under favorable propagation mechanisms, the spectral and energy efficiencies increase, resulting in a significant rise of the system performance. Moreover, in this chapter, the concepts and principles of BDMA as the key multiple access technology for 5G are discussed. It has been shown that enhanced mobile data services can be achieved for a large number of simultaneous users.

References

1. H. Papadopoulos, C. Wang, O. Bursalioglu, X. Hou, and Y. Kishiyama, "Massive MIMO technologies and challenges towards 5G," *IEICE Transactions on Communications*, vol. E99.B, no. 3, pp. 602–621, 2016.
2. T. L. Marzetta, "Noncooperative cellular wireless with unlimited numbers of base station antennas," *IEEE Transactions Wireless Communications*, vol. 9, no. 11, pp. 3590–3600, 2010.
3. F. Rusek et al., "Scaling up MIMO: Opportunities and challenges with very large arrays," *IEEE Signal Processing Magazine*, vol. 30, no. 1, pp. 40–60, Jan. 2013.

4. E. G. Larsson, O. Edfors, F. Tufvesson, and T. L. Marzetta, "Massive MIMO for next generation wireless systems," *IEEE Communications Magazine*, vol. 52, no. 2, pp. 186–195, 2014.

5. T. L. Marzetta, "Massive MIMO: An introduction," *Bell Labs Technical Journal*, vol. 20, pp. 11–22, Mar. 2015.

6. J. Gozalvez, "Tentative 3GPP timeline for 5G [mobile radio]," *IEEE Vehicular Technology Magazine*, vol. 10, no. 3, pp. 12–18, Sept. 2015.

7. J. G. Andrews et al., "What will 5G be?" *IEEE Journals on Selected Areas in Communications*, vol. 32, no. 6, pp. 1065–1082, Jun. 2014.

8. 3GPP, *Study on channel model for frequency spectrum above 6 GHz*, Technical Report 38.900, June 2016.

9. ITU-R, *Minimum requirements related to technical performance for IMT-2020 radio interface(s)*, Technical Report M.[IMT-2020.TECH PERF REQ], Oct. 2016.

10. J. Huang et al., "Multi-frequency mmWave massive MIMO channel measurements and characterization for 5G wireless communication systems," *IEEE JSAC*, vol. 35, no. 7, pp. 1591–1605, Jul. 2017.

11. K. Zheng, L. Zhao, J. Mei, B. Shao, W. Xiang, and L. Hanzo, "Survey of large-scale MIMO systems," *IEEE Communications Surveys & Tutorials*, vol. 17, no. 3, pp. 1738–1760, 2015.

12. M. J. Marcus, "5G and IMT for 2020 and beyond [spectrum policy and regulatory issues]," *IEEE Wireless Communications Magazine*, vol. 22, no. 4, pp. 2–3, Aug. 2015.

13. A. F. Molisch et al., "The COST259 directional channel model-Part I: Overview and methodology," *IEEE Transactions Wireless Communications*, vol. 5, no. 12, pp. 3421–3433, 2006.

14. L. Liu et al., "The COST 2100 MIMO channel model," *IEEE Wireless Communications*, vol. 19, no. 6, pp. 92–99, 2012.

15. M. Zhu, F. Tufvesson, and J. Medbo, "Correlation properties of large scale parameters from 2.66 GHz multi-site macro cell measurements," IEEE VTC, Yokohama, Japan, 2011, pp. 1–5.

16. S. Jaeckel et al., "Correlation properties of large and small-scale parameters from multicell channel measurements," *EuCAP 2009*, Berlin, Germany, 2009.

17. C. Shepard, H. Yu, N. Anand, E. Li, T. Marzetta, R. Yang, and L. Zhong, "Argos: Practical many-antenna base stations," *Proceedings 18th Annual International Conference on Mobile Computer and Networking*, Mobicom, pp. 53–64, 2012.

18. J. Jose, A. Ashikhmin, T. L.Marzetta, and S. Vishwanath, "Pilot contamination and precoding in multi-cell TDD systems," *IEEE Transactions Wireless Communications*, vol. 10, no. 8, pp. 2640–2651, Aug. 2011.

19. N. Fatema et al., "Massive MIMO linear precoding: A survey," *IEEE Systems Journal*, vol. 12, no. 99, pp. 1–12, 2017.

20. L. Lu, G. Y. Li, A. L. Swindlehurst, A. Ashikhmin, and R. Zhang, "An overview of massive MIMO: Benefits and challenges," *IEEE Journal of Selected Topics in Signal Processing*, vol. 8, no. 5, pp. 742–758, Oct. 2014.

21. O. Elijah, C. Y. Leow, Y. A. Rahman, S. Nunoo, and S. Z. Iliya, "A comprehensive survey of pilot contamination in massive MIMO 5G system," *IEEE Communications Surveys and Tutorials*, vol. 18, no. 2, pp. 905–923, 2016.

22. H. Yang and T. L. Marzetta, "Performance of conjugate and zeroforcing beamforming in large-scale antenna systems," *IEEE Journal on Selected Areas in Communications*, vol. 31, no. 2, pp. 172–179, Feb. 2013.

23. S. K. Mohammed and E. G. Larsson, "Per-antenna constant envelope precoding for large multi-user MIMO systems," *IEEE Transactions on Communications*, vol. 61, no. 3, pp. 1059–1071, Mar. 2013.

24. H. Huh, G. Caire, H. C. Papadopoulos, and S. A. Ramprashad, "Achieving "Massive MIMO" spectral efficiency with a not-so-large number of antennas," *IEEE Transactions Wireless Communications*, vol. 11, no. 9, pp. 3226–3239, 2012.

25. J. Hoydis et al., "Massive MIMO in the UL/DL of cellular networks: How many antennas do we need?" *IEEE Journals on Selected Areas in Communications*, vol. 31, no. 2, pp. 160–171, 2013.

26. H. Q. Ngo, E. Larsson, and T. Marzetta, "Uplink power efficiency of multiuser MIMO with very large antenna arrays," *49th Annual Allerton Conference on Communication, Control, and Computing*, Illinois, USA, pp. 1272–1279, Sept. 2011.

27. H. Q. Ngo, E. G. Larsson, and T. L. Marzetta, "Massive MU-MIMO downlink TDD systems with linear precoding and downlink pilots," *51th Allerton Conference on Communication, Control, and Computing*, Illinois, USA, pp. 293–298, 2013.

28. H. Q. Ngo, E. Larsson, and T. Marzetta, "Energy and spectral efficiency of very large multiuser MIMO systems," *IEEE Transactions on Communications*, vol. 61, no. 4, pp. 1436–1449, Apr. 2013.

29. L. Zhao, K. Zheng, H. Long, H. Zhao, and W. Wang, "Performance analysis for downlink massive multiple-input multiple-output system with channel state information delay under maximum ratio transmission precoding," *IET Communications*, vol. 8, no. 3, pp. 390–398, Feb. 2014.

30. L. Zhao, K. Zheng, H. Long, and H. Zhao, "Performance analysis for downlink massive MIMO system with ZF precoding," *Transactions on Emerging Telecommunications Technologies*, vol. 25, no. 12, pp. 1219–1230, Dec. 2014.

31. C. Sun, X. Gao, S. Jin, M. Matthaiou, Z. Ding, and C. Xiao, "Beam division multiple access transmission for massive MIMO communications," *IEEE Transactions on Communications*, vol. 63, no. 6, pp. 2170–2184, Jun. 2015.

32. Y. I. Choi, J. W. Lee, M. Rim, and C. G. Kang, "On the performance of beam division nonorthogonal multiple access for FDD-based large-scale multi-user MIMO systems," *IEEE Transactions Wireless Communications*, vol. 16, no. 8, pp. 5077–5089, Aug. 2017.

33. H. Jin, K. Peng, and J. Song, "A spectrum efficient multi-user transmission scheme for 5G systems with low complexity," *IEEE Communications Letters*, vol. 19, no. 4, pp. 613–616, Apr. 2015.

34. S. Timotheou and I. Krikidis, "Fairness for non-orthogonal multiple access in 5G systems," *IEEE Signal Processing Letters*, vol. 22, no. 10, pp. 1647–1651, Oct. 2015.

35. Z. Ding, Z. Yang, and P. Fan, "On the performance of non-orthogonal multiple access in 5G systems with randomly deployed users," *IEEE Signal Processing Letters*, vol. 21, no. 12, pp. 1501–1505, Dec. 2014.

36. Z. Yang, Z. Ding, P. Fan, and G. K. Karagiannidis, "On the performance of non-orthogonal multiple access systems with partial channel information," *IEEE Transactions on Communications*, vol. 64, no. 2, pp. 654–667, Feb. 2016.

37. Q. Sun, S. Han, C.-L. I, and Z. Pan, "On the ergodic capacity of MIMO NOMA systems," *IEEE Wireless Communications Letters*, vol. 4, no. 4, pp. 405–408, Aug. 2015.

38. Z. Ding, F. Adachi, and H. V. Poor, "The application of MIMO to nonorthogonal multiple access," *IEEE Transactions Wireless Communications*, vol. 15, no. 1, pp. 537–552, Jan. 2016.

39. A. Adhikary, J. Nam, J. Ahn, and G. Caire, "Joint spatial division and multiplexing - The large-scale array regime," *IEEE Transactions on Information Theory*, vol. 59, no. 10, pp. 6441–6463, Oct. 2013.

40. D. Kim, G. Lee, and Y. Sung, "Two-stage beamformer design for massive MIMO downlink by trace quotient formulation," *IEEE Transactions on Communications*, vol. 63, no. 6, pp. 2200–2211, Jun. 2015.

4

Distributed MIMO Network for 5G Mobile Communication

Rachit Jain
ITM Group of Institutions (ITM GOI)

Robin Singh Bhadoria
Birla Institute of Applied Sciences (BIAS)

Neha Sharma
CSIR-Central Electronics Engineering Research Institute (CSIR-CEERI)

Yadunath Pathak
Indian Institute of Information Technology (IIIT)

Varun Mishra
ASET, Amity University of Madhya Pradesh

CONTENTS

4.1 Introduction

In 5G mobile communication, it is quite difficult to achieve high-speed Internet with limited sources and bandwidth. Thus, in recent years, multiple-input multiple-output (MIMO) has become very popular because of its realistic performance. During communication in the wireless system, channels get disturbed by multipath fading, which occurs due to reception of transmitted signals through different angles and delay in time, resulting in scattering of radio waves. Fluctuation also occurs in the received power signal known as fading, hence degrading the quality and reliability of the received signal. Thus, MIMO is the best suitable technology for 5G mobile communication to achieve high-speed connectivity [1].

MIMO is advantageous over a conventional communication system in a high-speed wireless communication system because MIMO exploits information about the spatial dimension with the help of multiple antennas at the transmitter and the receiver.

MIMO technology helps in improving various performance gains [1,2]:

- **Array gain**. Array gain is also known as power gain that is achieved by using multiple antennas at the transmitter and/or receiver side, with respect to a single-input single-output case. In a broadside array, the array gain is almost precisely proportional to the length of the array. Array gain improves the range of the wireless network, resulting in better coverage with better resistance to noise.

- **Spatial diversity gain**. As we know, during the reception of the signal in a conventional wireless system, the signal fades or fluctuates. Spatial diversity gain makes fading less vulnerable, which is later realized by the receiver when it receives multiple individual copies of a transmitted signal, thus improving the reliability and quality of the received signal.

- **Spatial multiplexing gain**. In the MIMO system, different streams of data are transferred in different spatial dimensions from the same source and received by the receiver using multiple antennas. Furthermore, capacity is effectively enhanced by a multiplicative factor that is equal to the number of streams.

- **Interference avoidance and reduction**. When many users use wireless channels simultaneously for communication purposes, interference occurs. To reduce this interference, MIMO exploits the spatial dimension and increases separation between the users. Another way to minimize interference is by directing signals to authentic users. Interference avoidance and reduction increase the range, capacity and coverage of a wireless network.

4.2 Building Block for Multiple-Input Multiple-Output

It provides a high-speed wireless communication system over a conventional communication system. As depicted in Figure 4.1, data bits pass as input to a conventional encoder and interleaved block. Later, this encoder maps with data symbols with the use of the symbol mapping block. Such symbols enter into the space–time encoder, thereby generating output in one or more spatial data streams, which further goes to the transmitter antennas with the help of space–time precoding block. Signals travel through the medium from the transmitter end and are then received at the arrays of receiver antenna. At the receiver end, the receiver assembles the signals to decode the data through space–time processing, space–time decoding, symbol demapping, deinterleaving and decoding blocks. In this way, the encoded data are decoded. Furthermore, we see that MIMO provides high-speed connectivity and marvelous capacity gains.

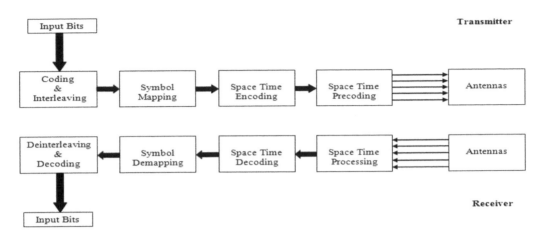

FIGURE 4.1
Block diagram of MIMO. MIMO, multiple-input multiple-output.

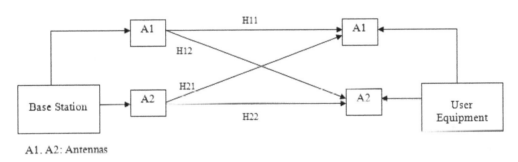

A1. A2: Antennas

FIGURE 4.2
Singleuser MIMO. MIMO, multiple-input multiple-output.

Single-user MIMO. In this, multitransmitter and receiver are used for wireless communication, which provides the bandwidth of wireless access point to a single device, as shown in Figure 4.2. Single-user MIMO transmits data streams to a single device at a time and also provides the highest data rates for the single-user equipment. In terms of MIMO, channel capacity is easier to obtain for single users rather than for multiple users. In fact, single-user MIMO capacity results are known for many cases, whereas in multiuser MIMO, many problems remain unsolvable [1,3].

Multiuser MIMO. Unlike the single-user MIMO system, multiuser MIMO provides access to multiple users. MIMO provides higher throughput when signal-to-noise ratio (SNR) is high and also provides capacity gain because multiple users access the channels at the same time. MIMO also permits spatial multiplexing gain at the base station without the need for multiple antennas at the user's equipment. MIMO is a type of extension of space-division multiple access (SDMA). Capacity benefits of multiuser MIMO are more than the single-user MIMO. [1,3]. The best example to understand multiuser MIMO is as follows: When people order any product from an online shopping website, then to ensure the proper delivery of a product to an intended recipient, seller uses courier service that

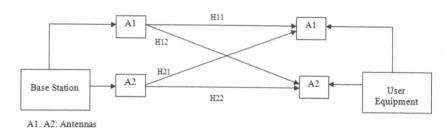

FIGURE 4.3
Multiuser MIMO. MIMO, multiple-input multiple-output.

means sharing the same service. This sharing mechanism saves expenses, provides better utilization of resources and improves experience and services [1]. Multiuser representation is shown in Figure 4.3 for better understanding.

4.3 Evolution of Mobile Network in India

In recent years, online ticket booking, online cab booking, online shopping, online food ordering and many more e-commerce activities are done through mobile applications. All these applications need 4G smartphones and high-speed Internet connectivity. Although this technological advancement saves time, energy, and money but still there is a requirement of updations and improvements at regular intervals of time. Evolution of mobile networks in India is represented in Figure 4.4.

5G technologies will offer more options and opportunities in this sector, which will optimize the whole infrastructure as a single unit system. For example, we can establish a vehicle platooning/flocking system in the transportation system. This technology enables vehicles to travel closer to each other safely with proper coordination and acceleration. This helps to increase road capacity with limited infrastructure facilities. In industries, 5G is a trending topic in the context of industrial automation. In the past few years, robots have

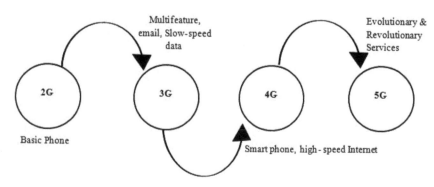

FIGURE 4.4
Evolution of mobile networks in India.

become general-purpose tools for automating production processes with precise operations and safety measures for human colleagues working in a factory or an industry. Apart from this, robots also help in tracking the products from beginning to its outcome. In farming, 5G can provide precision agriculture with proper monitoring of water, soil and humidity.

In the power sector, smart grids are popular to regulate control of power generation. The smart grid is an advanced technology that is useful for the automatic collection of behavioral data from customers and energy providers to improve the efficiency of the electricity supply. Smart meters are also helpful in providing greater information to the consumers about their electricity usage, improving the system reliability and quality with proper security.

In the medical field, connected health has the potential to improve patients' health by providing them the liberty to interact with their doctors and satisfy themselves. The health of a patient can also be improved through wireless monitoring of patient's health by automatic transfer of information at regular time intervals or in case of an emergency. Nowadays, doctors involve robots in critical surgeries known as robot-assisted surgeries. These robots help surgeons in terms of flexibility and precision. Robotic surgery allows doctors to perform many types of complex procedures with less complication, pain and blood loss. This is reliable and safe with 5G technology to treat patients even when the doctor is present in another country [4,5].

4.4 Enabling Technology and Network in 5G Mobile Communications

The International Telecommunications Union – Radio Communication Sector (ITU-R) has introduced criteria for promoting new technologies and applications in International Mobile Telecommunications-2020 (IMT 2020) for the benefit of society. Besides, the 3rd Generation Partnership Project (3GPP), which develops mobile telephony protocols, suggested three types of technologies and their 5G uses [4,6].

- **Enhanced mobile broadband (eMBB)** was firstly defined by the 3GPP as part of its SMARTER (Study on New Services and Markets Technology Enablers) project. eMBB provides higher data rates and speed connections with enhanced service quality at every cell site corner. This type of service also supports users when their mobility is very high and also provides high capacity Internet access in populated areas such as stadiums, conference rooms and so on. With this technology, consistent service can be achieved through enhanced connectivity Internet access at every place. Apart from these benefits, 360-degree video streaming is also available with smooth augmented and virtual reality support.

- **Massive machine-type communications (mMTC).** It supports IOT devices so that many devices can be connected in a small area for high-speed communication. Smart grids and smart cities are two major concepts that are supported by mMTC. The main objective of this technology is to connect users to a single base station living in highly dense areas.

- **Ultrareliable low-latency communications (uRLLC).** It is a game changer for many applications such as automation in factories, connected health, robotics automation and intelligent transportation system. All these applications need less end-to-end delay and more reliability with good network connectivity.

4.4.1 Distributed Computing in Multiple-Input Multiple-Output

Cloud computing provides many benefits by centralized servers with the advantages of resource sharing. Centralized servers are located at remote locations where running and maintenance costs are affordable, but time delays or response time increases, which is a big disadvantage for the communication system. Distributed computing solves this disadvantage by locating shared resources much closer to the service endpoints such as network functions virtualization (NFV) servers or 5G base stations to improve response time. In this way, efficiency and coverage of a high-speed broadband network are improved manifolds.

A distributed antenna system improves the performance of a communication system such as coverage, speed, diversity gain, capacity performance and so on. These benefits can be achieved by placing antennas location-wise through proper management. Good management can be established through the proper distribution of transmitting antennas and the central placement of the receiving antennas. This uniform distribution of antennas increases channel capacity to a greater extent.

In today's scenario, neither we think about the size of any file before downloading nor do we think about the data consumed by the file during the downloading process we just download even when signals are poor. Apart from this, we set up smooth video calls, watch movies or stream music online without thinking twice. All these services can be unlocked because of 5G new radio massive MIMO.

MIMO becoming enormous. Since MIMO systems are dependent on how the antennas and complicated algorithms expand, they are useful for wireless communication through the implementation of multiple antennas that enhance connectivity, offer better user experience and speed up with better performance. MIMO algorithms work out how the data transmitted or received through multiple antennas with proper management. For the smooth working of MIMO, there is a need of synchronizing way with great coordination between network and mobile devices. MIMO becomes massive MIMO after adding more number of antennas at the base station; this type of settlement helps in focusing energy, which will improve the throughput as well as efficiency, with the increment of complexity in the system which can be maintained by proper coordination with the operations [7].

Building blocks of MIMO systems. They are spatial diversity, spatial multiplexing and beamforming [8].

Spatial diversity. Spatial diversity is also known as space diversity. The main objective is to improve the reliability and quality of a wireless connection by linking two or more antennas. This scheme does not use additional bandwidth and increases the average SNR.

Spatial multiplexing. Spatial multiplexing is a transmission technique in which multiple data streams are simultaneously used by both transmitter and receiver to carry data within the same frequency band to maximize information capacity. This multiplexing technique also improves performance since multiple messages are transmitted simultaneously without any interference with each other as they are separated.

Beamforming. Beamforming is also known as spatial filtering. It is used in sensor arrays for directional signal transmission or reception. This means that wireless signals focus in a particular direction than broadcasting them. This is a signal processing technique that helps to improve the performance and capacity of the system. The phenomenon of directional transmission and broadcast transmission is applicable in laser light and flashlight. Flashlight gives broad light, whereas a laser pinpoints the location like beamforming [8,9].

In Figure 4.5, it is depicted that beamforming can give direction to the radio energy with the help of a radio channel toward a specific receiver. It is also seen that by varying

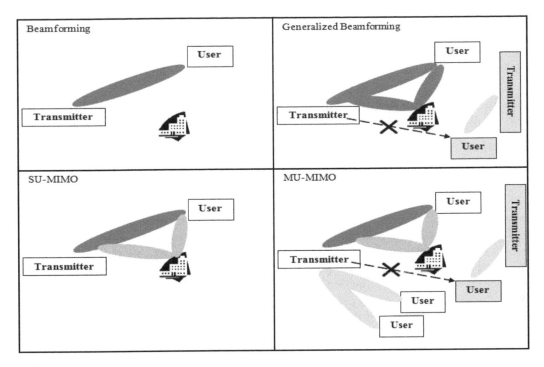

FIGURE 4.5
Beamforming in MIMO. MIMO, multiple-input multiple-output; MU, multiuser; SU, single-user.

the amplitude and phase of the transmitted signals, the constructive addition of the corresponding signals at the user equipment receiver can be achieved, which increases received signal strength and thus the end-to-end throughput. In the same way, during the reception of signals, beamforming is the skill to assemble the signal energy from a specific and precise transmitter. Beamforming in MIMO is best shown in Figure 4.5.

Mobile generates a feedback signal to identify its current position in the network; if a feedback signal is used to offer services to the mobile user, then it can reduce interference between beams directed in differing directions. Since mobile users are moving rapidly, these types of focused beams provide uninterrupted better services to the user.

4.4.2 Massive Multiple-Input Multiple-Output

Massive MIMO helps 5G to achieve faster data rates and enhance the potential level of 5G. Some advantages of Massive MIMO are as follows:

Improved coverage. Coverage is improved drastically with massive MIMO. Users get constant coverage without disturbance across the whole network even at the edges of a cell site so that users can get higher data rates in each corner of a cell site. Moreover, beamforming enables dynamic coverage for mobile users traveling in cars and also adjusts the coverage to suit the user, even in locations with the weaker network coverage.

Increased network capacity. Network capacity is defined as the ratio of total data volume provided to a user and the maximum number of users who can be served with some expected service. Massive MIMO contributes to increasing network capacity by enabling 5G NG (new ration) in the higher frequency range and by employing MU-MIMO where multiple users are served with equal resources such as time and frequency.

Better performance. Both the aforementioned benefits provide a better experience to users concerning sharing big data files with ease. Secondly, enabling users to download high-definition movies quickly. In addition, the applications that consume the Internet constantly for their operations also work smoothly.

4.4.3 In Terms of Power Consumption in Massive Multiple-Input Multiple-Output [10]

The total power consumed P is the sum of uplink and downlink transmissions in a massive MIMO system, which can be represented as:

$$P = P_A + P_c + P_s$$

where P_A represents the total uplink and downlink power consumed by the power amplifiers (P_A) at the base station and the user equipment, P_c indicates the total uplink and downlink circuit power expenses and P_s indicates various system hardware components used.

In comparison with the conventional system (LTE), massive MIMO systems achieve more energy-efficient gains that depend on the size of the system. As we know, the conventional system uses large signal processing techniques that require more energy to process; thus on increasing the size of the system, large computational power is consumed.

Energy efficiency (E.E), represented by

$$E.E = R/P$$

where R is represented by system throughput and P is represented by the power spent in achieving R.

After looking at the above equation, it is found that the massive MIMO system can achieve maximum throughput performance with low power consumption. In this direction, many approaches have been introduced for designing energy-efficient massive MIMO that includes the building of low-complexity algorithms for a base station in which multiuser detection and user scheduling can be implemented to minimize the power expenses. Other methods are also developed like redesigning of transmitters and receivers, selection of antenna, resource utilization and flexible hardware selection [11–13].

Many other technologies such as full-duplex, millimeter wave (mmWave), energy harvesting (EH), heterogeneous networks (HetNets) and cloud-based radio access are also accelerating as potential enablers for 5G. Each different technology offers a different set of performance benefits such as for enhancing throughput, mmWave operations are done with larger transmission bandwidths, and for less battery power expenses, EH is considered with the use of renewable energy. In a dense network, HetNets are used to achieve large throughput gains. In future, hybrid massive MIMO systems can be established depending upon 5G architectures enabled by massive MIMO and other 5G technologies with less power consumption, i.e., energy-efficient hybrid massive MIMO systems [14].

User requirements are increasing every day that leads to high-pressure demands on the radio access network to provide good capacity, increased coverage and throughput. To manage an exponential increase in data rates, mobile network operators (MNOs) must develop the radio access network in a way so that it can achieve a low cost per bit for end user performance. For developing such networks, there is a need for advanced antenna systems (AASs) that provide superior performance in both uplink and downlink and are also cost-effective. This AAS advances in the integration of radio, baseband and antennas to provide a reduction in the digital processing cost of advanced beamforming and MIMO.

AASs play a vital role for MNOs, as they offer better user performance, improved coverage and capacity on the existing network sites. This is essential for implementing 5G on existing cell sites and grids.

4.4.4 Advanced Antenna System

A distributed MIMO system is made up of many transmission points connected to a central server. Multiple antennas are used to transmit different MIMO streams at different transmission points. Optimized channel capacity is obtained with more number of spatial streams as a whole system, which acts as a virtual antenna array [11]. AAS with distributed MIMO is shown in Figure 4.6.

In Figure 4.6, coverage and throughput with one transmission point, with dual transmission point and with distributed MIMO are shown.

MIMO transmission is limited to rank two in a single transmission point but increased when distributed MIMO transmission principle is used, thus resulting in improved spatial multiplexing, transmission capacity and throughput. Distributed MIMO plays a vital role in improving the performance of future 5G communication systems.

Distributed MIMO (D-MIMO) provides the benefits of beamforming gain and suppression of spatial interference present in conventional massive MIMO. Hence, distributed MIMO is far better than conventional massive MIMO in terms of performance. D-MIMO also boosts the transmission and reception diversity and increases the degree of freedom and overall efficiency. But hardware complexity also increases with more energy consumption during transmission due to more signal processing. When point-to-point

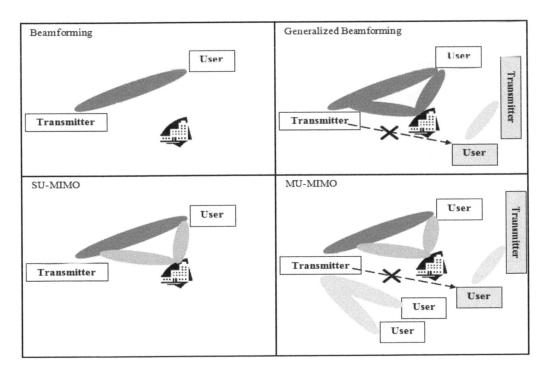

FIGURE 4.6
MIMO with single transmission point, MIMO with dual transmission point and distributed MIMO. MIMO, multiple-input multiple-output; MU, multiuser; SU, single-user.

communication links are established for error-free transmission, the complexity of the receiver system is more concerned rather than transmitter system; thus, proper synchronization and good control of interference are required.

An AAS is an arrangement of an AAS radio and a combination of AAS characteristics. An AAS radio is generated by an antenna array that is closely integrated with the software and hardware required to receive and transmit radio signals and algorithms for signal processing to enable the execution of AAS features. AAS characteristics are also referred to as multiantenna techniques which include beamforming and MIMO. Such types of characteristics are already available with conventional systems in present LTE networks. Applying AAS characteristics to an AAS radio results in considerable performance gains with the higher degrees of freedom given by the more number of radio chains, also termed as massive MIMO. The beams created by an AAS are continuously modified to the environment to give high performance in both uplink and downlink [15]. As compared with conventional systems, the combination used in the AAS provides much greater steerability, adaptivity and flexibility in correspondence with the propagation of multipath signals and the radiation pattern of the antenna with time-varying traffic.

These types of developments have made AASs a feasible option in existing 4G and future 5G mobile networks for large-scale deployments. The AAS ensures high-tech beamforming and MIMO techniques that are prominent and have an influential impact on improving user experience, coverage and capacity. This system will affect the performance of a network in a broad and better way in both uplink and downlink. This system equipped with beamforming in horizontal and vertical directions provides a better option for end users dwelling in the environment, which has tall buildings and dense area. In less dense areas, we do not need beamforming in the vertical direction, which will be cost-effective. High AAS performance can be obtained without the requirement for many MIMO layers, thus enabling network with better coverage, speed, connectivity and capacity in futuristic mobile radio networks.

The futuristic aim of 5G cellular communication is to address the high demands of capacity and data rates by the users. To fulfill this purpose, large bandwidth coverage is needed as compared with the conventional cellular system. Available solution for this purpose is utilizing higher frequency bands (30–300 GHz) of millimeter wavelengths.

4.5 Challenges in Distributed Multiple-Input Multiple-Output

There are several challenges in D-MIMO explained as follows:

Tracking and channel update. The major requirement of distributed MIMO is to obtain knowledge of available channels. Prior information about downlink channels is required to apply techniques such as beamforming. But fetching knowledge of downlink channels and tracking them are tedious job roles because both add additional load to the system.

Power control. Proper channel information is a key requirement for any MIMO multiuser network. Similarly, in wireless receivers, we need an extra controller to amplify and scale the received signals as per the requirement due to their weak strength. Now as we know that distributed MIMO utilizes multiple transmitters and receivers working together as a system such that each unit needs a power supply that is synchronized by the controller so that they all work coordinately. Timing and frequency synchronization are the core challenges to achieve to make the system work synchronously.

System complexity. Implementation of MIMO on the existing network is a complex scenario since MIMO requires multiple antennas. Hence, managing all of them together is a tedious job.

Antennas concerns. Designing antennas with proper interspacing and antenna length is a complex task for any distributed MIMO. The reason of complexity is that antennas are the main part of the system for signal transmission and reception and thus require precision failing that leads to poor signal transmission and reception. The number of elements in an antenna and right elevation angle to mount antenna are also prime concerns.

4.6 Conclusions

It is observed that various parameters for D-MIMO will increase the complexity of the system, but on the other hand, it also increases the capacity that provides higher data speed and coverage of radio links in correspondence to 5G mobile communication. Beamforming, spatial diversity and spatial multiplexing techniques improve the system's performance for making 5G communication and advance antenna systems better. D-MIMO also provides good signal transmission and reception during busy hours. Hence, the distributed MIMO network for 5G mobile communication will enhance the performance in terms of capacity and coverage.

References

1. Biglieri, E., Calderbank, R., Constantinides, A., Goldsmith, A., Paulraj, A., & Poor, H. V. (2007). *MIMO wireless communications.* Cambridge University Press.
2. Goldsmith, A., Jafar, S. A., Jindal, N., & Vishwanath, S. (2003). Capacity limits of MIMO channels. *IEEE Journal on Selected Areas in Communications*, 21(5), 684–702.
3. Making India 5G ready, Report of the 5G High level forum, prepared by The Steering Committee. (2018). http://dot.gov.in/whatsnew/making-india-5g-ready-report-5g-high-level-forum
4. Popovski, P., Trillingsgaard, K. F., Simeone, O., & Durisi, G. (2018). 5G wireless network slicing for eMBB, URLLC, and mMTC: A communication-theoretic view. *IEEE Access*, 6, 55765–55779.
5. Zhao, X., Jia, P., Zhang, Q., Li, Y., Xing, R., & Liu, Y. (2019). Analysis of a distributed MIMO channel capacity under a special scenario. *EURASIP Journal on Wireless Communications and Networking*, 2019(1), 189.
6. Xu, J., & Zhang, J. (2013). A spatial multiplexing MIMO scheme with beamforming and space-time block coding for downlink transmission. In 2013 *IEEE 24th Annual International Symposium on Personal, Indoor, and Mobile Radio Communications (PIMRC)* (pp. 1271–1275). IEEE.
7. Prasad, K. S. V., Hossain, E., & Bhargava, V. K. (2017). Energy efficiency in massive MIMO-based 5G networks: Opportunities and challenges. *IEEE Wireless Communications*, 24(3), 86–94.
8. Bogale, T. E., & Le, L. B. (2016). Massive MIMO and mmWave for 5G wireless HetNet: Potential benefits and challenges. *IEEE Vehicular Technology Magazine*, 11(1), 64–75.
9. Halvarsson, B., Karam, E., Nyström, M., Pirinen, R., Simonsson, A., Zhang, Q., & Ökvist, P. (2016, June). Distributed MIMO demonstrated with 5G radio access prototype. In 2016 *European Conference on Networks and Communications (EuCNC)* (pp. 302–306). IEEE.
10. Hamed, E., Rahul, H., Abdelghany, M. A., & Katabi, D. (2016, August). Real-time distributed MIMO systems. In *Proceedings of the 2016 ACM SIGCOMM Conference* (pp. 412–425). ACM.

11. Pathak, Y., Arya, K. V., & Tiwari, S. (2019). Feature selection for image steganalysis using levy flight-based grey wolf optimization. *Multimedia Tools and Applications*, 78(2), 1473–1494.

12. Pathak, Y., Arya, K. V., & Tiwari, S. (2019). An efficient low-dose CT reconstruction technique using partial derivatives based guided image filter. *Multimedia Tools and Applications*, 78(11), 14733–14752.

13. Bagwari, A., Bagwari, J., Tomar, G. S., & Bhadoria, R. S. (2018). New spectrum sensing technique for advanced wireless networks. In *Advanced Wireless Sensing Techniques for 5G Networks* (pp. 91–108). Chapman and Hall/CRC.

14. Mandloi, M., Bhatia, V., & Bhadoria, R. S. (2016). Interference alignment in MIMO cognitive radio networks. In *Introduction to Cognitive Radio Networks and Applications* (pp. 243–258). Chapman and Hall/CRC.

15. Jain, R., Jha, A. K., Bhadoria, R. S., & Arya, K. V. (2016). Analysis for cognitive radio sensor network architecture and its role in dynamic spectrum management. In *Introduction to Cognitive Radio Networks and Applications* (pp. 275–294). Chapman and Hall/CRC.

5

An Operation-Reduced Fast Modeling in 5G Communication Systems

Vladimir Mladenovic
University of Kragujevac

Sergey Makov
Don State Technical University

Yigang Cen
Beijing Jiaotong University

Asutosh Kar
Indian Institute of Information Technology

CONTENTS

5.1 Introduction

Modern research in the engineering field of sciences usually begins with theoretical analysis. The next step may be a simulation to prove the theoretical assumptions. The final step is practical implementation and measurements. Typically, theoretical performances are performed using ideal parameter values as an infinite number of samples or known signal values for the range from $-\infty$ to $+\infty$. The simulation is performed to gain insight into the functioning of the system or process. The result can be expected to agree well with the simulation for a finite number of measured values. The first critical step is adjusting the simulation parameters and writing code that fits the theory. The second key step is to select the number of iterations or the number of input samples. Symbolic processing (SP) can help to run the simulation code without errors, as well as to find errors in some published results. Also, SP can be used to find processing errors as a closed-form expression to calculate the number of iteration steps required or error functions due to a finite length word. Therefore, SP can help to gain insight into how the system works, which prefers to experiment with numerical simulations. SP corrects for some discrepancies between theoretical properties and numerical simulations. Numerical processing can produce unsatisfactory results and lead researchers to erroneous conclusions. Using symbolic tools, it is easy to find and prove errors for better insight into the whole process of analysis, simulation and modeling.

The main tools, used in telecommunications for calculating, designing and analyzing, are based on the numerical-only algorithms. On the other hand, the visualization of big data, which is used for processing, needs to be done in order to gain a better understanding of the phenomena under study. For such a thing, a functional visual interface is necessary to provide where all parametric of interest can be seen. So many computer-based numerical algorithms are developed as a consequence of development of hardware and computer techniques in numerical mathematic, adapted to computer calculations, and the main task of numerical processing is construction of algorithms, but the drawbacks are that a tremendous amount of numerical data are generated, the user might easily lose insight into the phenomenon being investigated and numerical computation manipulates with numerical values, so very often it is important to take care about final results: accuracy, execution time of numerical algorithm, efficiency-based reasoning and so on.

It is well known that researchers and engineers use more with methods in calculations of systems and processes when compared with traditional information delivery through the presentation. This is especially obvious for the upcoming technologies in the field of telecommunications, such as 5G networks. Their rapid development and standardization bring a large number of complex techniques to transfer information. Also, the modern 5G communication systems require near real-time estimation of error probability. Today, the application of the level crossing rate (LCR) and the average duration of a fade (ADF) is becoming increasingly popular because it depends on time and distance, which will be discussed in more detail in this chapter. The complexity of the system will be reflected in an increase in the number of customers, services increasing and a comprehensive integration of individual *smart* systems. It will be especially interesting challenges to the saving of the frequency space, device-to-device channels, fixed-to-vehicle channels, mobile-to-mobile communications, vehicle-to-vehicle channels, vehicular communications and moving scatterers. This will result needs for finding various algorithms and methodologies to speed up the calculation as an integral step in the design of these systems. On the other side, the calculations during design necessity use of mathematical integrals include iterative processes to get error probability. In this sense, this chapter presents a method that contributes to the calculation tools of complex techniques in these networks called the operation-reduced calculation method [1]. The presented method helps engineers and researchers to accelerate their calculations by reducing the number of operations. It will greatly help with a numerical calculation that can be considered as a link between theoretical and experimental analysis. An original procedure is developed using Wolfram language (WL) for efficiently solving the optimal result. The main motivation for the development and application of this algorithm is the question whether a system exists that must be extremely fast and accurate enough to calculate error probability. Such a response is provided in this chapter.

Theoretical, experimental and computational approaches are the basis for each study of the observed phenomenon in general. Nowadays, it is intended that every scientific and experimental result should be put into a function for the use. So the commercial use of products and services and many engineering uses emanate from a scientific to an engineering approach.

Searching for large databases [2,3], solving the complex processes described by mathematical models and analyzing phenomena in communications in the information space (such as the transmission of wireless signals in urban environments) [4–6], continuous delivery information [7] at a high speed without stagnation in software engineering poses the challenges of emerging technologies in information technology. The common feature of these observations is to obtain results directly for further processing or exploitation. For this purpose, the complex mathematical tools are used to perform analyses and simulations of performances of the observed processes and systems. Most often, the classical mathematical analysis does not provide with final answers in closed form in such complex phenomena, so special functions are used to solve them. And when we cannot get results here, we use numerical methods. Most of these numerical tools include complex calculations, such as differential and integral equations, algebraic structures, using numerical mathematics algorithms such as Newton-Cotes, Romberg integration, Gauss-Christophe, trapezium rule, Gauss formulas and so on [8,9]. For these reasons, students may not understand the complete process or system or cannot perform the method performance analysis to the end in education. Engineers and researchers may have not a good insight into the impact of the important parameters necessary for the good investigation or design. Even more, the numerical computation generates a large amount of data, which may sometimes lead to erroneous results [10,11]. They may be the result of the finite word length in the records or errors during shortenings of numbers in fractions, for example. These ways do not provide a possibility to manipulate with analytic expressions. These issues can be overcome introducing a new method that keeps variables and parameters as symbols. The method is named the iteration-based simulation method (IBSM) and is described in detail in Refs. [11,12]. In addition to the IBSM, which provides symbolic analysis, we give the possibility that the analysis to be observed partially through the concept of microsimulation analysis, which additionally enhances the better viewpoint for the influence of parameters and variables. Also, we provide a strong method for fast computation that, together with the operation reduction, gives a very accurate result in a very short time. Computer algebra systems (CASs) are very important tools for all of these analyses, developments and research, which allow us a completely new approach to understand and solve very complex cases. This chapter presents the visual scientific tool; the aforementioned methods are directly applied to two examples. Both examples are very complex to analyze and get symbolic closed-form expressions, and its numerical analysis takes a long time.

5.2 Microsimulation Semisymbolic Analysis

The first step to solving many problems in the engineering sciences is to understand engineering fields such as control theory, signals and systems, energy systems and electronics. The next step is to transfer knowledge to the appropriate software, where the most commonly used procedures are when solving problems. Finally, the last step is to encapsulate expert knowledge into some programming code. According to our assuredness, one of the most appropriate software is WL [13]. Starting with the assumption that the reader has the elementary knowledge of using Wolfram, it will be shown how it is possible to use SP tools to quickly calculate second-order statistics in wireless communications, especially in 5G technologies. Finally, a GUI (graphical unit interface) will be represented that shows the knowledge implementation, developed in the WL, in the Matlab environment. From a user viewpoint, it can be considered as a displayed method, and the GUI contains the elements necessary to calculate the response, in the frequency or time domain, automated generation of implementation code, SP and data visualization of the system. The complete procedure, which includes rapid modeling algorithms and GUIs that have been developed, is called the WL5G tool, which involves the application of WL in 5G technologies. The examples in this chapter are partly taken from published results [1,12,14,15], with extended published results that first time have been integrated into the visual environment of the GUI.

A large number of simulations, without a guarantee that tolerances might not be exceeded and exact, is one of the numerous drawbacks of numerical-based tools. We have a focus on a few goals. The first one is solving any analysis in closed form and providing for further simplification and manipulation. We do it with an IBSM. The second is development algorithm for fast computation of the aforementioned method. And finally, we reduce the number of operations of algorithm preparing it for implementation. All phases of development and testing are observed by microsimulation semisymbolic analysis (MSSA).

The IBSM is developed using CAS to simplify complex algebraic expressions, expressed as a closed-form expression, and offers acceptable analytic and reduced form for further manipulation or simulation as a closed-form solution published as in Ref. [12]. Basically, there are integrals in the most of the analyses; we approach them by elementary calculating when the integrals are presented using Riemann sum. The method makes low-complexity implementation into the high-complexity structure. Such approach provides the implementation in the hardware environment.

The CAS is the field of mathematics and computer science that performs symbolic mathematical operations. The CAS is based on algebraic calculations and manipulations in the same way that we perform the expressions by hand. It exclusively includes working with symbols, and the numerical calculation is a special case for CAS. Since symbols are used as variables, it indicates that it deals with SP. SP concerns the development, implementation and application of algorithms that manipulate and analyze mathematical expressions. CAS provides deeper understanding and prepares students to learn and engineers to simulate and design. The programming language that is suitable as CAS is the WL. It shows an ability to manipulate symbolic expressions in a way similar to traditional manual derivation [13]. The WL has the well-characterized high-performance computing, and program codes can be generated in a compact and short form.

The basic idea of the IBSM is to introduce a new parameter to get a closed-form expression. Since iteration is a new parameter, we use a transformation to turn the integral into a sum, i.e., series. For this purpose, we use Riemann sum transformation, with respect the features of the improper integrals. In this way, we enable the obtained closed-form expressions to be used for further manipulation, simplification, fast computation, reduction of operations, testing and verification of results. The general form of the Riemann integral transformation into a series is given as follows [16,17]:

$$\int_a^b f(x)dx = \lim_{\|\Delta x\| \to 0} \sum_{i=1}^n f(x_i)\Delta x_i \tag{5.1}$$

By observing the integrals we performed in the previous session, we define two types of Riemann sums. One is a single sum, and the other one is a double sum. The first one is to solve the single and the other to solve the double integrals. So respecting expression (5.1), WL5G code is given in Figure 5.1.

```
RiemannSumSingle[f_,{x1_,a1_,b1_},q_]:=
```
$$\text{Sum}[\text{Evaluate}[(f/.\{x1\to(a1+(b1-a1)k/q)\})],\{k,1,q\}]\frac{(b1-a1)}{q}$$

```
RiemannSumDouble[f_,{x2_,a2_,b2_},{x3_,a3_,b3_},q_]:=
```
$$\text{Sum}[\text{Evaluate}[(f/.\{x2\to(a2+(b2-a2)k/q),x3\to(a3+(b3-a3)k/q)\})],\{k,1,q\}]\frac{(b2-a2)}{q}$$

$$\frac{(b3-a3)}{q}$$

FIGURE 5.1
WL5G for Riemann sum.

The mark q is value of iteration in defined transformation in Figure 5.1.

Microsimulation is a method to mimic a complex phenomenon through the description of its microcomponents. Essentially, it leaves the system free to develop without too many constraints and simplifying assumptions [18]. But when we use microsimulation with symbolic-only contain, and we change with particular numerical values in the final stage, it becomes MSSA. Furthermore, we will observe each element of the symbolic calculation through MSSA, which will further provide a better testing and verification of the fast computation as well as reduction of operations [19,20]. Also, MSSA provides calculation directly in the first run without the need for more simulation attempts.

For previously mentioned reason, we introduce the next step, which is the development of an algorithm for fast computation. For that purpose, we observe the expression as series. As a slight reminder, we introduce a short description to better explain the concept of fast computation. It is said that the series converges slowly if a large number of members of the series needs to be added in order to determine the sum with the required accuracy. During the addition of members of series by the technique of a member by member, the process takes place automatically and is interrupted when a selected criterion for error evaluation is fulfilled. Due to the final summing up, the absolute value of the relationship between the last member and until the calculated sums is most often used. This criterion is not always reliable, especially in the case of the addition of trigonometric series. An error caused by an interrupted summing is always higher than estimated. On the other hand, contemporary computing machines can quickly add a huge number of members of the series. But, due to the format limitation of the records in registers, a certain number of decimal places are cut, which in the process of summing leads to the accumulation of errors and not least to completely absurd results. Of interest, therefore, are procedures for speeding up the convergence of series such as Kummer, Aitken, Cesar, Euler and so on. This session presents a very effective method for accelerating the convergence of the series based on Kummer's transformation.

We adhere two essential theorems. The first one says if there are $\sum_{k=1}^{n} a_k$ convergences, then $\lim_{k\to\infty} a_k = 0$. The second says if $\sum_{k=1}^{n} a_k$ and $\sum_{k=1}^{n} b_k$ are positive series and if $\lim_{k\to\infty} \frac{a_k}{b_k} = \rho$, $(b_k \neq 0)$, then convergence or divergence occurs at same time. Kummer's transformation (more famous as Kummer's acceleration method) provides speeding up the convergence of many series. The basic concept is to subtract from a given convergent series $\sum a_k$, and another equivalent series $\sum b_k$ whose sum $C = \sum\limits_{k \geq 0} b_k$ is well known and finite. Kummer's transformation is described with expression:

$$\sum_{k=0}^{\infty} a_k = \rho \sum_{k=0}^{\infty} b_k + \sum_{k=0}^{\infty}\left(1 - \rho\frac{b_k}{a_k}\right)a_k = \rho C + \sum_{k=0}^{\infty}\left(1 - \rho\frac{b_k}{a_k}\right)a_k \qquad (5.2)$$

The convergence of the right-hand side of equation (5.19) is faster because $1 - \rho \cdot b_k/a_k$ tends to 0 as k tends to ∞ [21].

The complete procedure is shown in Figure 5.2.

Reduction of operations is done by counting all math operations and functions contained in final expressions. WL5G allows to directly perform counts. Mathematical operations and functions in WL5G can be viewed both symbolically and as commands. Operations are recognized using the `FullForm` command, and the counting is done using the `StringPosition` command. Since we have sums to show that the numbers are repeated q times, it follows the WL code that completely counts the operations:

```
InnerOperations=q*FullForm[aK[z,q];]
StringPosition[InnerOperations,{"Times","Power","Plus","Rational",
"BesselI","Log","Exp"}];
```

The orders of `Times`, `Plus`, `BesselI`, `Log`, `Exp` are used functions in close-form expressions. Similarly, changing s with position of $a_k[z,q]$, we get the number of operations in accelerated algorithm.

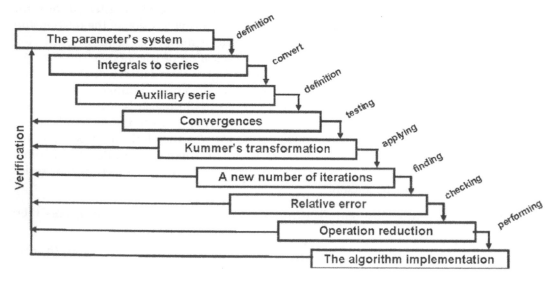

FIGURE 5.2
Steps of speeding up procedure and operation reduction.

5.3 Speeding Up Procedure and Operation Reduction

In this section, two applications of the operation reduction using fast computation of an IBSM with MSSA to problems of processing interest are presented to illustrate the fast computations required in the steps of the algorithm and also to demonstrate the wide applications to which it may be applied. The example in noncoherent ASK (amplitude-shift keying) with shadowing, interference and correlated noise illustrates the case with complex calculation. The second example in this section treats second-order statistics of SC macrodiversity system operating over gamma-shadowed Nakagami-m fading channels [22].

5.3.1 Noncoherent Amplitude-Shift Keying with Shadowing, Interference and Correlated Noise

Noncoherent ASK is a modulation scheme used to send digital information between digital equipment. The data are transmitted by the noncoherent system without carrier in a binary manner (Figure 5.3).

One of the most common models used in wireless communications to describe the phenomenon of multiple scattering is shadowing with interference as in Refs. [23–26]. The basic blocks of the system are shown in Figure 5.3. Both shadowing and interference cause strong fluctuations in the amplitude of the useful signal. Such case is present in urban areas and described as a log-normal distribution. In our analysis, we performed an outage probability. Transmitting signals by two symbols are observed in the noncoherent ASK system as in Refs. [27,28]. The noise, as a narrow-band stochastic process, is correlated, and this coefficient of correlation is marked by R ($R{\neq}1$). Mathematically, it can be described in the form of $n_i(t) = x_i(l){\cdot}\cos(\omega t) - y_i(t){\cdot}\sin(\omega t)$. The receiver is sheltered; there is no optical visibility toward transmitter, but there is interference $i_1(t) = A_1{\cdot}\cos(\omega t)$. If the system sends logical zero, then the signal $s_0(t) = a_0{\cdot}\cos(\omega t)$ has been sent, but if the system sends logical unit, then the signal $s_1(t) = a_1{\cdot}\cos(\omega t)$ has been sent. The parameters a_0 and a_1 are the signal elements from which the codewords are formed. Receiver detects information signal $b_0{\cdot}\cos(\omega t)$ and $b_1{\cdot}\cos(\omega t)$ with envelopes z_0 and z_1 after passing through a transmitting channel. The b_m ($m = 0, 1$) are elements of detected signals. The receiver system consists of filter and detector envelope. In the receiver input, the signal is in the form:

$$r_m(t) = b_m{\cdot}\cos(\omega t) + A_1{\cdot}\cos(\omega t) + x_m{\cdot}\cos(\omega t) - y_m{\cdot}\sin(\omega t) = z_m{\cdot}\cos(\omega t + \varphi_m), m = 0,1 \qquad (5.3)$$

with envelopes z_0 and z_1, and phases φ_0 and φ_1, respectively.

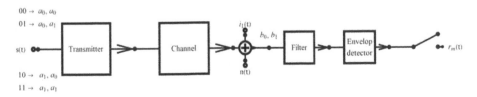

FIGURE 5.3
Noncoherent ASK system with interference $i_1(t)$. ASK, amplitude-shift keying.

The general form of the condition joint probability density function (JPDF) is:

$$p(x_0, x_1, y_0, y_1) = \frac{1}{4\pi^2 \sigma^4 \sqrt{1-R^2}} \exp\left\{-\frac{x_0^2 + x_1^2 + y_0^2 + y_1^2 - 2R(x_0 x_1 + y_0 y_1)}{2\sigma^2(1-R^2)}\right\} \quad (5.4)$$

The R is a coefficient of correlation and σ is variance. In order to ensure the continuation of the solving of the set of expressions, it is necessary to use the polar coordinates as follows:

$$x_0 = z_0 \cos\phi_0 - b_0 - A_1$$
$$y_0 = -z_0 \sin\phi_0$$
$$x_1 = z_1 \cos\phi_1 - b_1 - A_1 \quad (5.5)$$
$$y_1 = -z_1 \sin\phi_1$$

The next step is determining the condition JPDF. Substituting Equation (5.5) in Equation (5.4), it is obtained:

$$p(r_0, r_1/b_0, b_1, \varphi_0, \varphi_1, A_1) = p(x_0, y_0, x_1, y_1) \cdot |J| \quad (5.6)$$

Here, $|J|$ is Jacobian. A JPDF has log-normal distribution described as in Ref. [24]:

$$p_{ij}(b_0, b_1/A_1) = \frac{1}{2\pi\sigma^2 (b_0 + A_1)(b_1 + A_1)\sqrt{1-R^2}}$$
$$\times \exp\left\{-\frac{\left(10\log(b_0 + A_1) - a_i\right)^2 + \left(10\log(b_1 + A_1) - a_j\right)^2}{2\sigma^2(1-R^2)}\right\} \quad (5.7)$$
$$\times \exp\left\{\frac{2R\left(10\log(b_0 + A_1) - a_i\right)\left(10\log(b_1 + A_1) - a_j\right)}{2\sigma^2(1-R^2)}\right\}$$

For $i = j = 0$, the codeword 00 was sent; for $i = 1$ and $j = 0$, the codeword 01 was sent; for $i = 0$ and $j = 1$, the codeword 10 was sent; for $i = 1$ and $j = 1$, the codeword 11 was sent. So:

$$p(z_0, z_1/b_0, b_1, \phi_0, \phi_1, A_1) = \frac{z_0 z_1}{(2\pi)^2 \sigma^4 \sqrt{1-R^2}}$$
$$\times \exp\left\{-\frac{z_0^2 + z_1^2 - 2R(b_0 + A_1)(b_1 + A_1) - 2Rz_0 z_1 \cos(\phi_0 - \phi_1)}{2\sigma^2(1-R^2)}\right\} \times \quad (5.8)$$
$$\times \exp\left\{\frac{2(1+R)(b_0 + b_1 + 2A_1)(z_0 \cos\phi_0 + z_1 \cos\phi_1)}{2\sigma^2(1-R^2)}\right\}$$

The last expression can be transformed using modified Bessel function as in Ref. [29] before derivation of closed-form expression:

$$e^{x\cos\alpha} = \sum_{n=-\infty}^{\infty} I_n(x)\cos n\alpha \tag{5.9}$$

and applying trigonometric transformation:

$$\cos\alpha\cos\beta\cos\gamma = \frac{1}{4}\Big[\cos(\alpha+\beta+\gamma)+\cos(\alpha-\beta+\gamma)+\cos(\alpha+\beta-\gamma)+\cos(\alpha-\beta-\gamma)\Big] \tag{5.10}$$

expression (5.8) becomes

$$p\big(z_0,z_1/b_0,b_1,\phi_0,\phi_1,A_1\big)$$

$$= C\cdot e^{C_0} \sum_{n=-\infty}^{\infty}\sum_{m=-\infty}^{\infty}\sum_{k=-\infty}^{\infty} I_n(C_1)I_m(C_2)I_k(C_3)\times \cos[n\phi_0]\cos[m\phi_1] \tag{5.11}$$

$$\cos\big[k(\phi_0-\phi_1)\big]$$

Where,

$$C = \frac{z_0 z_1}{(2\pi)^2 \sigma^4 \sqrt{1-R^2}}$$

$$C_0 = -\frac{z_0^2 + z_1^2 - 2R(b_0+A_1)(b_1+A_1)}{2\sigma^2(1-R^2)}$$

$$C_1 = \frac{(b_0+A_1)(b_1+A_1)}{\sigma^2(1-R)}z_0 \tag{5.12}$$

$$C_2 = \frac{(b_0+A_1)(b_1+A_1)}{\sigma^2(1-R)}z_1$$

$$C_3 = \frac{R\cdot z_0 z_1}{\sigma^2(1-R^2)}$$

And using Bessel identity $I_n(x)=I_{-n}(x)$, it follows

$$p\big(z_0,z_1/b_0,b_1,A_1\big) = 2\pi^2 C \cdot e^{C_0(b_0,b_1,A_1)} \times \sum_{n=0}^{\infty} I_n\big[C_1(b_0,b_1,A_1)\big]\cdot I_n\big[C_2(b_0,b_1,A_1)\big]\cdot I_n[C_3] \tag{5.13}$$

The present interference is with the Rayleigh distribution described over probability density function (PDF) as [25,26]:

$$p(A_1) = \frac{A_1}{\sigma^2}\exp\left\{-\frac{A_1^2}{2\sigma^2}\right\}; \; 0 \le A_1 \le \infty \tag{5.14}$$

In order to eliminate the interference, it is necessary to perform averaging for all values of interference A_1.

$$p_{ij}(z_0,z_1/b_0,b_1) = \int_0^\infty p(z_0,z_1/b_0,b_1,A_1)\cdot p(A_1)dA_1 \tag{5.15}$$

The integral (5.15) is solved using integral:

$$\int_{-\pi}^{\pi}\int_{-\pi}^{\pi}\cos n\phi_0\cos m\phi_1\cos k(\phi_0-\phi_1)d\phi_0\,d\phi_1 = \begin{cases} \pi^2, & |n|=|m|=|k| \\ 0, & n\neq m\neq k \end{cases} \tag{5.16}$$

Distribution is obtained by averaging with φ_0 and φ_1 for all values between $-\pi$ and π.

$$p(z_0,z_1/b_0,b_1) = \int_{-\pi}^{\pi}\int_{-\pi}^{\pi}p(z_0,z_1/b_0,b_1,\phi_0,\phi_1)d\phi_0\,d\phi_1 \tag{5.17}$$

Distributions of envelopes for all combinations of codeword are obtained by the integration for all values b_0 and b_1. So, when the codeword $|ij|$ ($i=0, 1; j=0, 1$) has been sent (marked with H_iH_j in expression (5.18)) and when the same are detected on the input of receiver (marked with D_iD_j), the detection of the signals is described in the form:

$$P(D_iD_j/H_iH_j)=p_{ij}(z_0,z_1) = \int_0^\infty\int_0^\infty p(z_0,z_1/b_0,b_1)\cdot p_{ij}(b_0,b_1)db_0\,db_1 \tag{5.18}$$

The outage probability is obtained as:

$$P_{outage} = 1-\sum_{i=0}^{1}\sum_{j=0}^{1}P(H_iH_j)P(D_iD_j/H_iH_j) \tag{5.19}$$

where $P(H_iH_j)=P(H_i)\cdot P(H_j)=\frac{1}{2}\cdot\frac{1}{2}=\frac{1}{4}$, $i=0,1; j=0,1$. Expression (5.19) for the outage probability represents closed-form expression and very often cannot be present in the form of closed-form solution. Closed-form expression represents an implicit solution that is contained in a mathematical expression [12]. Let us define the difference between these two terms. Closed-form solution provides solved problem in terms of functions and mathematical operations from a given generally accepted set [30]. In other words, closed-form solution provides explicit solution of observed problem, whereas closed-form expression shows implicit or insufficient solution.

The JPDF in Equation (5.7) is described as in Figure 5.4.

After manipulations of expression (5.8) and substituting of expressions (5.12), it follows Figure 5.5.

Since interference A_1 is present (Equation 5.14) (Figure 5.6).

Averaging by all A_1 values is necessary to perform, according to expression (5.15). The general form of the condition JPDF defined in Equation (5.14) is described as in Figure 5.7.

$$JPDF[x\_,y\_,s\_,R\_,A1\_]:=\frac{1}{2*\pi*(s)^2*(b0+A1)*(b1+A1)*\sqrt{1-R^2}}*$$

$$Exp[-\frac{(10*Log[10,(b0+A1)]-x)^2+(10*Log[10,(b1+A1)]-y)^2}{2(s)^2(1-R^2)}]*$$

$$Exp[\frac{2*R*(10*Log[10,(b0+A1)]-x)*(10*Log[10,(b1+A1)]-y)}{2(s)^2(1-R^2)}]$$

FIGURE 5.4
Conditional joint probability density function for shadowing and interference in WL5G.

$$Cc=\frac{z0*z1}{(2\pi)^2(s)^4\sqrt{1-R^2}};$$

$$C0=-\frac{(z0)^2+(z1)^2-2R*(b0+A1)*(b1+A1)}{2(s)^2(1-R^2)};$$

$$C1=\frac{(b0+A1)*(b1+A1)}{(s)^2(1-R)}z0;$$

$$C2=\frac{(b0+A1)*(b1+A1)}{(s)^2(1-R)}z1;$$

$$C3=\frac{R}{(s)^2(1-R^2)}z0*z1;$$

FIGURE 5.5
Changing of coefficients for simplification.

$$LogNormalP[v\_,s\_,R\_,A1\_]:=$$

$$2\pi^2*Cc*Exp[C0]*\sum_{k=1}^{v}(BesselI[k,C1]*BesselI[k,C2]*BesselI[k,C3])$$

FIGURE 5.6
Rayleigh distribution for interference coded by WL5G.

$$pARayleght[A\_,s\_]:=\frac{A}{(s)^2}*Exp[-\frac{A^2}{2(s)^2}]$$

FIGURE 5.7
Log-normal distribution for noncoherent ASK in presence of shadowing and interference. ASK, amplitude-shift keying.

$$\frac{1}{q^2}(h_0 - h_1)^2$$

$$\sum_{k=1}^{q} \frac{1}{q}(h_1 - h_0) \sum_{k=1}^{q} z_0 z_1 \left(\frac{(h_1 - h_0)k}{q} + h_0 \right) I_1\left(\frac{R z_0 z_1}{(1-R^2)\sigma^2} \right) I_1\left(\frac{\left(2h_0 + \frac{2k(h_1-h_0)}{q} \right)^2 z_0}{(1-R)\sigma^2} \right) I_1\left(\frac{\left(2h_0 + \frac{2k(h_1-h_0)}{q} \right)^2 z_1}{(1-R)\sigma^2} \right)$$

$$\exp\left[-\frac{\left(\frac{10\log\left(\frac{2(h_1-h_0)k}{q} + 2h_0 \right)}{\log(10)} - a_0 \right)^2 + \left(\frac{10\log\left(\frac{2(h_1-h_0)k}{q} + 2h_0 \right)}{\log(10)} - a_1 \right)^2}{2(1-R^2)\sigma^2} + \right.$$

$$\frac{R\left(\frac{10\log\left(\frac{2(h_1-h_0)k}{q} + 2h_0 \right)}{\log(10)} - a_0 \right)\left(\frac{10\log\left(\frac{2(h_1-h_0)k}{q} + 2h_0 \right)}{\log(10)} - a_1 \right)}{(1-R^2)\sigma^2} - \frac{-2R\left(\frac{2(h_1-h_0)k}{q} + 2h_0 \right)^2 + z_0^2 + z_1^2}{2(1-R^2)\sigma^2} -$$

$$\left. \frac{\left(\frac{(h_1-h_0)k}{q} + h_0 \right)^2}{2\sigma^2} \right] \Bigg/ \left(4\pi(1-R^2)\sigma^8\left(\frac{2(h_1-h_0)k}{q} + 2h_0 \right)^2 \right)$$

FIGURE 5.8
Closed-form solution of PDF_{outage} of noncoherent ASK system. ASK, amplitude-shift keying; PDF, probability density function.

where s is marked variance σ, the R is correlation coefficient and v is order of the iterations. The finalization of IBSM is obtaining closed-form expressions of PDF and outage probability in terms of iterations. So:

The closed form of PDF_{outage} in Figure 5.8 provides the next parameters: the iteration q, h_0 and h_1 are the resolution of iteration, z_0 and z_1 are envelopes, R is coefficient of correlation and σ is variance.

Obviously, such expression cannot be obtained by hand and using numerical tools.

This resulting closed-form solution of P_{outage} is an expression that is ready for further processing. Accordingly, the viewpoint is an insight into the parameters and variables that participate in finally obtaining all the features of this case study.

It is now possible to draw characteristics, but this calculation would take too long time regardless of the accuracy we would use. On the other hand, for greater accuracy, we need a number of iterations, which again is not favorable for this form of expression.

Finally, the closed-form solution of P_{outage} is shown in Figure 5.9.

In our case, a member a_k represents a general member of the series in P_{outage} from closed-form solution and general term in in series of P_{outage} marked as a_k is shown Figure 5.10.

Convergence testing of the a_k verified that:

$$\lim_{\substack{k \to q \\ q \to \infty}} a_k = 0 \tag{5.20}$$

$$\frac{1}{q^2}\left\{\frac{\displaystyle\sum_{k=1}^{q}\frac{\displaystyle\sum_{k=1}^{q} q z^2 I_1\left(\frac{Rz^2}{(1-R^2)\sigma^2}\right)I_1\left(\frac{4k^2 z}{q^2(1-R)\sigma^2}\right)^2 \exp\left[-\frac{2z^2-\frac{8k^2R}{q^2}}{2(1-R^2)\sigma^2}-\frac{k^2}{2q^2\sigma^2}+\frac{R\left(\frac{10\log\left(\frac{2k}{q}\right)}{\log(10)}-1\right)^2}{(1-R^2)\sigma^2}-\frac{\left(\frac{10\log\left(\frac{2k}{q}\right)}{\log(10)}-1\right)^2}{(1-R^2)\sigma^2}\right]}{16\pi k(1-R^2)\sigma^8}}{q}\right. +$$

$$\sum_{k=1}^{q}\frac{\displaystyle\sum_{k=1}^{q} q z^2 I_1\left(\frac{Rz^2}{(1-R^2)\sigma^2}\right)I_1\left(\frac{4k^2 z}{q^2(1-R)\sigma^2}\right)^2 \exp\left[-\frac{2z^2-\frac{8k^2R}{q^2}}{2(1-R^2)\sigma^2}-\frac{k^2}{2q^2\sigma^2}+\frac{R\left(\frac{10\log\left(\frac{2k}{q}\right)}{\log(10)}+1\right)^2}{(1-R^2)\sigma^2}-\frac{\left(\frac{10\log\left(\frac{2k}{q}\right)}{\log(10)}+1\right)^2}{(1-R^2)\sigma^2}\right]}{16\pi k(1-R^2)\sigma^8}}{q} +$$

$$2\sum_{k=1}^{q}\frac{1}{q}\left\{\sum_{k=1}^{q}\frac{1}{16\pi k(1-R^2)\sigma^8} q z^2 I_1\left(\frac{Rz^2}{(1-R^2)\sigma^2}\right)I_1\left(\frac{4k^2 z}{q^2(1-R)\sigma^2}\right)^2 \exp\left[-\frac{2z^2-\frac{8k^2R}{q^2}}{2(1-R^2)\sigma^2}-\right.\right.$$

$$\left.\left.\frac{k^2}{2q^2\sigma^2}+\frac{R\left(\frac{10\log\left(\frac{2k}{q}\right)}{\log(10)}-1\right)\left(\frac{10\log\left(\frac{2k}{q}\right)}{\log(10)}+1\right)}{(1-R^2)\sigma^2}-\frac{\left(\frac{10\log\left(\frac{2k}{q}\right)}{\log(10)}-1\right)^2+\left(\frac{10\log\left(\frac{2k}{q}\right)}{\log(10)}+1\right)^2}{2(1-R^2)\sigma^2}\right]\right\}\right)$$

FIGURE 5.9
Closed-form solution of outage probability P_{outage} of noncoherent ASK system with shadowing and interference. ASK, amplitude-shift keying.

Convergence testing is performed with assumptions that $0 \le R < 1$, $\sigma > 0$, $z \ge 0$ and $q \ge 1$.

The selection of auxiliary function is one of the most important points of the MSSA [31]. Testing many series, the authors of this chapter have highlighted the series that shows the best performance to accelerate convergence, i.e., fast computation with the optimum number of iteration. Comparative analysis of different auxiliary series can be the subject of

$$-\frac{1}{\pi q^2 (R^2-1)\sigma^8} z^2\, 2^{-\frac{20}{(R+1)\sigma^2 \log(10)}-2}\, I_1\!\left(\frac{R\,z^2}{(1-R^2)\sigma^2}\right)$$

$$\left(e^{\frac{2R}{(R^2-1)\sigma^2}}\, 2^{\frac{20}{(R+1)\sigma^2 \log(10)}+1}\left(\frac{k}{q}\right)^{\frac{20}{(R+1)\sigma^2 \log(10)}}+2^{\frac{40}{(R+1)\sigma^2 \log(10)}}\left(\frac{k}{q}\right)^{\frac{40}{(R+1)\sigma^2 \log(10)}}+1\right)$$

$$\left(\frac{k}{q}\right)^{-\frac{20}{(R+1)\sigma^2 \log(10)}-1} I_1\!\left(\frac{k^4\,z}{q^4(1-R)\sigma^2}\right)^2 \exp\!\left(\frac{\frac{2q^2(-R+z^2+1)-k^2(R^2+8R-1)}{q^2(R-1)}-\frac{200\log^2\left(\frac{2k}{q}\right)}{\log^2(10)}}{2(R+1)\sigma^2}\right)$$

FIGURE 5.10
General term in series of P_{outage} marked as a_k.

particular surveys, and it is provided to the reader(s) to do it. So, in our case, the auxiliary series is:

$$C=\sum_{k=1}^{\infty}\frac{1}{k^5 2^{k-1}} \tag{5.21}$$

The series converges to 2·log2. For the full force of the formula in expression (5.2), we make a small modification of the member b_k, respecting the convergence theorems which have been aforementioned. A new member becomes $b_k \to a_k + c_k$, so

$$s=\sum_{k=0}^{\infty}a_k=\rho\sum_{k=0}^{\infty}a_k+\sum_{k=0}^{\infty}\left(1-\rho\frac{a_k+c_k}{a_k}\right)a_k=\rho C+\sum_{k=0}^{\infty}\left(1-\rho\frac{a_k+c_k}{a_k}\right)a_k \tag{5.22}$$

where c_k is general term in Equation (5.21). Following the next step in MSSA, we derive the term ρ (Figure 5.11).

We check that the value ρ tends to 1 after convergence testing. The fast computation is performed by assuming how much iteration is required to calculate the outage probability P_{outage} obtained by the IBSM. Otherwise, a large number of iterations are required to calculate the closest exact values of P_{outage}, but computation takes time-consuming. Then, such the resulting P_{outage} equalizes with a new series obtained by the Kummer's transformation and performs point matching for the various values of the envelopes, followed by a new reduced number of iterations. After that, the verification of the obtained results is performed by checking the relative error, which determines the degree of adjustability of the algorithm [31]. Finally, we check the number of operations of calculations in the expression in Figure 5.9, and then, we obtain a reduced number of operations with a new decreased number of iterations.

After all symbolic derivations, we use closed-form solutions to get results directly in the first attempt. To obtain concrete numerical results, we need to set the initial parameters. Let us suppose that the closest exact value is obtained after 500 iterations by using the outage probability P_{outage} in Figure 5.9, and resolution of iteration is $h_0=0$ and $h_1=1$. Also, we take $z_0=z_1=z$ to simplify analysis.

$$\left(2^q z^2 \left(\exp\left(\frac{R\left(100\log^2(2) + 7\log^2(10) + \log^2\left(\frac{512}{5}\right)\right) + \log^2(10\,240)}{(R^2-1)\sigma^2\log^2(10)}\right) + \right.\right.$$

$$\exp\left(\frac{R\left(100\log^2(2) + 7\log^2(10) + \log^2(10\,240)\right) + \log^2\left(\frac{512}{5}\right)}{(R^2-1)\sigma^2\log^2(10)}\right) +$$

$$\left.2\exp\left(\frac{R\left(8\log^2(10) + \log^2\left(\frac{512}{5}\right) + \log^2(10\,240)\right) + \log^2(10) + 100\log^2(2)}{(R^2-1)\sigma^2\log^2(10)}\right)\right)$$

$$\left.I_1\left(\frac{z}{(1-R)\sigma^2}\right)^2 I_1\left(\frac{Rz^2}{(1-R^2)\sigma^2}\right)\exp\left(\frac{(2z^2+1)\log^2(10) - 2R(100\log^2(2) + 7\log^2(10))}{2(R^2-1)\sigma^2\log^2(10)}\right)\right)\Bigg/$$

$$\left(2^q z^2\, e^{\frac{2z^2+1}{2(R^2-1)\sigma^2}}\left(\exp\left(\frac{R\log^2\left(\frac{512}{5}\right) + \log^2(10\,240)}{(R^2-1)\sigma^2\log^2(10)}\right) + \exp\left(\frac{R\log^2(10\,240) + \log^2\left(\frac{512}{5}\right)}{(R^2-1)\sigma^2\log^2(10)}\right) + \right.$$

$$\left.2\exp\left(\frac{R\left(-100\log^2(2) + \log^2(10) + \log^2\left(\frac{512}{5}\right) + \log^2(10\,240)\right) + \log^2(10) + 100\log^2(2)}{(R^2-1)\sigma^2\log^2(10)}\right)\right)$$

$$I_1\left(\frac{z}{(1-R)\sigma^2}\right)^2 I_1\left(\frac{Rz^2}{(1-R^2)\sigma^2}\right) - 8\pi q(R^2-1)\sigma^8$$

$$\left.\exp\left(\frac{R\left(R\log^2(10) + 2\left(4\log^2(10) + \log^2\left(\frac{512}{5}\right) + \log^2(10\,240)\right)\right)}{2(R^2-1)\sigma^2\log^2(10)}\right)\right)$$

FIGURE 5.11
The element Kummer's transformation μ in Section 5.3.1.

The next step is calculation of the new values of the number of iterations that are reduced for various values of the envelope z. This is performed by the command of FindRoot[s==Poutage,{q,1}]. The s is a new expression obtained by Kummer's transformation in Equation (5.22), and P_{outage} is the closed-form solution in Figure 5.8. We take the range of values $z=\{1,15\}$ for a concrete case [31]. The experiments are performed for various values of the coefficient of correlation R ($R=7/10, 8/10$) and the variance σ ($\sigma=2, 3$). All calculations were performed with a precision of 10^{-6}. All tests are performed on PC Intel® Core™ i5-6500 CPU@ 3.2GHz, 8 GB RAM, 64-bit Operating System, Windows 10 and Mathematica Wolfram 11.1. The reduced number of iterations is shown in Table 5.1.

In Figure 5.12, the changing of the iteration values in term of envelope z is shown for the accelerated algorithm. It can be noted that the reduced number of iterations is not the same for each value of the envelope. The minimum number of iterations is $z=10$ where it is considered to be a value that provides a true detection. However, the number of iterations is in range of 9–35 if we observe the total range of the envelope, which is a significant reduction compared with 500 iterations.

Since the absolute error is not precisely characterized by accuracy (Figure 5.13), the relative error is used as:

$$\delta = \frac{s - P_{outage}}{P_{outage}} \tag{5.23}$$

TABLE 5.1

Reduced Number of Iterations

z	q_1 $R = 7/10; \sigma = 2$	q_2 $R = 7/10; \sigma = 3$	q_3 $R = 8/10; \sigma = 2$	q_4 $R = 8/10; \sigma = 3$
1	25	32	21	30
2	20	27	16	25
3	19	25	14	22
4	18	24	12	21
5	17	23	11	20
6	17	23	10	20
7	18	23	9	19
8	18	23	9	19
9	20	24	9	20
10	21	25	9	19
11	23	26	10	20
12	27	27	11	21
13	29	28	12	21
14	32	30	14	22
15	36	31	16	23

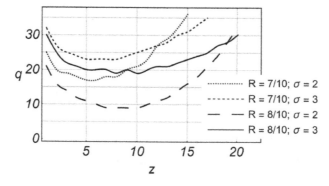

FIGURE 5.12
The number of iterations in terms of envelope z.

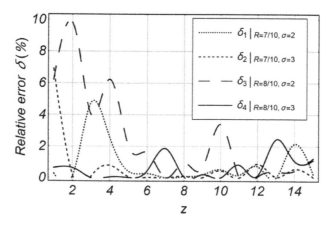

FIGURE 5.13
Relative error functions in terms of the envelope z.

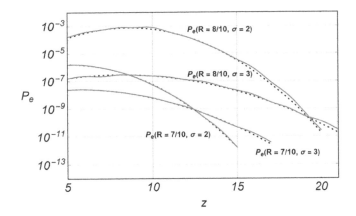

FIGURE 5.14
Comparative characteristics of P_{outage} and accelerated outage probability s.

Relative errors do not exceed more than 10% of the value. This indicates that the algorithm is quite accurate. In Figure 5.14, the comparative characteristics of P_{outage} and s are shown. The original characteristics are shown with solid lines, and the approximate values are dashed lines. The accelerated algorithm s is marked as $P_{e,approx}$.

The total calculation of formula P_{outage} consumes time of 1193.97 seconds (19 minutes, 54 seconds), so, the average time per iteration is 70.2335 seconds. Using of speeding up algorithm, total calculation time of accelerated formula is 1.25 seconds, so the average time per iteration is 0.0735294 seconds! WL5G for time-consuming is:

`Table[Timing[N[Poutage]],{z,15}]//Total`. Command Table provides a calculation for any value of envelope z, and command Timing provides the exact time of calculation. Command Total summarizes of all times per envelope. Similarly, changing the parameter Poutage with s (that is accelerated algorithm) in previous WL command line, we get time-consuming for fast computation. Our algorithm is accelerated:

$$\text{Ratio} = \frac{\text{time}(P_{outage})}{\text{time}(s)} = \frac{1193.97}{1.25} \approx 955\,\text{times} \tag{5.24}$$

In Figure 5.15, the number of operations in terms of number of iterations q for fast computation is shown. The number of iterations is fixed with $q=500$ for P_{outage} because we assumed on the beginning that this number of iteration is satisfied for closest exact value of P_{outage}. We can see that the number of operations for fast computation of IBSM is pretty much less than P_{outage}. For 500 iterations, we count 120,000 math operations for P_{outage}. The number of math operations changes in the range of ≈ 9000 to $\approx 34,000$, which is the result of a variety of the number of the iteration for fast computation.

5.3.2 Second-Order Statistics in Wireless Channels

The LCR and the ADF are important second-order statistical characteristics for description of the fading channel in mobile communications. These values are suitable for designing mobile radio communication systems and for their performance analysis. In digital telecommunications, a sudden drop in the value of the received signal directly leads to a drastic increase in the probability of error. For optimization of the coding system required for correct errors, it is not enough to know only how many times the received signal passes

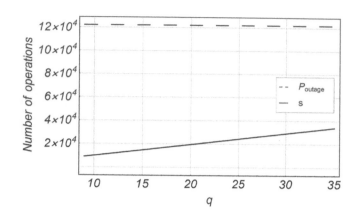

FIGURE 5.15
The number of operations in terms of the number of iterations q for fast computation. The number of iterations is fixed with $q=500$ for P_{outage}.

through the given level in time and already also how long on average that signal is below the specified level. The LCR and ADF are exactly the appropriate measures closely related to the quality of the received signal [26].

The LCR of the signal $Z(t)$, marked as $N_Z(z)$, is defined as the signal speed crossing through the level z with a positive derivative at the intersection point z. The ADF, marked as $T_Z(z)$, represents the mean time for which the signal overlay is below the specified z level.

The LCR at envelope z is mathematically defined by formula [24]:

$$N_Z(z) = \int_0^{\infty} \dot{z} p_{z\dot{z}}\left(z, \dot{z}\right) d\dot{z} \tag{5.25}$$

In expression (5.25), z is the envelope of received signal, $\dot{z}$ is its derivative in the time and $p_{z\dot{z}}\left(z, \dot{z}\right)$ is JPDF. The average fade duration (AFD) is determined as [24]:

$$T_Z(z) = \frac{F_Z(z \leq Z)}{N_Z(z)} \tag{5.26}$$

In expression (5.26), $F_Z(z \leq Z)$ represents the probability that the signal level $Z(t)$ is less than the level z. Evaluation and calculation of LCR and ADF are trivial in an environment where there are no large reflections, a large number of transmission channels and shadowing, which makes the mathematical description of the distribution of the signal simpler. However, in complex environments, obtaining LCR and ADF characteristics is very exhausting and long term. One such example of a complex environment is described in Ref. [22].

Namely, in this example, the LCR and ADF expressions were obtained. Their analytical shapes are closed forms, but complexity shows long-term computation. Thus, the LCR value is normalized by the Doppler shift frequency f_d [22, eq. 15]:

$$\frac{N_Z(z)}{f_d} = \frac{z^{M_1-1}}{\Gamma(M_1)\Gamma(c_1)\Gamma(c_2)}\left(\frac{N_1 m_1}{r_1}\right)^{M_1-1}\sqrt{\frac{2\pi}{m_1}}$$

$$\times \sum_{k=0}^{\infty} \frac{\left(N_1 m_1 z/r_1 \left((1/\Omega_{01})+(1/\Omega_{02})\right)\right)^{\frac{M_1-c_1-c_2+k-1/2}{2}}}{c_2 (1+c_2)_k \, \Omega_{01}^{c_1} \Omega_{02}^{k+c_2}} K_{(M_1+c_1+c_2+k-1/2)} \left(2\sqrt{\frac{N_1 m_1 z(\Omega_{01}+\Omega_{02})}{r_1 \Omega_{01}\Omega_{02}}}\right)$$

(5.27)

$$+ \frac{z^{M_2-1}}{\Gamma(M_2)\Gamma(c_1)\Gamma(c_2)} \left(\frac{N_2 m_2}{r_2}\right)^{M_2} \sqrt{\frac{2\pi}{m_2}}$$

$$\times \sum_{k=0}^{\infty} \frac{\left(N_2 m_2 z/r_2 \left((1/\Omega_{01})+(1/\Omega_{02})\right)\right)^{\frac{M_2-c_1-c_2+k-1/2}{2}}}{c_2 (1+c_2)_k \, \Omega_{01}^{c_1} \Omega_{02}^{k+c_2}} K_{(M_2+c_1+c_2+k-1/2)} \left(2\sqrt{\frac{N_2 m_2 z(\Omega_{01}+\Omega_{02})}{r_2 \Omega_{01}\Omega_{02}}}\right)$$

where the parameters are as follows: $\Gamma(x)$ denotes the gamma function, M_i is defined as $\left(m_i N_i^2\right)/r_i$, m_i is well-known Nakagami-m fading severity parameter, N_i denotes the number of identically assumed channels at each microlevel, the parameter r_i is related to the exponential correlation ρ_i, c_i denotes the order of gamma distribution, Ω_{0i} is related to the average powers of the gamma long-term fading distributions and $K_v(x)$ is the modified Bessel function of the second kind. Similarly, the AFD is obtained as [22, eq. 16]:

$$T_Z(z) = \frac{F_Z(z \le Z)}{N_Z(z)}$$

$$P_z(Z) = \frac{2\left(N_1 m_1/r_1\right)^{M_1}}{\Gamma(M_1)\Gamma(c_1)\Gamma(c_2)M_1} \times$$

$$\sum_{k=0}^{\infty}\sum_{l=0}^{\infty} \left(\frac{N_1 m_1}{r_1}\right)^k \frac{\left(N_1 m_1 z/r_1 \left((1/\Omega_{01})+(1/\Omega_{02})\right)\right)^{\frac{c_1+c_2+l-k-M_1}{2}}}{c_2 (1+c_2)_k \, \Omega_{01}^{c_1} \Omega_{02}^{l+c_2}} K_{(c_1+c_2+l-k-M_1)}$$

(5.28)

$$\left(2\sqrt{\frac{N_1 m_1 z(\Omega_{01}+\Omega_{02})}{r_1 \Omega_{01}\Omega_{02}}}\right) + \frac{2(N_2 m_2/r_2)^{M_2}}{\Gamma(M_2)\Gamma(c_1)\Gamma(c_2)M_2} \times$$

$$\sum_{k=0}^{\infty}\sum_{l=0}^{\infty} \left(\frac{N_2 m_2}{r_2}\right)^k \frac{\left(N_1 m_1 z/r_1 \left((1/\Omega_{01})+(1/\Omega_{02})\right)\right)^{\frac{c_1+c_2+l-k-M_2}{2}}}{c_2 (1+c_2)_k \, \Omega_{01}^{c_1} \Omega_{02}^{l+c_2}} K_{(c_1+c_2+l-k-M_2)}$$

$$\left(2\sqrt{\frac{N_1 m_1 z(\Omega_{01}+\Omega_{02})}{r_1 \Omega_{01}\Omega_{02}}}\right) +$$

As in the previous example, we define a general term a_k from expression (5.27) (Figure 5.16).

Using the expression in Figure 5.17, we derive the term ρ that tends to 1 when $q \rightarrow \infty$.

In this case, expressions (5.27) and (5.28) have already been provided in advance in a closed form where the iteration parameter q is present, so it is applied the IBSM. For computation of closest exact values of LCR and AFD, it takes about 100 iterations in Ref. [22].

$$\left(\sqrt{\pi} \sqrt{m} \, \Gamma(c+1) \Gamma(q+1) \, \Omega^{-2c-q} \left(-\frac{L\,m\,(R-1)^2}{L\,(R^2-1)-2\,R\,(R^L-1)} \right)^{-\frac{L^2\,m\,(R-1)^2}{L\,(R^2-1)-2\,R\,(R^L-1)}} \right.$$

$$2^{\frac{1}{20}z\left(-\frac{L^2\,m\,(R-1)^2}{L\,(R^2-1)-2\,R\,(R^L-1)}-1\right)+\frac{5}{4}} \, 5^{\frac{1}{20}z\left(-\frac{L^2\,m\,(R-1)^2}{L\,(R^2-1)-2\,R\,(R^L-1)}-1\right)}$$

$$\sqrt[4]{\frac{L\,m\,(R-1)^2\,\Omega\,10^{z/20}}{2\,R\,(R^L-1)-L\,R^2+L}} \left(-\frac{L\,m\,(R-1)^2\,\Omega\,2^{\frac{z}{20}-1}\times 5^{z/20}}{L\,(R^2-1)-2\,R\,(R^L-1)} \right)^{c+\frac{L^2\,m\,(R-1)^2}{2\,L\,(R^2-1)-4\,R\,(R^L-1)}+\frac{q}{2}}$$

$$K_{\frac{L^2\,m\,(R-1)^2}{L\,(R^2-1)-2\,R\,(R^L-1)}+2c+q+\frac{1}{2}}\left(2\sqrt{2}\sqrt{-\frac{10^{z/20}\,L\,m\,(R-1)^2}{(-2\,R^{L+1}+L\,R^2+2\,R-L)\,\Omega}}\right) \Bigg) \Bigg/$$

$$\left(c\,q!\,\Gamma(c)^2\,\Gamma(c+q+1)\,\Gamma\left(-\frac{L^2\,m\,(R-1)^2}{L\,(R^2-1)-2\,R\,(R^L-1)}\right) \right)$$

FIGURE 5.16
General term of LCR marked as a_k. LCR, level crossing rate.

Using Kummer's transformation, both LCR and AFD are calculated in the first iteration. All computations are performed taking the values of $m=1$, $L=2$, $\Omega=1$, $c=2$ and $R=1/5$. As an auxiliary series is used:

$$C = \sum_{k=1}^{\infty} e^{-k^2} \tag{5.29}$$

The series C converges to $(1/2)\cdot(\theta_3(0, e^{-1})-1)$, where $\theta_a(u,x)$ $(a=1,...,4)$ is the theta function, defined as [32]:

$$\theta_3(u, x) = 1 + 2\sum_{k=1}^{\infty} x^{k^2} \cos(2k \cdot u) \tag{5.30}$$

Figure 5.18 shows the comparative characteristics of LCR and accelerated LCR.

The deviation of the accelerated series is small in relation to the original series, and the relative error is shown in Figure 5.19, in the specified range of envelope $-35 \le z \le 30$.

The total calculation of formula LCR consumes time of 30.6563 seconds, so the average time per iteration is 0.437946 seconds. Using speeding up algorithm, total calculation time of accelerated formula is determined as 1.53125 seconds, so the average time per iteration is 0.021875. Our algorithm is accelerated:

$$\text{Ratio} = \frac{\text{time}(\text{LCR}_{\text{orig}})}{\text{time}(\text{LCR}_{\text{accelerated}})} = \frac{30.6563}{1.53125} \approx 20 \tag{5.31}$$

$$\left(\sqrt{\pi}\ \sqrt{m}\ e^{q^2}\ \Gamma(c+1)\Gamma(q+1)\ 2^{\frac{1}{20}z\left(-\frac{L^2\,m\,(R-1)^2}{L\,(R^2-1)-2\,R\,(R^L-1)}-1\right)+\frac{5}{4}}\ 5^{\frac{1}{20}z\left(-\frac{L^2\,m\,(R-1)^2}{L\,(R^2-1)-2\,R\,(R^L-1)}-1\right)}\right.$$

$$\sqrt[4]{\frac{L\,m\,(R-1)^2\,\Omega\,10^{z/20}}{2\,R\,(R^L-1)-L\,R^2+L}}\left(-\frac{L\,m\,(R-1)^2\,\Omega\,2^{\frac{z}{20}-1}\times 5^{z/20}}{L\,(R^2-1)-2\,R\,(R^L-1)}\right)^{c+\frac{L^2\,m\,(R-1)^2}{2\,L\,(R^2-1)-4\,R\,(R^L-1)}+\frac{q}{2}}$$

$$K_{\frac{L^2\,m\,(R-1)^2}{L\,(R^2-1)-2\,R\,(R^L-1)}+2\,c+q+\frac{1}{2}}\left(2\sqrt{2}\ \sqrt{-\frac{10^{z/20}\,L\,m\,(R-1)^2}{(-2\,R^{L+1}+L\,R^2+2\,R-L)\,\Omega}}\right)\right/$$

$$\left(c\,q!\,\Gamma(c)^2\,\Omega^{2\,c+q}\,\Gamma(c+q+1)\left(-\frac{L\,m\,(R-1)^2}{L\,(R^2-1)-2\,R\,(R^L-1)}\right)^{\frac{L^2\,m\,(R-1)^2}{L\,(R^2-1)-2\,R\,(R^L-1)}}\right.$$

$$\Gamma\left(-\frac{L^2\,m\,(R-1)^2}{L\,(R^2-1)-2\,R\,(R^L-1)}\right)+\sqrt{\pi}\ \sqrt{m}\ e^{q^2}\ \Gamma(c+1)\Gamma(q+1)$$

$$2^{\frac{1}{20}z\left(-\frac{L^2\,m\,(R-1)^2}{L\,(R^2-1)-2\,R\,(R^L-1)}-1\right)+\frac{5}{4}}\ 5^{\frac{1}{20}z\left(-\frac{L^2\,m\,(R-1)^2}{L\,(R^2-1)-2\,R\,(R^L-1)}-1\right)}$$

$$\sqrt[4]{\frac{L\,m\,(R-1)^2\,\Omega\,10^{z/20}}{2\,R\,(R^L-1)-L\,R^2+L}}\left(-\frac{L\,m\,(R-1)^2\,\Omega\,2^{\frac{z}{20}-1}\times 5^{z/20}}{L\,(R^2-1)-2\,R\,(R^L-1)}\right)^{c+\frac{L^2\,m\,(R-1)^2}{2\,L\,(R^2-1)-4\,R\,(R^L-1)}+\frac{q}{2}}$$

$$\left.K_{\frac{L^2\,m\,(R-1)^2}{L\,(R^2-1)-2\,R\,(R^L-1)}+2\,c+q+\frac{1}{2}}\left(2\sqrt{2}\ \sqrt{-\frac{10^{z/20}\,L\,m\,(R-1)^2}{(-2\,R^{L+1}+L\,R^2+2\,R-L)\,\Omega}}\right)\right)$$

FIGURE 5.17
The element *Kummer*'s transformation ρ in Section 5.3.2.

In Figure 5.18, the number of operations of LCR (N_Z) in terms of number of iterations q for fast computation is shown. The number of iterations is fixed with $q = 100$ for LCR$_{orig}$ because we assumed on the beginning that this number of iteration is satisfied for closest exact value of LCR$_{orig}$. For 100 iterations, we count 20,200 math operations for LCR$_{orig}$. The number of math operations is 1184 for LCR$_{accelerated}$ calculated in the first iteration for fast computation.

In the same way, the AFD is obtained by applying expression (5.22).

In Figure 5.20, the comparative characteristics of AFD and accelerated AFD are shown. There is a small deviation in range of $-35 \le z \le -28$, and this is perceived through a relative error in Figure 5.21.

The total calculation of formula AFD$_{orig}$ consumes time of 19553.1 seconds (5 hours 25 minutes), so the average time per iteration is 279.33 seconds (4 minutes 19.33 seconds). Using speeding up algorithm, total calculation time of accelerated formula is determined

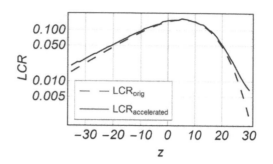

FIGURE 5.18
Comparative characteristics of LCR and accelerated LCR. LCR, level crossing rate.

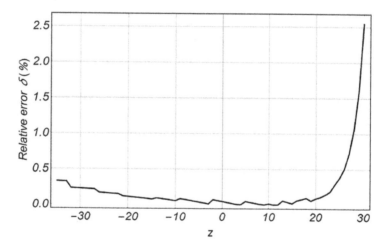

FIGURE 5.19
Relative error function of LCR in terms of envelope z. AFD, average fade duration; LCR, level crossing rate.

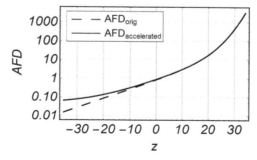

FIGURE 5.20
Comparative characteristics of AFD and accelerated AFD. AFD, average fade duration.

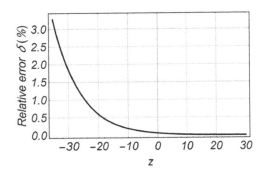

FIGURE 5.21
Relative error function of AFD in terms of envelope z. AFD, average fade duration.

as 1.29688 seconds, so the average time per iteration is 0.0185268! There is an obvious difference in the time calculation because the number of sums of AFD is increased in expression (5.27), where we have sums by k, l and q. In this case, our algorithm is accelerated:

$$\text{Ratio} = \frac{\text{time}\left(\text{AFD}_{\text{orig}}\right)}{\text{time}\left(\text{AFD}_{\text{accelerated}}\right)} = \frac{19553.1}{1.29688} \approx 15,000 \tag{5.32}$$

For 100 iterations, we count the imposing $344 \cdot 10^6$ math operations for AFD_{orig}. The number of math operations is 5619 for $\text{LCR}_{\text{accelerated}}$ calculated in the first iteration for fast computation. These methods can be modified to additionally speed up fast computation. We make an additional reduction of the number of iterations and get a new expression of the accelerated expression in the closed form shown in expression 5.22. This expression is used to test various values of coefficient correlation R (7/10 and 8/10) for various values of variance σ (2 and 3).

The new general term in the resulting sum, as well as the appearance of the new term ρ, is represented to test the convergence [15].

$$\rho^{(2)} = \lim_{k \to \infty} \frac{a_k^{(2)} + b_k}{b_k} \neq 0 \tag{5.33}$$

$$s^{(2)} = \sum_{k=0}^{\infty} \underbrace{\rho^{(1)} \cdot b_k + \left(1 - \rho^{(1)} \cdot \frac{a_k^{(1)} + b_k}{a_k^{(1)}}\right) a_k^{(1)}}_{a_k^{(2)}} \tag{5.34}$$

Exponential indices (1) and (2) denote the *Kummer's* transformation of the first and second order, respectively. By obtaining new closed-form expressions, it is possible to present the general term for the *Kummer's* transformation for the n order of repetition:

$$\rho^{(n)}(q) = \lim_{k \to \infty} \frac{a_k^{(n)}(q) + b_k}{b_k} \neq 0 \tag{5.35}$$

Since these parameters depend on the number of iterations q, the following expression shows the full form:

$$s^{(n)} = \rho^{(n)}(q) \cdot C + \sum_{k=0}^{\infty} \left(1 - \rho^{(n)}(q) \cdot \frac{a_k^{(n)}(q) + b_k}{a_k^{(n)}(q)}\right) a_k^{(n)}(q) \qquad (5.36)$$

By applying the same auxiliary function that is used in Ref. [6] (Figure 5.22)

$$C = \sum_{k=1}^{\infty} \frac{1}{k^5 2^{k-1}} \qquad (5.37)$$

Figure 5.23 shows the characteristics of reduction of the number of iterations of the first and second order of the *Kummer*'s transformation. The z is parameter of the envelope. It is clearly seen that there is a further decrease in the number of iterations and the relative error of the second order of the *Kummer*'s transformation.

The development of algorithms for fast computations that reduce the number of operations can be implemented in a GUI. In this case, a Matlab environment was used where the GUI was linked to the WL5G by exporting closed-form expressions. The GUI shows some details in Figure 5.24. The left side of the window contains basic initialization parameters, such as mobile user (MU) speed, number of scatterers, the magnitude of fast Fourier transform, number of samples and sampling rate relative to wavelength. Frequency, power, a radius around the MU and distance from the base station (BS) were selected for

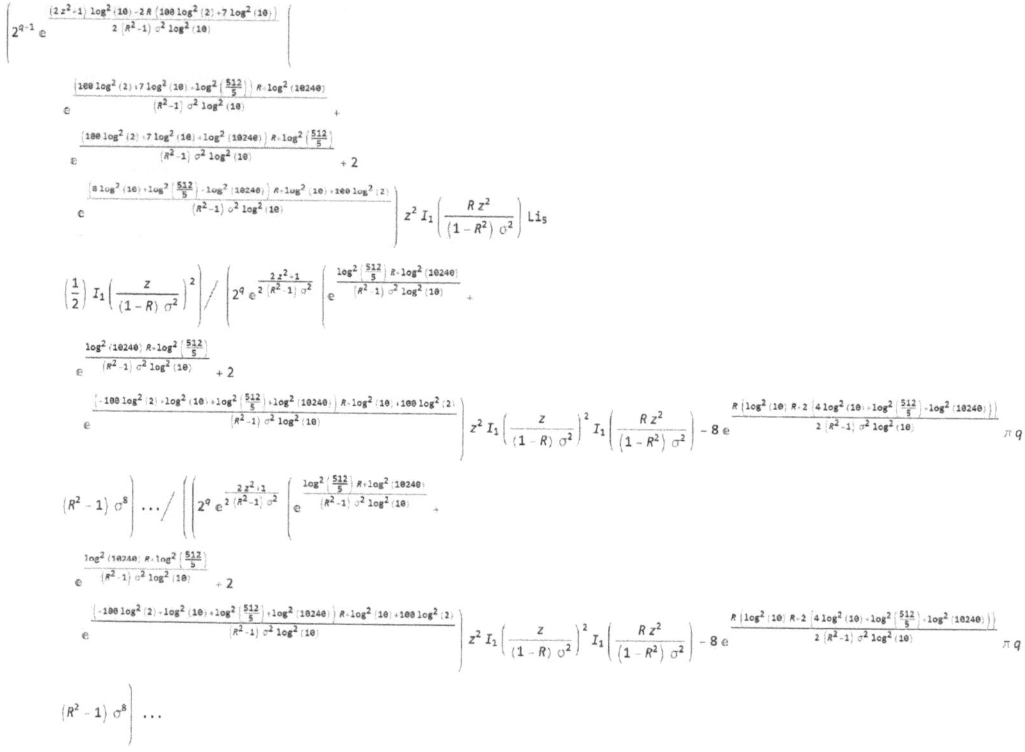

FIGURE 5.22
The fragment of closed-form expression of IBSM using second order of Kummer's transformation. IBSM, iteration-based simulation method.

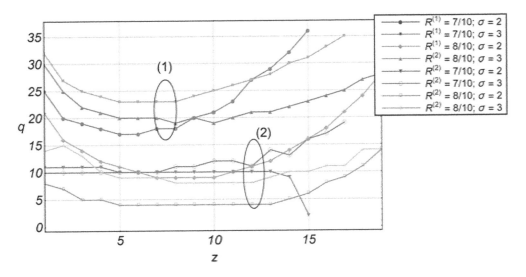

FIGURE 5.23
The comparative view of the number of iterations.

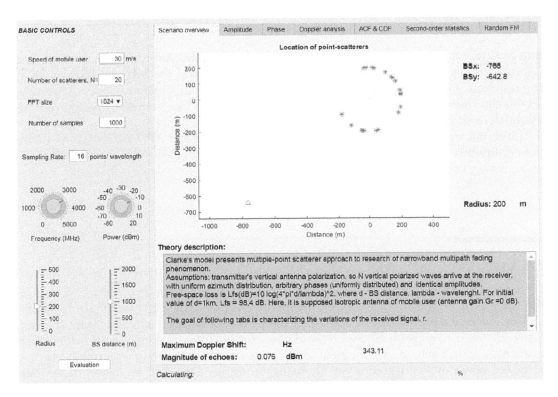

FIGURE 5.24
The parameter initialization implemented into GUI. GUI, graphical user interface.

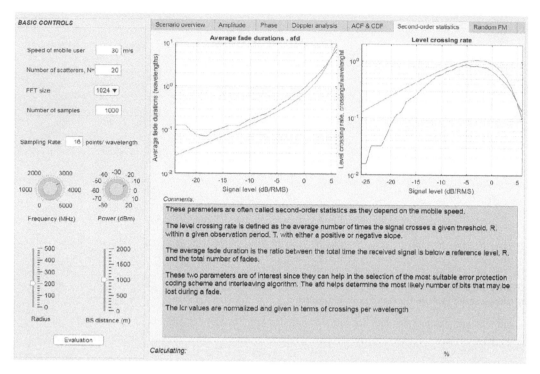

FIGURE 5.25
LCR and AFD implementation into GUI. AFD, average fade duration; LCR, level crossing rate.

fine-tuning. The results in Ref. [33] were partly used as the basis for the scenario. The first window shows a scenario overview showing the positions of the MU and BS, as well as the number of scatterers. The entire model is run via the Evaluation button. After that, all the algorithms presented in this chapter are started, and in a very short period of time, all the desired responses are obtained.

The characteristics of second-order statistics such as LCR and AFD are shown as a window of interest in Figure 5.25. Dynamic changes of all values in the initial part of the GUI make it very easy to track all the characteristics.

5.4 Concluding Remarks

Upcoming 5G technology will provide powerful development of new services based on the Internet of things (IoT), artificial intelligence (AI), big data and blockchain technologies. In the ocean of new services, new processes and phenomena will be present in rural, especially in urban areas. All aspects and behavior of different scenarios of information transmission in such an environment have not yet been sufficiently examined, but it is already known that very complex mathematics is required for their analysis. All measuring devices will be based on very sophisticated algorithms for measuring and detecting new parameters, it will be necessary for the results to be obtained in real time. Certainly, fast computation algorithms will be part of all new software tools that will facilitate fast simulations, and these methods will make a huge contribution to beyond.

References

1. V.Mladenovic, D.Milosevic, M.Lutovac. An Operation – Reduced Calculation Method for Solving Complex Communication Systems, *Studies in Informatics and Control*, Vol. 26, No. 2, pp. 213–218, 2017, ISSN 1220-1766.
2. S.C.Kan, Y.G.Cen, Y.H.Wang, V.M.Mladenovic. SURF Binarization and Fast Codebook Construction for Image Retrieval, *Journal of Visual Communication and Image Representation*, Vol. 49, pp. 104–114, Nov. 2017.
3. F.Zhanga, Y.G.Cen, R.Zhaoa, S.Hua, V.Mladenovic. Multi-Separable Dictionary Learning, *Signal Processing*, Vol. 143, pp. 354–363, 2017.
4. L.Lin, X.Ma, C.Liang, X.Huang, B.Bai. An Information-Spectrum Approach to the Capacity Region of the Interference Channel, *Entropy*, Vol. 19, p. 270, 2017,
5. P.H.Lin, E.A.Jorswieck. Multiuser Channels with Statistical CSI at the Transmitter: Fading Channel Alignments and Stochastic Orders, an Overview, *Entropy*, Vol. 19, p. 515, 2017.
6. J.Kakar, A.Sezgin. A Survey on Robust Interference Management in Wireless Networks, *Entropy*, Vol.19, p. 362, 2017.
7. L.Chen, Continuous Delivery: Overcoming Adoption Challenges, *Journal of Systems and Software*, Vol.128, pp.72–86, June 2017
8. G.Ferrari, G.Colavolpe, R.Raheli. *Detection Algorithms for Wireless Communications*, John Wiley & Sons, Ltd, UK, 2004.
9. I.A.Glover, P.M.Grant, *Digital Communications*, 3E, Harlow:Prentice Hall, 2010.
10. V.Pavlovic, V.Mladenovic, M.D.Lutovac, Computer Algebra and Symbolic Processing in Modern Telecommunication Applications: A New Kind of Survey, In Telecommunications: Applications, Modern Technologies and Economic Impact, Barringer, J.P.; Nova Science Publishers, pp. 29–116, 2014.
11. V.Mladenovic, S.Makov, V.Voronin, M.Lutovac. An Iteration-Based Simulation Method for Getting Semi Symbolic Solution of Non-Coherent FSK/ASK System by Using Computer Algebra Systems, *Studies in Informatics and Control*, Vol. 25, No. 3, pp. 303–312, 2016.
12. V.Mladenovic, D.Milosevic. A Novel-Iterative Simulation Method for Performance Analysis of Non-Coherent FSK/ASK Systems over Rice/Rayleigh Channels Using the Wolfram language, Serbian, *Journal of Electrical Engineering*, Vol. 13, No. 2, pp. 157–174, June 2016.
13. S.Wolfram, *An Elementary Introduction to the Wolfram Language*, Wolfram Media, Inc., Campaign, IL, 2017.
14. V.Mladenovic, D.Milosevic, M.Lutovac, Y. Cen, M.Debevc. An Operation Reduction Using Fast Computation of an Iteration-Based Simulation Method with Microsimulation-Semi-Symbolic Analysis, *A Special Issue of Entropy*, Vol. 20, No. 1, 62, 2018.
15. V. Mladenović, D. Milosević, M. Greconici, M. Lutovac. Multiple Procedures for Fast Computation of an Iteration-Based Simulation Method, in *Proceedings 5th International Conference on Electrical, Electronic and Computing Engineering*, IcETRAN, pp. 798–801, 2018.
16. H.Anton, *Calculus: A New Horizon*, 6E, New York:Wiley, pp. 324–327, 1999.
17. E.W.Weisstein, Riemann Sum, Wolfram Web Resource, Available online: http://mathworld. wolfram.com/RiemannSum.html (accessed on 01–09–2017).
18. F.Castiglione, *Microsimulation of Complex System Dynamics: Automata Models in Biology and Finance*. PhD thesis, Universität zu Köln, Koln, Germany, 2001, Available online: http://kups. ub.uni-koeln.de/487/ (accessed on 08-08-2017).

19. D.Grabowski, C.Grimm, E.Barke. Semi-Symbolic Modeling and Simulation of Circuits and Systems, in *Proceedings IEEE International Symposium on Circuits and Systems*, pp. 983–986, 2006.
20. L.Gil, M.Radetzki. Semi-Symbolic Operational Computation for Robust Control System Design, in *Proceedings 22nd International Conference on Methods and Models in Automation and Robotics (MMAR)*, pp. 779–784, 2017.
21. M.Abramowitz, I.Stegun. *Handbook of Mathematical Functions*, National Bureau of Standards, Dover, New York, 1964.
22. D.Stefanovic, S.Panic, P.Spalevic. Second-Order Statistics of SC Macrodiversity System Operating over Gamma Shadowed Nakagami-m Fading Channels, *International Journal of Electronics and Communications (AEÜ)*, Vol. 65, No. 5, pp. 413–418, 2010.
23. A.Radwan, J.Rodriguez. *Energy Efficient Smart Phones for 5G Networks*, Cham:Spinger, 2015.
24. A.Borhani, M.Patzold. Modeling of Vehicle-to-Vehicle Channels in the Presence of Moving Scatterers, *Vehicular Technology Conference* (VTC Fall), 2012 IEEE, pp. 1–5, Sept. 2012.
25. M.Patzold, *Mobile Radio Channels*, 2nd ed., Chichester:John Wiley & Sons, 2011.
26. M. Patzold, *Mobile Fading Channels*, New York: John Wiley & Sons, Ltd., 2003.
27. M.Lutovac, V.Mladenovic. M.Lutovac. Development of Aeronautical Communication System for Air Traffic Control Using OFDM and Computer Algebra Systems, *Studies in Informatics and Control*, Vol. 22, No. 2, pp. 205–212, 2013.
28. W.Martinez, A.Martinez. *Computational Statistics Handbook with MATLAB*, New York: Chapman & Hall/CRC, 2015.
29. G.Andrews, R.Askey, R.Roy. *Special Functions (Encyclopedia of Mathematics and Its Applications)*, Cambridge:Cambridge University Press, pp. 212–222, 2001.
30. J.M.Borwein, R.E.Crandall. Closed Forms: What They Are and Why We Care, *Notices of the American Mathematical Society*, Vol. 60, pp. 55, Jan. 2013.
31. V.Mladenovic, S.Makov, Y.G.Cen, M.Lutovac. Fast Computation of the Iteration-Based Simulation Method - Case Study of Non-Coherent ASK with Shadowing, *Serbian Journal of Electrical Engineering*, Vol. 14, No. 3, pp. 415–431, Oct. 2017.
32. http://reference.wolfram.com/language/ref/EllipticTheta.html
33. F.P.Fonta, P.M.Espineira. *Modeling the Wireless Propagation Channel a Simulation Approach with MATLAB*, Chichester:John Wiley & Sons, Aug. 2008.

6

5G ML-Based Networks toward 6G: Convergence and Key Technology Drivers

Dragorad A. Milovanovic and Zoran S. Bojkovic
University of Belgrade

Dragan D. Kukolj
University of Novi Sad

CONTENTS

6.1 Introduction

5G service-driven network efficiently and flexibly supports mobile broadband, reliable low-latency and massive machine-type communications (MTCs). Major inflection points of 5G deployment are radio standard, cloud network virtualization and artificial intelligence (AI) innovative technique [1]. AI introduced in 5G is limited to the operation, management and maintenance of the network, with the goal of smart management and maintenance. In 6G, AI is a key enabler. Core requirements for 6G create immersive communications, industry revolutions and intelligent driving. Moreover, mobile operators straggle to extend service coverages and increase network capacity. As a consequence of various requirements and heterogeneity in devices and applications, the mobile networks are complex systems [2]. Traditional approach for networks planning, control, operation and optimization is centrally managed and reactive; novel proactive, adaptive and predictive networking is necessary. The origin of 6G key drivers are the technology-driven evolution of wireless networks and the challenges and performance limits imposed by 5G [3,4].

We explore in this chapter an evolution of machine learning (ML) research in mobile networking. Key AI technology drivers and solutions that, at present, have not been fully addressed in 5G mobile networks and that are attracting further research efforts toward 6G are pointed out in the second part of this chapter.

6.2 Convergence of Machine Learning and Communications

5G networks are at advanced stage of technical development and standardization. As the specifications have been developing quickly and commercially deployed worldwide, mobile operators straggle with network complexity and quality of user experience. 5G cellular networks are expected to increase the bit rate significantly, up to 20 Gbps. These bit rates, end-to-end latencies down to 1 ms, ultrareliability (packet error rate 10^{-5} or less) and massive multiple access will foster services such as enhanced mobile broadband, device-to-device (D2D) communication, ultrareliable and low-latency Internet of Things (IoT) and MTC, e-health, augmented reality and tactile Internet, industrial control for the Industry 4.0, automated driving and flying.

ML techniques have been used in the network field for a long time for prediction and classification [5,6]. The evolution of sensing, mining, prediction and reasoning modules in mobile networks is shown in Table 6.1.

Network complexity refers to deployment of densely distributed 5G base stations, configuration of large-scale antenna arrays and global scheduling of virtualized cloud networks. Technical specifications enhance massive MIMO technology, introduce the non-orthogonal multiple access (NOMA), network virtualization and network slicing based on cloud computing. Networks are maintained and operated in a smarter and more agile manner. Recently, applied ML provides new concepts and possibilities for research in academia, industry and standardization organizations. The partnership project 3GPP eNA (*Enablers for Network Automation*) and focus group International Telecommunication Union ITU ML5G (*Machine Learning for Future Networks*) have proposed research projects on ML communications. The standardization increases the interoperability and modularity of a systems. Current activities of ETSI-ENI (experimental network intelligence) group are in industry specification for automating network configuration and monitoring. Experts group ISO/IEC MPEG-NN standardizes interoperable format for digital compression of trained neural networks. In July 2018, the ITU established the Network 2030 working

TABLE 6.1

The Evolution of ML in 5G Systems

	4G	5G	ML	Modules
Services	MBB	eMBB, mMTC, uRLLC	Service-aware	
Radio resource management	Granted	• Granted or grant-free • Flexible bandwidth • Flexible symbol length	UE-specific on-demand	Sensing, mining, reasoning
Mobility management	Unified	• On-demand	Location tracking/ awareness	Sensing, prediction, reasoning
Management and orchestration	Simple	• Operator-tailored	Enhanced self-organization and trouble-shooting capability	Sensing, mining, prediction, reasoning
Service provisioning management	Unified	• End-to-end network slicing	Network slice autoinstantiation	Mining, prediction, reasoning

eMBB, enhanced mobile broadband; ML, machine learning; mMTC, massive machine-type communication; uRLLC, ultrareliable low-latency communication; UE, user equipment.

group with the goal to explore the 6G system technologies for 2030 and beyond [7]. AI-enabled network management is responsible for monitoring real-time network key performance indicators and adjust network parameters to provide optimal quality of experience.

ML is highly suitable for complex system modeling. The workflow in 5G systems is similar to the classical ML.

- **Problem formulation** (classification, clustering, decision-making). A given problem is classified into one of the ML categories. An improper problem formulation results in model with unsuitable structure and unsatisfactory performance.
- **Data collection** (traffic traces, performance logs). In this step, huge quantity of data are measured and aggregated. According to the application needs, tie data are recorded in offline/online phases from different network layers.
- **Data analysis** (preprocessing, feature extraction). Networks are affected by many factors, but usually only several characteristics have the significant impact on performance metric. The goal is to find the most influencing features for model training.
- **Model construction** (training and tuning). Model selection and training according to the size of the dataset and the problem category are crucial.
- **Model validation** (error analysis). Offline validation evaluates the learning algorithm. Cross-validation tests the overall accuracy of the model against overfitting or underfitting. The results are guidance for optimization of the model, reducing model complexity as well as increasing the data volume.
- **Deployment and inference** (trade-offs). The trade-off between accuracy and the computational complexity is significant for the system performance under limited computation or energy resources.
- **Model life cycle**. The maintenance of ML model is important to identify the model update cycle duration considering the computational and validation complexity of updated ML model.

There is no doubt that the complexity of 5G network systems could be managed with ML applications. ML provides solutions for many wireless network technology components such as software-defined network (SDN), network function virtualization (NFV), network slicing (NS), mobile edge computing (MEC), massive multiple-input multiple-output (mMIMO), new radio access technology (RAT), millimeter wave (mmWave) access and green communications (Tables 6.2 and 6.3). ML techniques for efficient operation, control and optimization are necessary in fully operative and efficient networks.

6.3 Toward 6G Networks

ML has been evolving to the point that nowadays this technique enables 5G mobile networks. ML is great to get insights about complex networks that use large amounts of data, and for predictive and proactive adaptation to dynamic networks. When 5G is applied commercially at the beginning of 2020, 6G research becomes to be underway. AI is becoming a crucial technology toward new 6G service classes of ubiquitous mobile ultrabroadband

TABLE 6.2

Nonfunctional Challenges in ML-Based Solutions

Solutions	Algorithm	Data Model	Challenges
Energy modeling	*Regression analysis POMDP (Markov decision process)*	• Regression function • Generalization of *Markov* decision	• Energy demand prediction • Recover simultaneous wireless transmissions • Transmission power control in energy harvesting • Base station association under the unknown energy status
Anomaly/fault/ intrusion detection	*PCA/ICA (component analysis) DL NN*	• Linear mixtures of variables	• Detectors act as classifiers • Recover statistically independent source • Signals from their linear mixtures • Analysis of network flows
Users' behavior classification	*Unsupervised ML*	• Clustering and decision tree classification	• Context extraction and profiling
Network planning	*Supervised ML*	• Classification and prediction	• Network capacity planning and operation
Network deployment, maintenance/ optimization, management	*Deep learning*	• Self-organizing network	• Adaptive and autonomous functions • Monitoring network slicing • Network configuration, optimization, healing

ML, machine learning.

TABLE 6.3

Functional Challenges in ML-Based Solutions

Solutions	Algorithm	Data Model	Challenges
mMIMO (massive multiple-input multiple-output)	*Regression analysis*	• Regression function • Majority vote • Nonlinear mapping • Gaussians mixture model, expectation maximization, hidden Markov models	• Channel estimation and data detection • Designing pilot patterns • Estimating or predicting radio parameters • Optimal handover solutions • Interfering links of the adjacent cells • User location/behavior learning/ classification
mmWave	*KNN (K-nearest neighbors)*	• Online learning	• Beam selection with environment awareness • Model of propagation channel • Clustering and power allocation • Spectral white state estimation, prediction and handoff decisions in CR networks
Cognitive radio (CR)	*SVM (support vector machine)*	• Majority vote • Nonlinear mapping	• Spectrum sensing and white space detection • Cognitive spectrum sensing/detection • Cooperative wideband spectrum sensing • Resource management

(Continued)

TABLE 6.3 (*Continued*)

Functional Challenges in ML-Based Solutions

Solutions	Algorithm	Data Model	Challenges
HetNets (heterogeneous network)	*Bayesian learning*	• Mixed integer programming • Model-free reinforcement • Model resource allocation problem • Multiparametric modeling	• Cell clustering in cooperative ultradense small-cell networks • Clustering to avoid interference • Network selection/association problems • Self-configuration/optimization femtocells • Cell outage management and compensation • Allocation problems in wireless scenarios • Load balancing • User association (cell selection) mechanisms
SDN NVF/NS	*DNN, SOM, Q-learning, decision trees, reinforcement learning*	• Deep learning	• Achieving dynamic 6G network orchestration and slice management • Providing on-demand dynamic 6G network configuration • Achieving autonomous 6G network management and maintenance

NVF, network function virtualization; NS, network slicing; SDN, software-defined networking.

(uMUB), ultrahigh-speed with low-latency communications (uHSLLC) and ultrahigh data density (uHDD). AI brings agility and flexibility to the air interface, while efficiency becomes improved. For example, AI-based cognitive radio faces a huge gap between building modules and network. Next, photonics-defined radio represents a key evolution in 6G. Recent advances make it possible to apply ML to RF signal processing, spectrum mining and mapping. By combining low latency and high reliability, scalable AI can be achieved in 6G infrastructures [8–10].

The 6G mobile network should be an intelligent system architecture. Each network node is intelligent, which supports AI from the application layer to the physical one. The main idea is to build an AI-based end-to-end PHY architecture, which would transform and replace aspects that require expert knowledge and modules. The data link layer includes sublayers for the service data adaptation protocol, packet data convergence protocol, radio link control and medium access control, each of which can be enhanced by AI. The role of AI is to enable resource allocation and can choose the most suitable scheduling for users. The network layer provides user functions, such as user radio resource control connectivity and mobility management together with BS-specific functions such as load balancing. With AI, users choose optimal serving cells, manage multiple connection and choose optimal handover target cells to guarantee service continuity. On the other hand, BSs with AI can optimize system parameters, such as mobility parameters, to achieve load balancing and enhance network robustness.

Conventionally, wireless networks are designed using statistical models. The signal processing flows from transmission to reception into several independent modules, such as coding, modulation and detection. These modules are designed independently to match specific channel models and overcome performance degradation over wireless channels. In practical networks, the signal is distorted due to nonideal hardware components, as well as perfect system characteristic information cannot always be captured.

TABLE 6.4

A Comparison of Key Performance Indicators (KPI)

KPI	5G	6G
Traffic capacity	10 Mbps/m^2	~ 1–10 Gbps/m^3
Data rate: downlink/uplink	20 Gbps/10 Gbps	1 Tbps/1 Tbps
Uniform user experience	50 Mbps, 2D everywhere	10 Gbps, 3D everywhere
Latency (radio interface)	1 ms	0.1 ms
Jitter	*Not specified*	1 µs
Reliability (frame error rate)	1–10–5	1–10–9
Energy/bit	*Not specified*	1 pJ/b
Localization precision	10 cm in 2D	1 cm in 3D

AI can predict and make decision with big data learning. This permits to transform and replace the models in wireless networking or to enhance the performance of networks with imperfect system characteristic information. Therefore, AI provides a new way to design wireless networks and will be an innovative technology for 6G, leading to superior performance (Table 6.4).

6.3.1 Technology Drivers

In general, in the evolution and development of mobile networks, emerging new transmission schemes, such as code-division multiple access (CDMA) in 3G, multiple-input multiple-output (MIMO) and orthogonal frequency-division multiplexing (OFDM) in 4G, the mmWave technique and massive MIMO in 5G boil down to the question of how to innovatively exploit the degrees of freedom (DoFs) of time, frequency and spatial resources to satisfy the diverse and refined requirements of society. Following on the innovation directions of previous mobile networks, 6G will be superflexible in terms of the utilization of time–frequency–space resources in order to provide higher speed, greater capacity and ultralow latency for future faster than 5G applications. In general, the time–frequency–space resource utilization is interrelated [11].

- **Frequency dimension.** 6G will utilize higher-frequency spectrum than previous generations in order to improve the data rates. On one hand, high-frequency bands, such as mmWave band, terahertz band and even visible-light frequency band, will be used for the 100 Gb/s transmissions in 6G.

- **Space dimension.** For further taking advantage of multipaths, the number of antennas equipped on both the transmitter and receiver will be increased. Postmassive MIMO (PM-MIMO) techniques, such as ultramassive MIMO (UM-MIMO) for terahertz communications, may support hundreds to thousands of transmit/receive antennas.

- **Time dimension.** 6G will deliver the low latency and architectural shift that 5G is promising. Even more, the basic time slot unit in 6G may be more compressed to more efficiently use the high-frequency bands and satisfy latency-sensitive services. The flexibility and versatility of the networks will be improved and hence facilitate their downward compatibility for 2G to 5G.

To deal with increased traffic, the mobile network will become more intelligent, with learning mechanisms to autonomously modify itself based on users' experience and situation-aware decision-making and networking.

- As of today, about 1 million terabytes (1 exabyte=1018 bytes) of data per day are exchanged over the mobile networks all over the world. The amount of data exchanged by mobile users will continue to increase, on one side due to the increasing number of nonhuman devices connected (including vehicles, UAVs, and autonomous systems) and to the enhanced quality 3D video/holographic-type communication that will be used by humans.
- Network intelligence will be used to allow fast and flexible spectrum allocation/reallocation, with consequent large bitrates available to the users.
- Other human senses will be communicated to improve the quality of the teleinteraction, including 3D/holographic-type communication, taste, smell, touch.
- Users will not necessarily need to bring a smartphone but will benefit of wireless-devices-as-a-service, with distributed devices available to anyone. All information being in the cloud, users will just need to be authenticated and then access the network by using any available device.
- The need to put devices on recharge will be dramatically reduced, so that the battery life will be substantially extended.
- We will see the appearance of quantum computers, capable to solve problems that are not solvable with nonquantum computers. Also, quantum communication and networks will be available, e.g., for cryptographic key exchange, also from satellites.
- The need for privacy and security for a proper management of personal data will be of paramount importance.
- The IoT and the Industry 4.0 will bring the network very close to the real infrastructures. Therefore, a security breach on the networking side may quickly become a very important safety issue in the real life.
- The NFV-SDN technologies and the related slicing capabilities will boost the emergence of virtual operators. This may lead to a significant innovation also in the business models and commercial strategies in the field.

SDN, NFV and NS were introduced to design the 5G network architecture. Today, they are still an important for designing 6G. The connection with AI can achieve the evolution to autonomous 6G networks with orchestration, optimization and management. AI-enabled network orchestration is used to satisfy the demands of constantly changing services and applications. Three typical AI design patterns for wireless communication systems can be available: layer-free, layered, and cross-layer. As a comparison of these design patterns, the cross-layer AI pattern is more attractive.

6.3.2 Future Research Directions

Some issues and challenges that, at present, have not been fully addressed in 5G mobile networks and that are attracting further research efforts are [12] as follows:

- Coverage issues and site densification to increase network capacity pose economic issues that may slow down considerably spatial and temporal 5G deployment.
- Emerging applications will push the required quality of service to extreme levels that appear very challenging for currently 5G architectures.

- D2D communication challenges and vulnerability have not yet been addressed in a totally satisfactory manner in 5G.
- MEC at the edge of the radio access network (RAN) poses concerns.
- RAN virtualization and RAN intelligence should be based on the use of commercial off-the-shelf hardware and standardized open interfaces.
- Network orchestration and slicing based on NFV-SDN technologies promise to support the implementation of a large variety of services. Softwarized functionalities are hosted into cloud computing platform over data centers implemented with standard hardware. In the slicing concepts, different subsets of customers may subscribe a service contract with different operators (either real or virtual) and share in the end the same infrastructure, which paves the way to novel business models and opportunities for the network providers.

The research in 6G is still in its infancy; hence, there are numerous open issues to resolve.

- Subterahertz and terahertz communications. The availability of large bands beyond 100 GHz (D-band, 110–170 GHz) will lead to transmission systems at high data rate over short distances, so releasing the lower band radio spectrum for long range uses.
- Massive use of multiple antenna systems. The increase in the frequency will require multiple antenna systems able to exploit the multirays propagation, not only for larger throughput but also for precise localization and for energy transfer.
- Dynamic spectrum allocation. The precious radio spectrum in the lower bands will be used more effectively by allocating the frequencies every second or so, based on the context.
- Free-space optical communication. Optical free-space communication will allow high data rate, for both outdoor and indoor scenarios, so releasing the lower band radio spectrum for long-range uses.
- High accuracy indoor localization. Context awareness needs a precise user's localization.
- Access schemes for massive wireless networks. New access schemes will be needed to handle a massive number of nonorthogonal users (more than 10 devices/m²) in an efficient and scalable way.
- ML and AI. The complexity of the network and the number of connected devices will lead to a network, which will learn from the experience to modify itself and accommodate new services.
- Wireless energy transfer. This could be in some situations a viable way to extend the battery life, avoiding frequent recharges.
- Cybersecurity. The possibility to use nonpersonal devices for personal communication will impose new challenges on biometric authentication and privacy.

6.3.3 Standardization Activities

The 5G specifications have already been prepared, and even though it has already been launched in some parts of the world, the full phase of 5G will be deployed in 2020. Research activities on 6G are in their initial stages. From 2020, a number of studies will be performed worldwide on the standardization of 6G; 6G communication is still in its infancy. Many researchers have defined 6G as B5G or 5G+. Preliminary research activities have already started in the United States. China has already started the concept study for the development and standardization of 6G communications in 2019. The Chinese are planning for active research work on 6G in 2020. Most European countries, Japan, and Korea are planning several 6G projects. The research activities on 6G are expected to start in 2020.

For example, in September 2017, the European Union launched a 3-year research project on the basic 6G technologies. The main task is to study the next-generation forward error correction coding, advanced channel coding and channel modulation technologies for wireless terabit networks (https://futurecomresearch.eu). At the end of 2017, China began to study the 6G mobile communication system to meet the inconstant and rich demands of the IoT in the future, such as medical imaging, augmented reality and sensing. In April 2018, the *Academy of Finland* announced an 8-year research program, 6Genesis, to conceptualize 6G through a joint effort of the University of Oulu and Nokia. Research is organized into four unified planned parts: wireless connectivity, distributed computing, services and applications. Scientific innovations will be developed for important technology components of 6G systems. More recently, the U.K. government has invested in some potential techniques for 6G and beyond; some universities in the United States have launched research on terahertz-based 6G wireless networks, and South Korea Telecom (SKT) has started 6G research based on the cell-free and nonterrestrial network techniques. *Samsung Electronics* has opened an R&D center for the development of essential technologies for 6G mobile networks.

International Telecommunication Union (ITU) standardization activities were based on IMT-2020. Consequently, ITU-R will probably release IMT-2030, which will summarize the possible requirements of mobile communications in 2030. In July 2018, the newly established ITU-T NET-2030 Focus Group on 6G technologies considers the long-term evolution of 5G and develop future ICT use cases. Immersive communications, industry revolutions and intelligent driving create core requirements for 6G. Emerging Internet of everything (IoE) applications will require a convergence of communication, sensing, control and computing functionalities. Although 5G cellular system supports ultrareliable, low-latency communication, it has short-packet, sensing-based functions that limit the delivery of high-reliability low-latency services with high data rates. Starting from the point of the challenges performance limits of 5G, technology trends and network evolution, 6G use cases are in a position to be offered. The seven representative use cases (holographic-type communications, tactile Internet for remote operations, intelligent operation network, network and computing convergence, digital twins, space-terrestrial integrated network and Industrial IoT [IIoT] with cloudification) have been evaluated according to five abstract network requirement dimensions (Table 6.5) [7].

TABLE 6.5

ITU-T NET-2030 Key Performance Indicators

	Relevant Network Requirements
AI	Data computation, storage, modeling, collection and analytics, autonomy, programmability
Bandwidth	Bandwidth, capacity, QoE, QoS, flexibility, adaptable transport
Time	Latency, synchronization, jitter, accuracy, scheduling, coordination, geolocation accuracy
Security	Security, privacy, reliability, trustworthiness, resilience, traceability, lawful intercept

QoE, quality of experience; QoS, quality of service.

TABLE 6.6

Work in Progress to Specify Major Technical Solutions Pertinent to the CSR Targets

Technology Area	Evolution Trend	Reference Roadmap(s)
Circuit and device	Nanometers level with node scaling targets of power–performance–area–cost breaking through the limits of Moore's law	ITRS (International Technology Roadmap for Semiconductors) 2.0
Radio transceiver	RF front-end and baseband design to support extreme requirements (Tbps data rate, sub-ms latency, sub-mWatt power)	3GPP 5G NR Evolution WiFi 802.11 Evolution NFC 2.0
Radio system	Integrating licensed and unlicensed, terrestrial and nonterrestrial, comms and non-comms, in a volumetric space with fluid topologies	IEEE Future Networks Networld2020 SRIA
Network	Protocols catering for the requirements of next-generation Internet including determinism, time sensitivity and automation	3GPP 5G Core Evolution IRTF RGs ITU-T NET2030 FG Networld2020 SRIA
Data and AI	Data-driven E2E optimizations with pervasive collaborative intelligence distributed across terminals, edge, fog and cloud	ITU-T ML5G FG ETSI ENI ISG

AI, artificial intelligence; CSR, critical system requirement; RF, radio frequency.

The research in 6G is still in its infancy; hence, there are numerous open issues to resolve. We are in the phase of determination of critical system requirements (CSRs) for area of focus and definition corresponding targets. Next, we need to specify major technical solutions pertinent to CSR targets and estimate corresponding maturity timelines (Table 6.6) [12].

6.4 Concluding Remarks

Besides applying 5G commercially at the very beginning of 2020, 6G research becomes to be underway. The origin of key drivers are the challenges and performance limits imposed by 5G as well as the technology-driven evolution of wireless networks. AI is introduced in 5G, and the intelligence is limited to the operation, management and maintenance of the network, with the goal of smart management and maintenance. In 6G, AI is a key enabler. AI principles, which incorporate learning, reasoning and decision-making mechanisms, are natural choices for designing a future integrated network.

The areas of ML and communication technology are converging. After about 17 years of development, potential of AI in telecom application is stimulated. Since 2013, with the rapid development of deep learning and knowledge mapping technology, the trend is on a fast track with focus on the three basic levels of data and computing layer (base), algorithm layer (technology) and application layer. Next-generation wireless networks (5G and beyond), which will be extremely dynamic and complex, pose many critical challenges. Conventional approaches for network planning, operation and management that require complete and perfect knowledge of the systems are inefficient or even inapplicable in networks beyond 5G. Furthermore, recently developed procedures offer new ways to jointly optimize the components of a communication system. Also, in many emerging application fields of communication technology, ML methods are of central importance. We pointed out that the future development of complex networks should be driven by real-time ML technology.

References

1. K.David, H.Berndt. 6G Vision and requirements: Is there any need for Beyond 5G? *IEEE Vehicular Technology Magazine*, vol.13, no.3, pp.72–80, Sept. 2018.
2. Z.Zhan, et al. 6G Wireless networks: Vision, requirements, architecture, and key technologies. *IEEE Vehicular Technology Magazine*, vol.14, no.3, pp.28–41, Sept. 2019.
3. P.Yang, Y.Xiao, M.Xiao, S.Li. 6G Wireless communications - Vision and potential techniques. *IEEE Network*, vol.33, no.4, pp.70–75, July/Aug. 2019.
4. B.Zong, et al. 6G Technologies: Key drivers, core requirements, system architectures, and enabling technologies. *IEEE Vehicular Technology Magazine*, vol.14, no.3, pp.18–27, Sept. 2019.
5. W.Samek, S.Stanczak, T.Wiegand. The convergence of machine learning and communications. *ICT Discoveries, Special Issue: The Impact of AI on Communication Networks and Services*, vol.1, no.1, pp.49–59, Mar. 2018.
6. T.E.Bogale, X.Wang, L.B.Le. Machine intelligence techniques for next-generation context-aware wireless networks. *ICT Discoveries, Special Issue: The Impact of AI on Communication Networks and Services*, vol.1, no.1, pp.109–120, Mar. 2018.
7. ITU-T FG NET-2030, Representative use cases and key network requirements for Network 2030, Jan. 2020 (https://www.itu.int/en/ITU-T/focusgroups/net2030/)
8. E.C. Strinati, et al. 6G: The next frontier: From holographic messaging to Artificial Intelligence using sub-terahertz and visible light communication. *IEEE Vehicular Technology Magazine*, vol.14, no.3, pp.42–50, Sept. 2019.
9. Z.Xiong, et al. Deep reinforcement learning for mobile 5G and beyond: Fundamentals, applications, and challenges. *IEEE Vehicular Technology Magazine*, vol.14, no.2, pp.44–52, 2019.
10. Q.Mao, F.Hu, R.Hao. Deep learning for intelligent wireless networks: A comparative survey. *IEEE Communications Surveys Tutorials*, 4(20), pp.2595–2621, 2018.
11. M.Latvaaho, K.Leppänen (Eds.). 6G Research visions: *Key drivers and research challenges for 6G ubiquitous wireless intelligence*. 6G Flagship, University of Oulu, Finland, Sept. 2019
12. A.Mourad, R.Yang, P.H.Lehne, A.De La Oliva. A baseline roadmap for advanced wireless research beyond 5G. *Electronics*, vol.9, no.2, pp.1–14, MDPI, 2020

Section II

Multiservices Network

7

An Evolution of 5G Multimedia Communication: New Ecosystem

Dragorad A. Milovanovic and Zoran S. Bojkovic

University of Belgrade

CONTENTS

7.1 Introduction

5G systems are fundamentally transforming a radio network from wireless connectivity to the network for services. 5G wireless is a combination of evolved long-term evolution (LTE) network with new radio access technologies targeting for wide bandwidth, expanded spectrum, enhanced mobile broadband (eMBB), as well as mission critical connections. The 5G vision also goes beyond traditional mobile broadband applications. The new services are eMBB, massive machine-type communications (mMTCs) and ultrareliable low-latency communications (uRLLCs). An eMBB primary service category is oriented to the service in human user access to multimedia content. Key elements of the service are massive mobile connectivity, high data rates and connection density, mobility as well as access to multimedia content (4k/8k UHD [ultrahigh definition] streaming, 3D immersive experiences) [1,2].

The increasing demand for mobile broadband has led to eMBB improvements. eMBB is among the first services to be supported with early 5G deployment. The primary goal is increasing the end-user data rates and system capacity [3]. Enhanced performance enables usage cases in new application areas. In the case of high user density area, huge traffic capacity is needed, while the requirement for mobility is low. In the case of wide area coverage, seamless coverage and medium-to-high mobility are required. Minimum technical requirements and key performance indicators (KPIs) in ITU-R and 3GPP standardization frameworks are outlined in this chapter.

In the 5G era, applications such as UHD video and 3D video will be better served with data rates up to hundreds of megabits per second or even gigabits per second. In addition, the demand of uplink high data rate service is also emerging, for example, with HD video sharing. These service requirements, together with the anywhere anytime experience requirements with high user density and user mobility, define new limits for the eMBB scenario. It is not just about the consumption of multimedia content for entertainment purposes. It will support entire smart office where all devices are wirelessly and seamlessly connected. Ultimately, it will enable applications from fully immersive augmented reality (AR) and virtual reality (VR) to real-time video monitoring and virtual meetings with 360° video and real-time interaction [4].

The chapter is organized as follows. The 3GPP development framework of overall 5G system is presented in the first part. Performance requirements for eMBB, technical specification for mobile broadcast services, requirements for professional media services and quality of experience (QoE) evaluation in immersive applications are outlined. The research consortia and proof of 5G concepts of new multimedia communication ecosystem are presented in the second part.

7.2 5G Multimedia System and Services

The 3GPP completed the first 5G Release 15 in 2018, and initial rollouts largely focus on enhancing the MBB experience. As early 5G systems are being deployed, Release 16 is underway, and its work items are supposed to evolve and expand the 5G ecosystem bringing industrial IoT, private networks and V2X applications to the market. However, many items will need to be pushed out to meet the Q4 2020 deadline and deferring Release 17 and beyond. Figure 7.1 shows timeline for 5G over the next few years.

The fourth generation of mobile communication systems, formally referred to as International Mobile Telecommunications-Advanced (IMT-Advanced) systems, provided a versatile platform for enabling a wide range of MBB applications. Development of the next generation of mobile communication systems – the International Mobile Telecommunications-2020 (IMT-2020) commonly referred to as 5G – the fifth generation of mobile communication

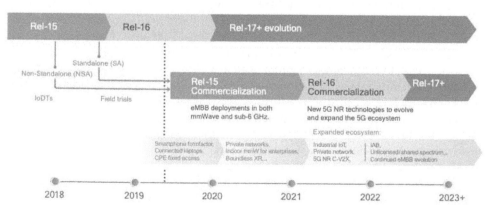

FIGURE 7.1

An evolution of 5G communication. IoDT, interoperability development testing.

systems is motivated on demand for eMBB services and the vast potential for mobile communications providing ultralow latency and ultrahigh reliability for higher frequencies.

Candidate IMT-2020 systems are undergoing a rigorous evaluation process to ensure they fulfill the requirements set out by the International Telecommunication Union (ITU) for IMT-2020 systems to meet the performance requirements of emerging 5G applications, commonly grouped into eMBB. The prime IMT-2020 candidate system, the 5G new radio (NR) system developed by the 3GPP, fulfills the IMT-2020 system requirements set out by the ITU. An ITU-R working group (WP 5D IMT systems) has published recommendations on global broadband multimedia IMT-2020 in liaison with ITU-T. The overall objectives of the IMT development as well as expected usage cases are described in the ITU-R M.2083 framework [5]. Associated capabilities cases are spectrum and bandwidth flexibility, reliability, resilience, security and privacy and operational lifetime. For the overall system, the requirements are in the form of KPIs for three main services:

- eMBB is about very high throughput rates to support applications such as UHD streaming or VR. This concept is direct successor to the high data rate services provided by LTE.
- uRLLC promises low-latency interfaces with very high reliability for mission-critical applications such as health monitoring drones or driverless automobiles.
- mMTC enables millions of devices communicating incorporating low data rates, less energy and lower costs. This approach opens the doors to major automation in every area of life and industries.

3GPP initiated the development of the 5G standard in September 2015. The first version of the standard was released in June 2018, under the Release 15, with a specification that covers the non-stand-alone (NSA) and stand-alone (SA) options. NSA implies that the 5G access node relies on a 4G core or on another 4G eNodeB to establish the connection. This is the deployment option selected for almost all the early deployment scenarios as it allows operators to capitalize on earlier 4G investments. The SA option relies on a complete 5G gNode B and a 5G core.

The fast development of this first release was pushed by the aggressive deployment agenda of operators. Consequently, the first release has focused on the eMBB use case and on the NR specification, in view of the high-rate high-capacity options contemplated by these operators. The standard covers notably the following:

- The NR functionalities for eMBB and uRLLC is defined in 3GPP TR 38.913 [6]. uRLLC focus has primarily been on low latency. NR specification considers frequency ranges up to 52.6 GHz. The NR functionalities have been designed to be forward compatible and allow for smooth introduction of additional technology components and support for new use cases. Backward compatibility of the NR to LTE is not required.
- User plane layer 1 and layer 2 specifications include a common part to all supported architecture options.
- Several connectivity options and scenarios as defined in 3GPP TR 38.801 [7] and corresponding to either NSA or SA options are considered for this normative work. These scenarios envisage several configurations to link (or not) to an existing 4G infrastructure, hence providing maximum flexibility to operators to implement 5G as a function of their existing 4G deployment status.

3GPP specifications are split into multiple studies and work items. TR represents a technical report, or study, and TS represents a technical specification. Study item RAN (radio access network) TR 38.912 covers NR deployment scenarios, physical layer, layer 2, architecture for next-generation (NG) RAN (NR), radio transmission/reception, higher layers and network procedures. Study item 3GPP TR 38.913 [6] describes the KPIs for the different deployment scenarios eMBB, uRLLC and mMTC. 3GPP standards Release 15 (phase 1) is being gradually developed and included different milestones for the various dual connectivity options (early drop for NSA EN-DC option in December 2017), main release for SA NR option in June 2018 and late drop for further dual connectivity options in March 2019. The work continued on Release 16 (Phase 2) is well underway until March/June 2020 as well as on Release 17 until June 2021.

3GPP is defining a new system architecture to support 5G requirements. The network needs to support the variety of 5G services, many different types of devices and varied traffic loads. The 5G core network must be flexible and efficient. Many operators are moving to software-defined networking (SDN) and network function virtualization (NFV). Distributed cloud, network slicing (NS) and self-optimizing networks are key enabling technologies. These technologies help virtualize the network architecture and management plane to create enhanced communication capabilities.

3GPP specification of both the 5G NR and 5G new core provides a whole picture on 5G end-to-end (E2E) system and key features. A major change in the core (5GC) architecture compared with evolved packet core (EPC) and the previous generations is the introduction of the service-based architecture (SBA) (Figure 7.2). In EPC architecture, the control plane functions communicate with each other via the direct interfaces (or reference points) with a standardized set of messages. In the SBA, the network functions (NFs), using a common framework, expose their services for use by other NFs. In the 5GC architecture model, the interfaces between the NFs are referred to as service-based interfaces (SBIs). The service framework defines the interaction between the NFs over SBI using a producer–consumer model. Such a service offered by an NF producer) could be used by another NF (consumer)

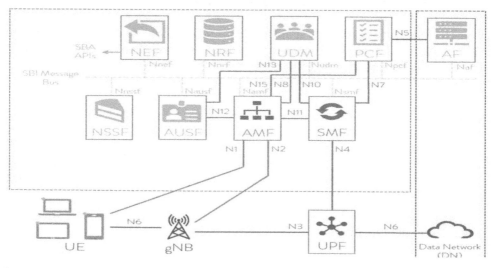

FIGURE 7.2
Service-driven 5G SBA network architecture. UPF, user plane function; AUSF, authentication server function; NEF, network exposure function; NRF, NF repository function; PCF, policy control function; UDM, unified data management.

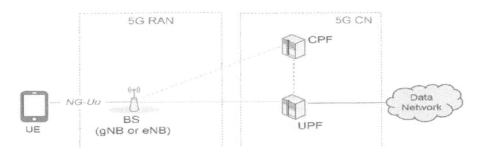

FIGURE 7.3
5G system end-to-end architecture. UE, user equipment; RAN, radio access network (5G NR gNB base station); CN, core network; UPFs, user plane functions; CPFs, control plane functions.

that is authorized to use the service. The services are generally referred to as NF service in 3GPP specifications [8].

5G system provides data connectivity and services. The overall system is defined as 3GPP system (5GS) consisting of core (5GC), RAN and user equipment (UE) (Figure 7.3). The 3GPP system architecture work identifies the features and functionality needed to deploy a services-based operational network for 5G system. These specifications are represented in 3GPP TS 23.xxx documents. System architecture specification covers interworking, mobility, quality of service (QoS), policy control and charging, authentication and in general 5G system-wide features 3GPP TS 23.501 [9]. The specification 3GPP TS 23.502 [10] defines procedures and NF services for the 5G system architecture as well as policy and charging control framework which 3GPP TS 23.503 [11]. Unlike previous mobile communication technologies, the core network for 5G is designed to work seamlessly with more than one access technology. Within the 3GPP NG-RAN, a new cellular access technology NR is introduced [12,13].

For mobility, devices must maintain connection as they travel through the network. The 5G RAN operates with both 5G NR (gNB) and LTE (eNB) base stations. The new RANs are interconnected through the Xn interfaces and connected to the 5G core through the NG interfaces. 5G NR can operate in NSA mode, where the UE requires a legacy eNB with connection to the EPC for control plane to support 5G NR communication. In SA mode, the 5G network can operate independently of the 4G core network. There are seven different connectivity options defined in the 5G NR specification, allowing network equipment manufacturers to plan upgrade paths to the NG core network.

5G NR specifications are covered by the RAN working groups. The output of the workgroup is public. All documents, meeting reports and published specifications are available on the 3GPP website. 5G NR documents are represented in the 38.xxx series. Technical reports begin with TR, and specifications begin with TS. For 5G NR, RAN study items and specifications define functions, requirements and interfaces to the networks (RAN1 radio layer 1, RAN2 radio layer 2 and radio layer 3, RAN3 radio network, RAN4 radio performance and protocol, RAN5 mobile terminal conformance tests). The radio interface between the UE and the network is described in layers 1, 2 and 3 of the communications stack. These are commonly known as the physical layer (RAN1 TR 38.802, TS 38.201-38.215), the data link layer and the network layer (RAN2 TR 38.804, TS 38.300-38.331). The physical layer represents the interface to the real world and includes the hardware and software to control this linkage. The data link layer enables data transfer between the different networks. Known as medium access control, it provides different logical channels to the radio link control in the network layer. Layer 3 connects with the nodes in the network so that the UE can travel through the network.

Service-driven 5G network architecture (SBA) provides a modular framework from which common applications can be deployed using components of varying sources and suppliers. The 3GPP defines a SBA, whereby the control plane functionality and common data repositories of a 5G network are delivered by way of a set of interconnected NFs, each with authorization to access each other's services [8]. SBA aims to flexibly and efficiently meet diversified mobile service requirements. With SDN and NFV supporting the underlying physical infrastructure, 5G comprehensively cloudifies access, transport and core networks. Cloud adoption allows for better support for diversified 5G services and enables the key technologies of E2E NS, on-demand deployment of service anchors and component-based NFs.

The 5G QoS model is based on QoS flows where the QoS parameters are negotiated by means of traffic flow template (Figure 7.4). The main idea is that data connection conveys multiple QoS flows where each flow is accompanied by mechanisms such as packet labeling and bearer mapping. A flow of this type can be considered as a granularity for QoS forwarding treatment in the overall 5G system. All traffic that is mapped to the same QoS flow will receive the same forwarding treatment (the scheduling policy, queue management, prioritization, rate allocation policy, protocol layer configuration) (3GPP TS 38.300) [14].

NS enables to steer traffic in fine granularity to enforce policy-based routing and meet E2E QoS requirements of various applications. E2E NS is a foundation to support diversified 5G services and is key to 5G network architecture evolution (Figure 7.5). Based on NFV and SDN, physical infrastructure of the future network architecture consists of sites and three-layer data centers (DCs). According to diversified service requirements, networks generate corresponding network topologies and a series of NF sets (network slices) for each corresponding service type using NFV on a unified physical

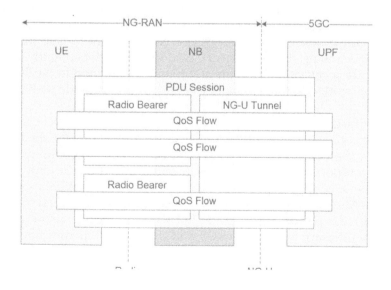

FIGURE 7.4
QoS flow architecture. NB, network-based; NG, next-generation; QoS, quality of service; UE, user equipment; UPF, user plane function.

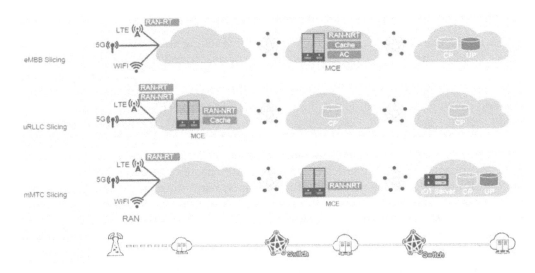

FIGURE 7.5
End-to-end network slices (function sets) generated for eMBB service on a unified physical infrastructure. CP, control plane; eMBB, enhanced mobile broadband; IoT, Internet of things; LTE, long-term evolution; mMTC, massive machine-type communication; NRT, non-real time; RAN, radio access network; RT, real time; UP, user plane; uRLLC, ultrareliable low-latency communication.

infrastructure. Each network slice is derived from a unified physical network infrastructure, which greatly reduces subsequent operators' network construction costs. Network slices feature a logical arrangement and are separated as individual structures, which allows for heavily customizable service functions and independent observations and measurements [15,16].

One operational network is a cost-efficient way to engineer, maintain and operate radio networks from the perspective of capital and operational expense. Conformance tests ensure a minimum level of performance in UEs and base stations. UE requirements are very extensive to ensure RF transmission and reception, radio access, signaling and demodulation. Base station tests are structured around RF parameters. Conformance tests only provide pass/fail results, offering no indication of how the device will perform when integrated into a wireless communications system. Vendors will test a wider set of parameters early using verification and regression testing to ensure quality and sufficient margins. Operators will also conduct device acceptance tests to evaluate whether a device has adequate performance on a specific network. 5G NR conformance tests are found in the following 3GPP documents: base stations (TS 38.141) and UE devices (TS 38.521, 38.522, 38.533, 38.508, 38.509, 38.523) [17–22].

3GPP conformance tests define measurement definitions and procedures necessary to achieve compliance against the core specifications. UE conformance tests involve connecting a device to a wireless test system and performing the required 3GPP tests. Conformance tests are conducted by third parties. The test systems used to perform conformance tests must be validated and calibrated to ensure that the conformance test is performed with known uncertainties and under controlled conditions. Once the UE passes conformance tests and meets specifications, the next step is to validate the device on a specific network. Device acceptance testing is used to evaluate whether the device has adequate performance and helps identify and resolve issues before a device is allowed on the network in the hands of consumers.

7.2.1 Performance Requirements for Enhanced Mobile Broadband

The first eMBB use case has the primary goal of increasing the end-user data rates and system capacity. The trends in end-user data rates growth and the key ingredients contributing to the capacity of wireless communications systems in the past decade and next decade are shown in Figure 7.6. As observed for the past 10 years, air-interface improvements and new spectrum acquisition have contributed to roughly 20 times and 25 times capacity increase, respectively. Technologies such as orthogonal frequency-division multiplexing (OFDM) and multiple-input-multiple-output (MIMO) have significantly improved the air-interface spectrum efficiency. Looking into the next decade, we should first observe that the air-interface spectrum efficiency has been approaching its capacity limit, and it is likely that there will be less possibility for new spectrum acquisition. Therefore, higher capacity gains would come from the network side. Network architecture improvements as well as information and communication technology convergence accompanied by the corresponding upgrades in devices are expected to be the key drivers for the wireless communications system capacity increase in the next decade [2,12].

The 5G vision defines an advanced MBB communication system that can support an ultra-connected society. 5G NR is very different from the previous generations of wireless standards. It introduces a new E2E network architecture that promises high data throughput and ultrareliable low-latency connections. With the introduction of 5G NR standards, networks need to support a diversity of 5G devices with eMBB usage scenarios.

The IMT-2020 framework includes feature performances for a more flexible and reliable system. The minimum technical requirements of IMT-2020 radio interface technologies are described in a report ITU-R M.2410 [23]. The report also provides the necessary background information concerning the individual requirements and the chosen operational values. The another report ITU-R M.2412 [24] prescribes methodology, test environments

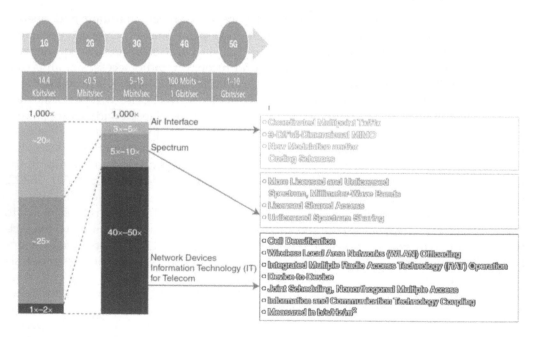

FIGURE 7.6
The trends in end-user data rates and network capacity growth. MIMO, multiple-input multiple-output.

and configurations of channel models used for evaluation for radio interface requirements. Within the development of detailed specifications, the key minimum technical performances are used in specification and evaluation of the radio interface technologies. These requirements are planned to provide the objectives and to set a specific level of performance. The report presents the technical performance requirements for the radio interface and specifies their values (Table 7.1).

The ITU sets the requirements for the NG telecommunication systems. The ITU requirements are addressed by 3GPP standardization committee. 3GPP has also specific requirements and target values before to their studies on 5G radio interface in ITU-R. In addition, ITU-R technical performance requirements include items not in the 3GPP, for example, bandwidth to be supported. 3GPP is finishing out 5G NR KPIs (3GPP TR 38.913) [25]. KPIs include target peak data rates, spectral efficiency, latency, reliability and more.

3GPP LTE standards Release 12, Release 13 and Release 14 set the foundation for 5G in the period March 2015–June 2017. Release 15 had great importance as it introduced NR technical specifications for the very first time, including the introduction and standardization of core network (5GC). Release 15 focused on MBB use cases with very high bitrate and low latency and, in addition, some features for uRLLC were standardized. Release 15 was completed in three steps: the early drop in December 2017 that included E-UTRA-NR dual connectivity and EN-DC, the focus on standalone NR in June 2018 and the late drop in March 2019 that introduced other architecture options.

5G Release 15 is the first full set of 5G phase 1 specification, driving commercial launches across the globe based on March 2019 version of specification 3GPP TR 21.915 [26], enabling 5G radio access technology that is flexible enough to support sub-6 GHz and

TABLE 7.1

Minimum eMBB Requirements for IMT-2020 Radio Interface in Comparison with IMT-Advanced

Technical Performance Requirement	Download Link Upload Link	IMT-Advanced Requirement
Peak data rate	DL: 20 Gbps UL: 10 Gbps	~6× LTE-A Release 10
Peak spectral efficiency	DL: 30 bit/s/Hz UL: 15 bit/s/Hz	~2× IMT-advanced
User experience data rate	DL: 100 Mbps UL: 50 Mbps	
User spectral efficiency	DL: 0.3 bps/Hz UL: 0.21 bps/Hz	~3× IMT-advanced
Average spectral efficiency	DL: 9 bps/Hz UL: 6.75 bps/Hz	~3× IMT-advanced
Area traffic capacity	10 Mbps/m^2 –	
Energy efficiency	High sleep ratio and long sleep duration under low load	
Mobility class with traffic channel link data rates	– Up to 500 km/h with 0.45 bps/Hz	1.4× mobility class 1.8× mobility link data rate
User plane latency	4 ms	>2× reduction
Control plane latency	DL: 20 ms UL: 20 ms	>5× reduction
Mobility interruption time	DL: 0 ms UL: 0 ms	Much reduced

millimeter-wave (mmWave) frequencies as well as new use cases and applications. Equally important, 5G system architecture must evolve to keep pace with the radio access changes.

- **NR air interface.** Like LTE, NR uses OFDM but makes it highly flexible. For example, variable subcarrier spacing, flexible radio frame structure including a self-contained slot and carrier bandwidth parts are introduced. Both sub-7 GHz spectrum and mmWave spectrum are supported. New high-performance channel coding techniques of LPCC (low-density parity check coding) and polar coding are defined. Spatial multiplexing techniques used in LTE, SU-MIMO and MU-MIMO are enhanced in 5G. NR is a beamformed air interface with fewer beams at low-frequency bands and more beams at high-frequency bands. 5G supports hybrid beamforming, where both digital beamforming (available in LTE) and analog beamforming are combined.

- **NG-RAN, NGC and SBA.** NG-RAN includes NR-based 5G base stations called NG node Bs or gNBs. A gNB can be decomposed or disaggregated into a central unit and a distributed unit. Such gNB architecture reduces infrastructure and transport costs and provides scalability. While LTE uses a limited number of nodes in the EPC, 5G defines more NFs that have fewer responsibilities. The overall 5G system is based on SBA, where NFs communicate with each other using SBIs. SBA facilitates design and deployment of the 5G system using virtualization and automation technologies such as NFV, SDN, opens stack and orchestration.

- **Deployment options.** Release 15 fully defines two deployment options for the network architecture, NSA NR and SA NR. NSA NR with the EPC uses the LTE eNB as the master node and makes use of additional NR radio resources of a gNB when possible. SA NR with the NGC does not rely on LTE eNB at all and allows direct communications between the UE and the gNB.

- **Network slicing.** 3GPP introduces the concept of NS, where different logical networks are created using the same physical network to cater to different services and different customer requirements for a given service. Three standard slices of eMBB, uRLLC and massive IoT are defined with the support for numerous operator-defined **network slices.**

- **Edge computing.** 3GPP supports edge computing, where the applications are located close to the UE. More specifically, 3GPP allows the selection of a gateway that is close to the gNB. Since the use traffic passes through a local gateway instead of a remote gateway located deep inside the core network, the E2E latency is reduced, and transport requirements are also reduced.

3GPP Release 16 is the major specification of 5G phase 2 comes to its freeze date March 2020 and completion June 2020 (3GPP TR 21.916) [27], because it will bring the specification organization's IMT-2020 RIT/SRIT submission (to ITU-R WP 5D) for an initial full 3GPP 5G system to its completion. Release 16 served to broaden the use cases where NR can be applied as well as improve capacity and performance of the system. These features and use cases included the following:

- Serving highly demanding critical industrial use cases as well as support of time-sensitive communications (TSCs) as part of a focus on uRLLC and industrial IoT, e.g., by providing accurate timing information from the network to the device.

- 5G NR provides more capacity for traditional and new operators to meet customer needs by utilizing the larger bandwidths at 5 and 6 GHz in unlicensed spectrum. In addition, the work regarding the introduction of nonpublic network (NPN) in SA or as a part of public networks served to enhance the support of vertical and LAN services.

- NR V2X solutions complement existing LTE V2X solutions for advanced automotive industry services. For this purpose, NR side link was introduced to allow vehicle-to-vehicle communication and vehicle communication to roadside units.

- MIMO features such as joint transmission from multiple TRPs and enhanced feedback from the terminals allow more extensive usage of advanced multiantenna schemes.

- Integrated access and backhaul (IAB) refers to the solution where the backhaul link of a node uses NR link. Such a relay node is called IAB node. This sort of solution is useful for operation in high frequencies with larger spectrum availability but poorer coverage.

7.2.2 Technical Specification for Mobile Broadcast Services

Broadcast and multicast represent a key opportunity in 5G for the massive consumption of multimedia services in the near future. These technologies permit to offload an important portion of this traffic in peak demand scenarios where users are consuming parallel content. The 3GPP defined multimedia broadcast/multicast service (MBMS) in 2005 to optimize the distribution of video traffic. This standard covers the terminal, radio, core network and user service aspects. MBMS is a point-to-multipoint (PTM) service in which data are transmitted from a single source entity to multiple recipients. This MBMS standard has evolved into enhanced MBMS (eMBMS) that builds on top of the 3GPP LTE standard. eMBMS evolution brings improved performance thanks to higher and more flexible LTE bit rates, single-frequency network operations and carrier configuration flexibility. The version further evolved multimedia broadcast multicast service (FeMBMS) is specified in June 2017, referred to as LTE broadcast. The latest version is available in 2019 for 5G NR broadcast (EnTV), such as media and entertainment (M&E), public warning systems (PWS), automotive and Internet of things (IoT).

The work on adding broadcast/multicast support to 3G networks began in 2002 when 3GPP created work items in GSM/UMTS and CDMA2000 [28–33]. Requirements are defined for optimization of the distribution of video traffic. The MBMS standard was introduced in the 3GPP specification version Release 6 and has evolved in subsequent releases [34–43].

- Release 6 covers the MBMS terminal, radio, core network and user service aspects and splits the feature into the bearer service and the user service. Release 7 improves support for a dual receiver UE receiving simultaneously MBMS and dedicated services on separate carriers, as well as SFN operation of MBMS in the physical layer.

- Originally, Release 8 LTE was planned to include enhanced MBMS (eMBMS); however, starting from 2008, it was shifted into Release 9. The specification is evolved to eMBMS (LTE broadcast) service with improved performance based on higher and more flexible Release 9 LTE-A bit rates, single-frequency network operations and carrier configuration flexibility. The most important improvements in LTE are the support of larger carrier bandwidths, use of OFDM modulation in the physical layer and efficient and flexible support for SFN configurations. A number of eMBMS enhancements have been included into 3GPP releases to support large area broadcast, with dedicated eMBMS carrier and longer cyclic prefix and agreed eMBMS enhancements in LTA-A.

- Release 10 and Release 11 support service continuity and enhancements (reception reporting aggregation, content schedule information, pre-FEC repair QoE metrics report, unicast file repair, location filtering). Release 11 also brings improvements in the areas of the service layer with, for example, a video codec for higher resolutions and frame rates and forward error correction (FEC), and the radio network with procedures to ensure MBMS reception in a multifrequency LTE network. eMBMS allows LTE network and backhaul offload. It creates new opportunities by enabling delivery of premium video content to many users with QoS in defined areas and by enabling pushed content services via UE caching.

- Examples of enhancements for LTA-A Release 12 are broadcast/unicast switching based on demand, enhancements to improve accuracy and utility, multilayer transmission (MIMO) to increase capacity, dedicated carrier design and emergency alert in eMBMS. Release 13 includes technologies to support new services. A representative example is enhancement of device-to-device (D2D) proximity services that was specified in Release 12 for the support of peer discovery, indoor positioning enhancements and single-cell point-to-multipoint (SC-PTM) as a complementary tool for support of eMBMS. The 22 MBMS RAN requirements and six notification mechanism(s) have been identified in Release 13.

- LTE Release 14 includes some key technology areas and components, referred to as enhancement-for-television (EnTV) services: enhanced operation in unlicensed spectrum, machine-type communication, massive multiantenna systems, latency reduction, intelligent transportation systems, enhanced positioning and enhanced multimedia broadcasting. It includes all high-level broadcaster requirements: free to air linear TV without SIM cards, SA eMBMS network (operation using entire capacity of a carrier), predictable and sustained QoS, large coverage areas, fixed and mobile reception. The most important enhancement is support for high-power high-tower (HPHT) transmitters alongside cellular masts, to create overlay networks able to provide required coverage way for popular linear content in both rural and urban areas.

- Release 15 is opened from June 2016 to June 2018 for further improvement in MBMS services. 3GPP has added features to enable delivering television services in new and improved ways. The standardized interface enables for television and content providers to directly offer their services. The system enhancement includes wider radio broadcast range, free-to-air services and transparent mode delivery of digital video signals. Television services for mobile devices and stationary TV sets are improved compared with eMBMS and unicast. A standardized interface xMB between mobile network operators and service providers as well as TRAPI between the network and application on the handset are supported.

- Release 16 joint efforts of broadcasters and mobile industry develop further LTE MBMS features to meet the requirements of 3GPP TR 38.913 scenarios and requirements for NG access technologies: new RAT shall support static and dynamic resource allocation between multicast/broadcast and unicast; new RAT shall in particular allow support of up to 100% of DL resources for multicast/broadcast

(100% meaning a dedicated MBMS carrier); new RAT shall support multicast/ broadcast network sharing between multiple participating MNOs, including the case of a dedicated MBMS network.

- Release 17 is important for more advanced use cases compared with the traditional TV broadcast. An example is the public safety scenario and broadcast in terms of group call and video. Furthermore, intelligent transportation systems (ITS) could utilize this technology for the distribution of road conditions and traffic sign/lights. The future common framework will enable single- and multicell (SFN) broadcast. By June 2021, the Release 17 standards are expected to be finalized and published.

There are three methods of pushing media from content servers to end-user devices: unicast, broadcast and multicast.

- In the unicast case, the media is transported from the content server to the end-user device on a dedicated radio and core network link. There are as many bidirectional links as there are end-user devices. Unicast is efficient to cover all services requiring a bidirectional link. Unicast is also efficient for services dealing with users spread out over multiple radio cells, consuming different content at different times such as on-demand video streaming.

- In the broadcast case, the media is transported from the content server to the end-user device on a single unidirectional link shared by several users in one radio cell, and any user can receive the broadcast service. Broadcast more efficiently covers all services for which many users located in the same areas consume the same content at the same time. Broadcast is efficient to quickly push the same content to many devices at the same time without interaction with the user.

- Multicast further allows to activate areas where a number of end-user devices have joined the service beforehand or to charge users based on the services actually received. In eMBMS, multicast is only used for the transport of data within the network.

Deployment of MBMS services requires definition of a set of media codecs, formats and transport/application protocols. To support UltraHDTV, Release 14 upgraded media codecs and formats to values specified for broadcast television. HDR (high dynamic range) format support was added in Release 15 (3GPP TS 26.346) [44]. The responsibilities of working group 3GPP SA WG4 (Codec) include the following items:

- Development and maintenance of specifications for media codecs, media transport and handling related session descriptions, as well as for presentation layer
- Guidance to other 3GPP groups related to required QoS parameters and other system implications, including channel coding requirements
- Multimedia quality evaluation including new methods, testing, verification, characterization and selection criteria
- E2E performance, including terminal characteristics of multimedia services

MBMS QoE metrics is an optional feature for both streaming server and client. It is necessary that MBMS server should activate the gathering of client QoE metrics with SDP and receive reporting procedure. The MBMS client has to measure the quality in accordance with the specification, aggregate QoE metrics and report to the specified server. The metrics are sampled in the transport layer after FEC decoding and at the same time in the application layer for higher accuracy. The measurement period corresponds to the streaming duration. The resolution of each reported metrics value is defined by the Measure-Resolution field. In the case of late joiners or early leavers, the measurement period may be less than the session duration. In addition, the measurement period should not include pause or any buffering or freeze/gap that impacts the actual play.

Given the current state of technology and future trends with unicast connectivity providing the default transport, the 5G multicast architecture, through efficient extensions, should benefit from the unicast architecture as much as possible. To cover a multitude of possible deployments with increased flexibility over existing multicast services, 5G multicast solutions are required. One 3GPP system requirement for 5G is a flexible broadcast/multicast service for three types of devices (eMBB, uRLLC and mMTC). The need for multicast/broadcast and PTM transmissions is also evident in the general system requirements, such as those for efficient content delivery and resource efficiency. In general, the requirements for 5G multicast/broadcast can be classified into two different operation modes: SA deployment of dedicated broadcast networks and mixed unicast/multicast mode. Although the existing solutions are capable of enabling existing multicast/broadcast services (downloads, streams, group communications, TV), the inclusion of new vertical applications is also needed, such as V2X, interactive media and entertainment with personalized content. These services are made available in areas where the number of users during popular events (in stadiums) can be high and the user distribution within the multicast area changes over the time.

7.2.3 Requirements for Professional Media Services

A 5G network can facilitate audio–visual (AV) production services by providing flexibility, reducing costs and reducing communications setup times. Media could be produced within or outside the premises of a production company (TV center). The 5G system supports media flows from open standard–based broadcast workflows and be agnostic to the data carried. 3GPP report TR 22.827 [45] describes relevant use cases and proposes respective potential service requirements for 5G systems to support production of AV content and services:

- Provision of pre-defined bandwidth capacity, E2E latency and other QoS requirements for larger live events or high-quality cinematic video production
- Time synchronization among all devices (cameras, microphones, in-ear monitors)
- Coverage-related issues dealing with nomadic and ad hoc production deployments
- Interoperability issues related to existing AV production standards and protocols
- Dependability assurance and related topics (network isolation, QoS monitoring)

Here are examples of the AVPROD use cases where 3GPP contribute:

- **Studio-based production.** Professional media could be produced in a studio using wireless microphones connected to a variety of sources. A 5G system can replace costly and inflexible fixed infrastructure.

- **Newsgathering.** This use case represents unplanned ad hoc production such as covering of an important event. A 5G system can be set up quickly to produce relevant media and supply such media to the central facility for further processing and distribution.

- **Planned outside broadcasts.** An elaborate infrastructure with numerous cameras and microphones can be installed for a planned event. A 5G system can facilitate media transmission from such event facilities to the central production base. Some media preprocessing could also be carried out locally. Sometimes, a large coverage sports area is needed, and an airborne 5G NG-RAN can be deployed.

- **Live immersive media service.** Multiple cameras can be installed on various sports stadium locations and players to create an immersive experience for the local audience and global audience. The real-time AV production for live immersive media involves three major processes, including media acquisition, media production and media distribution. An example is deployment of 30+ cameras of UE capability around the stadium to be one of the players and connected to 5G network via NG-RANs in local nonpublic network operated by third-party or mobile network operators.

The service and performance requirements for the operation of professional video, audio and imaging via a 5G system are described in 3GPP report TS 22.263 [46]. The report document introduces requirements related to professional video, imaging and audio services. Unlike other consumer multimedia applications envisioned for 3GPP systems, the applications in which this document focuses have more demanding performance targets and include user devices that are managed in different workflows when compared with typical UEs. The report focuses on services for the production of AV data for any area that requires high-quality images or sound. This may include AV production, medical or gaming applications. To enable devices such as professional cameras, medical imaging equipment and microphones to use the 5G network either directly or via the addition of a dedicated intermediate technology, certain key parameters are required [47,48].

AV production includes television and radio studios, live news-gathering, sports events, among others. Typically, numerous wireless devices such as microphones, in-ear monitoring systems or cameras are used in these scenarios. In the future, the wireless communication service for such devices could be provided by a 5G system. AV production applications require a high degree of confidence, since they are related to the capturing and transmission of data at the beginning of a production chain. This differs drastically when compared with other multimedia services because the communication errors will be propagated to the entire audience that is consuming the content on both live and recorded outputs. Furthermore, the transmitted data are often postprocessed with filters which could actually amplify defects that would be otherwise not noticed by humans. Therefore, these applications call for uncompressed or slightly compressed data and very low probability of errors. These devices will also be used alongside existing technologies which have a high level of performance, and so any new technologies will need to match or improve upon the existing workflows to drive adoption of the technology.

The performance aspects that are covered in the report TS 22.263 also target the latency that these services experience (Table 7.2). Since these applications involve physical feedback on performances that are happening live, the latency requirements are very strict. The document also refers to how the network structure of the 5G system is configured in order to accommodate these applications. Many of these are nomadic scenarios that

TABLE 7.2

Performance Requirements for Low-Latency in Video Production Applications

Profile	Active UEs	UE Speed	Service Area	E2E Latency	Packet Error Rate	Data Rate
Uncompressed UHD	1	0 km/h	1 km²	400 ms	UL: 10^{-10} DL: 10^{-7}	UL: 12 Gbps DL: 20 Mps
Uncompressed HD	1	0 km/h	1 km²	400 ms	UL: 10^{-9} DL: 10^{-7}	UL: 3.2 Gbps DL: 20 Mbps
Mezzanine compression UHD	5	0 km/h	1000 m²	1 s	UL: 10^{-9} DL: 10^{-7}	UL: 3 Gbps DL: 20 Mbps
Mezzanine compression HD	5	0 km/h	1000 m²	1 s	UL: 10^{-9} DL: 10^{-7}	UL: 1 Gbps DL: 20 Mbps
Tier 1 events UHD	5	0 km/h	1000 m²	1 s	UL: 10^{-9} DL: 10^{-7}	UL: 500 Mbps DL: 20 Mbps
Tier 1 events HD	5	0 km/h	1000 m²	1 s	UL: 10^{-8} DL: 10^{-7}	UL: 200 Mbps DL: 20 Mbps
Tier 2 events UHD	5	7 km/h	1000 m²	1 s	UL: 10^{-8} DL: 10^{-7}	UL: 100 Mbps DL: 20 Mbps
Tier 2 events HD	5	7 km/h	1000 m²	1 s	UL: 10^{-8} DL: 10^{-7}	UL: 80 Mbps DL: 20 Mbps

require simplified deployment often in different countries. For this reason, the 5G system enables nonpublic (private) networks that can be deployed in an agile ad hoc way. AV production also relies on a number of other technologies that will be deployed by a 5G system such as the use of UAV's to capture video and high bandwidth connectivity for file transfer. Some aspects of specific 5G specifications such as direct communications between devices or multicast/broadcast could also be used to enable future user cases.

7.2.4 Quality of Experience Evaluation in Immersive Applications

The increased bandwidth capacity and decreased latency of 5G networks will allow access to complex 3D immersive AV experiences over mobile networks [49]. The future of VR and AR depends on 5G. These cutting-edge technologies superimpose digital data onto the physical world and enable computers to generate simulations of 3D images or create 3D environments. Exploiting the immersion that it offers, VR technology is being increasingly used in fields such as entertainment, video games, training, military simulations, construction, architecture and design. Early AR applications focused mostly on the entertainment and gaming industries, but current advances in the field have led to major interest from sectors such as education, professional training, knowledge sharing and collaborative environments (Table 7.3).

- **VR** refers to a 3D computer-generated environment that users can explore and interact with, using specialized equipment. VR environments typically offer a 360° view capability and may either resemble actual physical environments or be completely artificial, thus providing enhanced flexibility to content creators and allowing the users to engage in novel interactive experiences. Head-mounted displays are the most common equipment for accessing VR content, due to low cost, availability of equipment, and the immersion levels that they can offer. Specially designed rooms equipped with multiple large screens can also be used but at a

TABLE 7.3

Example Applications of Virtual (VR) and Augmented Reality (AR)

Domain	Application	VR	AR
Archeology	Superimposing features onto modern landscapes	−	+
Archeology	Recreation of site	+	+
Architecture	Visualization of building projects	+	+
Commerce	Superimposing marketing material on the real world	−	+
Education	Superimposing multimedia content	−	+
Engineering	Visualization of new designs	+	+
Live events	Superimposing interesting information and commentary	−	+
Live events	Roaming through the environment and selecting view point	+	−
Medicine	Enhancing telemedicine capabilities	+	+
Military	Information overlay on maps and visual equipment	−	+
Navigation	Display of information on vehicle windscreens	−	+
Tourism	Objects of interest rendered on the landscape	−	+
Workplace	Facilitate collaboration among distributed team members	+	+

much larger cost and in a nonportable manner. Smaller devices that have the ability to present 3D images, such as smartphones, tablets and computers, may also be used, though offering a less immersive experience. Additionally, while VR is usually focused on auditory and visual feedback, it may incorporate various other types of sensory feedback, such as haptic or smell. There are a variety of devices and sensors that can contribute to the immersion of the user into the virtual environment, including force-feedback sensors, seats and other motion platforms, omnidirectional treadmills and nasal wearables for smell creation. Employing such a diverse set of sensory input allows the user to become more immersed into the virtual environment and offers a more realistic experience [50].

- **AR** and mixed reality (**MR**) entail the creation of an enhanced version of reality by superimposing computer-generated content over the user's view of the real environment, thus augmenting the user's perception of reality. The superimposed computer-generated content is not limited to images and video but can be expanded across multiple sensory modalities (visual, auditory, haptic) in order to create rich interactive experiences of environment. The addition of artificial components on the real view can either be constructive – adding information to the real environment, or destructive – masking (removing) information from the environment and replacing it with artificial information. The most important advantage of AR is that it does not change the user's perception of the real environment by just displaying data, but through providing immersive sensory stimuli that is perceived as an actual part of the environment.

- **360°** (360-degree) videos are video recordings that capture a view toward every direction at the same time. Omnidirectional cameras or multiple regular cameras are used for recording such videos. In the latter case, overlapping angles of the video are filmed simultaneously, and video stitching, color and contrast calibration methods are employed in order to merge the footage into one spherical video piece. Normal flat 2D displays, such as the ones found in computers, TV monitors, smartphones and tablets can be used for playback, by displaying only the video region within the field-of-view of the viewer. At the same time, the viewers are able to control the viewing direction similarly to panorama imagery. 360° video can also be displayed using multiple projectors projecting on spherical surfaces for a more cinematic experience or using HMDs for more portability and comfort. An equirectangular projection is typically used in order to encode 360° video using generic video compression algorithms, and the encoded video can either be monoscopic or stereoscopic – displaying the same image to both eyes or having a distinct image directed at each eye, respectively. 360° video technology expands video viewing to a more immersive experience that allows the audience to watch a video from multiple viewing angles. Furthermore, it can potentially revolutionize the way that remote viewers experience live events. By using multiple omnidirectional cameras and broadcasting 360° video of the events, the viewers will have the ability to virtually roam through the venue of the events and select their favorite viewpoint or join the acts [51].

NG VR and AR experiences will have six degrees of freedom (6DoF), the next level of immersion allowing users to move within and intuitively interact with the environment. 6DoF content is an order of magnitude richer in naturalness and interactivity than current three degrees of freedom (3DoF) video. 3DoF experiences, such as 360° video, allow the user to look around rotationally from a fixed position. 6DoF experiences, which are available in video games today, allow the user to move spatially through the environment just by walking or leaning their head forward. 6DoF head motion tracking is required to enjoy 6DoF content in an intuitive manner. Most components of the video delivery pipeline are currently ill-suited for 6DoF video (capture devices, production software, codecs, compression algorithms, network, players). 6DoF video also demands bit rates in the range of 200 Mbps to 1Gbps, depending on the E2E latency [52,53].

Based on requirements, 3GPP defined a consistent set of interoperability points for VR 360° streaming applications in the first release of 5G technical specification TS26.118 [54]. Specification defines three video media profiles: (1) a simple legacy version based on H.264/AVC, (2) a version that permits streaming and download based on existing MPEG DASH and file format clients and (3) an advanced profile that permits decoding on existing chipsets but requires innovative processing and streaming technologies to enable viewport-adaptive streaming and one audio operating point, the OMAF 3D Audio Baseline Media profile based on MPEG-H 3D Audio Low Complexity profile enabling the distribution of channel, object and scene-based 3D audio. The subjective and objective test methodologies for the evaluation of immersive audio systems have also been specified in TS26.259 [55] and TS26.260 [56], and characterization test results for VR streaming audio have been documented in TR26.818 [57].

- SA1 conversational VR use cases (spherical video calls, videoconferencing with 360° video) and user-generated live streaming and virtual world communication, which involve interactive real-time encoding, delivery and consumption of VR content, are documented in TS26.114 [58] and TS26.223 [59]. Furthermore, VR support over 5G conversational services was studied by 3GPP SA4 during the Release 15 time frame, and relevant gaps and potential solutions were documented in TR 26.919 [60]. Based on the findings, a new Release 17 work item was launched and is being currently progressed to specify VR capabilities in TS26.114 [61] and TS26.223 [62] to enable support of an immersive experience for remote terminals joining teleconferencing and telepresence sessions. More specifically, the normative work aims on specifying recommendations of audio and video codec configurations (profile, level and encoding constraints of HEVC, AVC as applicable) to deliver high-quality VR experiences.

- S4 study on media distribution and immersive delivery over the 5G system is completed in Release 15 TR26.891 [63]. 5G can support a wider range of QoS requirements including high-bandwidth low-latency needs of interactive VR/AR/XR applications, through a NR air interface as well as flexible QoS enabled via 5G core network architecture and NS. Moreover, the ability of the 5G system to leverage edge computing is essential to meet the performance requirements of immersive media, not only for better delivery performance via edge caching but also to offload some of the complex VR/AR/XR processing to the edge to perform various operations such as decoding, rendering, graphics, stitching, encoding, transcoding and so on, toward lowering the computational burden on the client devices. A relevant potential solution for offloading compute-intensive media processing to the edge is based on MPEG-I network-based media processing (NBMP) specification ISO/IEC 23090-8, which aims to specify video media/metadata formats and APIs for intelligent edge media processing. Benefits of other technologies for DASH streaming enhancements at the edge are also applicable for immersive media, such as those from server- and network-assisted DASH (SAND), which is already specified as part of DASH in TS26.247 [64] with the supported modes of proxy caching, consistent QoE/QoS and network assistance.

5G media streaming architecture work item TS26.501 [65] specifies the objectives to develop architectures for immersive media distribution, considering various aspects such as QoS framework, network assistance, QoE reporting and content rights management. Another Release 16 work item is also in progress to define 3GPP media codec profile(s) and network-based media processing functions (video stitching, media transcoding, content reformatting) and enablers (network APIs) for immersive media support over framework for live uplink streaming (FLUS) service in TS26.238 [66] and also recommend new QoS classes trading-off video quality for immersive media delivery latency.

The 3GPP has been conducting standardization work on immersive media since its launch of 5G-targeted standardization activities in 2015, starting with the completion of two study items SA1 (Requirements) and SA4 (Codecs and Media) working groups.

- SA1 developed a set of use cases on immersive media in Release 14 TR22.891 [67] and later in Release 15 TS22.261 [68] for 5G stage 1 development, which described requirements toward supporting VR and interactive conversation use cases, including relevant latency requirements for video and audio.
- SA4 technical report TR26.918 [69] documented a broad range of on-demand and live streaming, broadcast and conversational VR use cases, relevant VR technologies for audio and video and various subjective quality evaluations.

The relevance of AR/VR in the context of 3GPP services addressing in Release 16 3GPP TR26.928 [70] aspects such as use cases, relevant technologies, media formats, metadata, interfaces and delivery procedures, client and network architectures and APIs and QoS service parameters. Also, 3GPP initiated a study item on QoE metrics relevant to VR user experience TR26.929 [71]. The QoE parameters and metrics that may need to be reported by the client to the network for evaluation of user experience in VR services has been investigating. The study considers E2E VR delivery chain including content creation, network transmission and device capabilities. A key consideration is placed on QoE reference metrics for viewport-dependent delivery, which uses HEVC codec tiling concept to deliver content in tiles allowing streams to have a different quality or resolution for different areas/regions of the omnidirectional video. While the viewport-dependent video delivery approach helps optimize the quality-bandwidth trade-off, the interactivity performance in this case not only depends on latency at the rendering level but also on other metrics such as network-level latencies (since the high-quality tiles corresponding to the new viewport need to be fetched continuously from the network) such as network request delay, origin-to-edge delay (in case of cache miss), transmission delay (accounting for access network delay) and delays incurred in the client device due to buffering, decoding and rendering.

The immersive services such as AR/VR, UHD and 3D video have critical requirement, and the 5G network will be able to transfer these data traffic in a flexible and efficient manner. In order to better understand what happens to the user end, the network operators need to collect QoE metrics that are able to represent these features and events. It is necessary to take into account the following three aspects: content quality, the network constraints and the device limitations. The user's device plays an important role in the E2E user experience. The goal should be that the overall QoE is maximized within the given constraints and that a minimum QoE is ensured. Therefore, 3GPP is working on the following:

- Technical evaluation of the subjective quality of the relevant parts of E2E system needs to be performed to assess the overall QoE.
- Full technical description and either an implementation of these blocks and/or performance requirements are required to ensure the tested QoE.

Higher picture quality is a general trend for all services and devices. With the expected increase of personal HMDs and new content generation devices, researchers have recently started to quantify, model and manage QoE when the user-consumed content is beyond traditional audio and video materials. On the other hand, service providers seek to deliver as high quality as possible, despite the growing associated costs and the issues related to network capacity. Defining VR service–specific QoE metrics will allow 5G operators to understand and manage how end users are experiencing specific VR services. Based on these QoE metrics, operators may also perform problem analysis and trouble shooting. VR-oriented E2E network operation and management system become critical for understanding the user-perceived quality of immersive multimedia experiences. The VR video quality can be degraded by various faults which can be hard to separate or distinguish from. This necessitates the development of effective QoE-oriented E2E solutions at various points of VR video delivery system for real-time monitoring, detecting and demarcating faults. This would enable service providers to enhance VR video streaming service experience through a multisensing VR QoE solution that can model user experience and measure the perceivable media quality during future softwarized network transmission.

7.3 A New Media Ecosystem

5G is a technology enabler for a massive market and ecosystem, which could be created by all involved companies or organization. It rather requires standard organizations and industry consortiums to work collaboratively with their own expertise and requirements for 5G in order to develop a healthy ecosystem with a unified 5G standard. 3GPP as the core standard developing organization for 5G communication technology has been in the center stage of cross-industry effort and has been evolving to be more open and flexible.

A number of traditional industries, such as automotive, healthcare, energy and municipal systems, participate in the construction of the ecosystem. The expansion of service scope for mobile networks enriches the telecom network ecosystem. 5G is the beginning of the promotion of digitalization from personal entertainment to society interconnection. Digitalization creates tremendous opportunities for the mobile communication industry but poses strict challenges toward mobile communication technologies. With the continued growth in mobile traffic, there is a need for a more efficient technology, higher data rates and spectrum utilization. In terms of consumer digitalization and richer user experience, new applications based on VR/AR will require higher bandwidths and lower latency.

The first phase of 5G is to provide a massive improvement of the MBB service with eMBB service. There will be significantly improvement in the capacity, coverage and latency of the network. The end user finds his Internet experience faster and in more places. But perhaps the most significant improvement is in the reduction of the latency, which allows for the extension of the Internet to another human sensory system. eMBB aims to meet the people's demand for an increasingly digital lifestyle and focuses on services that have high requirements for bandwidth, such as high-definition (HD) videos, R and AR.

Several standards bodies are involved in the specification of the 5G ecosystem. The most important one is the 3GPP project, even though others, such as ETSI (European Telecommunications Standards Institute) and IEEE (Institute of Electrical and Electronics Engineers, are also working on defining some aspects of the 5G system. In September 2018, 3GPP has completed the definition of the first set of 5G features, called 5G phase 1, under the Release 15. This first set defines the new telecommunication architecture components and protocols and messages needed to make interwork the 5G new access technology, called 5G NR, with the legacy systems. From the usage scenarios point of view, 5G phase 1 has focused on mainly defining the eMBB services, whereas the other two main usage scenarios uRLLC and mMTC are in the focus of the ongoing 5G phase 2, under the Release 16, planned to be completed by the middle of 2020. The content of what will come afterward, most probably called 5G LTE, under the work of Release 17, is still a matter of development.

With the development of 5G NR standard, there is also 5G trial campaign ongoing to verify the candidate technology, test the system design and promote 5G industry. This facilitates the maturity of 5G industry and build a good foundation for commercialization. 5G NR standardization for NSA and SA had been frozen in December 2017 and June 2018, respectively, and the 5G industry is becoming mature in the aspects of network, chipset, device, etc. Currently, we are at the stage of deployment, both precommercial and commercial. More launches are planned at the beginning of 2019, but it is in 2020 that a real broad deployment of 5G services is expected to take place. Meanwhile, the experimentation phase and the first trials are going into the field with more and more real users, so to make mature the newly defined 5G technology. Such trials and first commercial launches are key to identify the most promising technical enhancement needed to make the 5G system proposition a commercial success.

7.3.1 Proof of Concepts

The implementation of the 5G vision set out by the research and innovation actions. The peak of proofs of concept (PoCs) and trials targeting verticals occurs with multimedia and entertainment. One reason for this is that eMBB is among the first services to be supported with early 5G deployment. 5GPP projects cover vertical clusters of automotive, energy, health, industry, media and entertainment, public safety, smart cities, transport and logistics. All 5G service classes are delivered over a scalable and cost-efficient network. 5G technological innovations transform the network into a secure, reliable and flexible orchestration platform across multiple technology.

5GPPP 5G-Xcast media delivery project has built-in unicast/multicast/broadcast and caching capabilities and enables media services to use any mix of the available mobile, fixed and broadcast networks. The project analyzes the commercial and technical requirements of 4K/8K UHDTV, HDR&WCG, HFR, object-based content, VR/AR/MR and 360° visual media as well as NG audio. The project defines system architectures, as well as the top-level specifications for the transport and application layers. 5G-Xcast takes a practical approach, PoC prototypes and demonstrations. 5G-Xcast is focusing on large-scale media distribution, as this use case is one of the most demanding requirements in terms of data rate (capacity), scalability (cost-effectiveness) and ubiquity (coverage). The 5G-Xcast media delivery solutions are built-in unicast/multicast/broadcast modes and caching capabilities. The project covers multiple disciplines from the radio interface to the transport and application layers, including protocols and APIs, as well as network and system architecture aspects. The development of the 5G-Xcast media delivery solution is focused on the media and entertainment vertical. PoC prototypes and technology demonstrators are pivotal tasks of the project. Special emphasis is being given to emerging new 3D immersive media services that cannot be efficiently delivered by existing technologies and networks. The 5G-Xcast consortium is a balanced combination of telecom and media entities, covering the complete ecosystem.

5GPPP *Superfluidity* media distribution. Working group WG11 (MPEG) investigating if and how compression standards can be affected by 5G. With the increasing complexity and sophistication of media services and the incurred media processing, offloading complex media processing operations to the 5G cloud/network are becoming critically important in order to keep receiver hardware simple and power consumption low. MPEG promoted ISO/IEC 23090-8 network-based media 0rocessing (NBMP) to final draft international standard (FDIS) in January 2020. The NBMP standard defines a framework that allows content and service providers to describe, deploy and control media processing for their content in the cloud by using libraries of prebuilt third-party functions. The framework includes an abstraction layer to be deployed on top of existing commercial cloud platforms and is designed to be able to be integrated with 5G core and edge computing (MEC). MEC additional opportunities for cost savings for mobile operators and wireless communication ecosystem partners by tight interworking of data communication, processing, storage and management capabilities very near the radio access nodes. MEC will run on general-purpose processors (GPPs) that are similar to those used in the data centers of SNDs and NFVs, but within the cloud MEC, they will be very close to the radio edge or even within it, typically inside the BS or in its close proximity. ETSI is working on the standardization of MEC building blocks such as platform services and APIs, virtual machine service-level agreements and the MEC application platform management interface [72].

5G-MiEdge project focus is on mmWave access/backhauling with MEC to enable eMBB services and mission-critical low-latency applications via cost-efficient RANs for 5G phase 2. To achieve this goal, an ultralean and interoperable control signaling RAN C-plane has defined in combination of eMBB and uRLLC into a new class services ultrahigh-speed low-latency communications. Acquisition of context information and forecasting of service requests are key steps to enable a proactive orchestration of radio and computation resources. The final target is to demonstrate a first 5G implementation through testbeds at the 2020/2021

Summer Olympics Games. Current research efforts on 5G RANs strongly focus on mmWave access for addressing a critical weakness of deployed cellular systems, i.e., the capacity to realize eMBB services, as discussed at the World Radiocommunication Conference 2015 (WRC-2015). Recently, mmWave technologies have reached a significant degree of maturity operated in the 60 GHz unlicensed band and are already in the market. However, there are many unsolved issues for an effective deployment of mmWave 5G RAN. The most critical issue is the impossibility of providing 10 Gigabit Ethernet backhaul everywhere. MEC is considered as a key technology to enable mission-critical (low-latency) applications by allocating storage and computation resources at the edge of the network, so to circumvent the backhaul networks' limited capacity. However, in the case of mobile networks, it is not easy to reallocate computational resources on demand, while meeting the strict latency constraints foreseen in 5G networks. In summary, the 5G MiEdge project develops a feasible 5G ecosystem by combining RAN C-plane and user/application-centric orchestration.

7.3.2 Research Consortia

The 5G communication system is officially launching worldwide in 2020. As 5G research is maturing toward a global standard, the research community must focus on the development of beyond-5G solutions. Around 2030, our society will become data driven, enabled by nearly instantaneous, unlimited wireless connectivity (Figure 7.7).

The next generation, 6G, will include relevant technologies considered too immature for 5G or which are outside the defined scope of 5G. However, research activities on 6G are in their initial stages. 6G communication is still in its infancy. From 2020, a number of studies will be performed worldwide on the standardization of 6G. Many researchers have defined 6G as B5G or 5G+. Preliminary research activities have already started in the United States. China has already started the concept study for the development and standardization of 6G communications in 2019. The Chinese are planning for active research work on 6G in 2020. Most European countries, Japan and Korea are planning several 6G projects. For example, in September 2017, the EU launched a 3-year research project on the basic 6G technologies. The main task is to study the NG forward error correction coding, advanced channel coding and channel modulation technologies for wireless terabit networks (https://futurecomresearch.eu). At the end of 2017, China began to study the 6G mobile communication system to meet the inconstant and rich demands of the IoT in the future, such as medical imaging, AR, and sensing (www.china.org.cn). In April 2018, the Academy of Finland announced an 8-year research program (6Genesis) to conceptualize 6G through a joint effort of the University of Oulu and Nokia. Research is organized into four unified planned parts: wireless connectivity, distributed computing, services and applications. Scientific innovations will be developed for important technology components of 6G systems. More recently, the U.K. government has invested in some potential

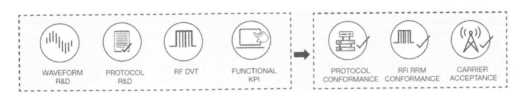

FIGURE 7.7
5G research and development span from simulation and design verification (DVT), to device acceptance and ultimately to manufacturing and deployment.

techniques for 6G and beyond, some universities in the United States have lunched research on terahertz-based 6G wireless networks and South Korea Telecom (SKT) has started 6G research based on the cell-free and nonterrestrial network techniques.

ITU standardization activities were based on IMT-2020. Consequently, ITU-R will probably release IMT-2030, which will summarize the possible requirements of mobile communications in 2030. Specifically, the newly established ITU Focus Group on Technologies for Network 2030 will consider the LTE of 5G and develop future use cases.

The new requirements of 6G will influence the main technology trends in its evolution process [73]. The success of 6G will have to leverage breakthroughs in novel technological concepts. Hence, there are numerous open issues to resolve [74–77]:

- Subterahertz and terahertz communications. The availability of large bands beyond 100 GHz (D-band, 110–170 GHz) will lead to transmission systems at high data rate over short distances, so releasing the lower band radio spectrum for long-range uses.

- Massive use of multiple antenna systems. The increase in the frequency will require multiple antenna systems able to exploit the multirays propagation, not only for larger throughput but also for precise localization and for energy transfer.

- Dynamic spectrum allocation. The precious radio spectrum in the lower bands will be used more effectively by allocating the frequencies every second or so, based on the context.

- Free-space optical communication. Optical free-space communication will allow high data rate, for both outdoor and indoor scenarios, so releasing the lower band radio spectrum for long-range uses.

- High accuracy indoor localization. Context awareness needs a precise user's localization.

- Access schemes for massive wireless networks. New access schemes will be needed to handle a massive number of nonorthogonal users (more than 10 devices/m²) in an efficient and scalable way.

- Machine learning and artificial intelligence. The complexity of the network and the number of connected devices will lead to a network, which will learn from the experience to modify itself and accommodate new services.

- Wireless energy transfer. This could be in some situations a viable way to extend the battery life, avoiding frequent recharges.

- Cybersecurity. The possibility to use nonpersonal devices for personal communication will impose new challenges on biometric authentication and privacy.

7.4 Conclusions

This chapter reviews the research and development of latest 5G mobile cellular systems, discusses multimedia communication issues and points out what we may expect beyond 5G ecosystem. The MBB human-to-human communication was the main base of the cellular ecosystems. Key elements of eMBB are massive mobile connectivity, high data rates, connection density and mobility and human-centric use cases for access to multimedia content. Although the eMBB service remains a significant segment of the wireless market, the capabilities of 5G allow it to move beyond the eMBB service and support vertical industries. This convergence of human-to-human communication with human-to-machine and machine-to-machine communications are a key enabler of the fourth industrial revolution. 5G constitutes a larger ecosystem, including more stakeholders than in the past, with more complex relationships, more heterogeneity and more dynamicity. Application sectors are more and more actively involved in the creation and provision of services, taking full part in the 5G value chain. 5G provides services not only to customers but also to industrial stakeholders. 5G is also an opportunity for network operators to return in the service creation and management arena.

5G is the new generation of the global telecommunication network. Aiming to become the new reference architecture for the global mobile and fixed telecommunication network, 5G is not only an evolution in terms of performance, but it also creates a breaking point with respect to previous generations. 5G is based on a cloud-native, softwarized, E2E architecture, encompassing the radio access, metro and core network sections, as well as the edge and cloud computing resources within the network. 5G supports diversified vertical services, targeting different types of users and including services not exclusively dedicated to human users. eMBB is service characterized by very high data rates and very high density of users, requiring extremely high-quality mobile video distribution, and support of the expected increases in immersive 3D video consumption.

5G is kick-off with eMBB as its first use case. With the continued growth in mobile traffic, there is a need for a more efficient technology, higher data rates and spectrum utilization. In terms of consumer digitalization and richer user experience, new applications based on VR and AR require higher bandwidths and lower latency. From the usage scenarios point of view, 5G phase 1 has focused on mainly defining the eMBB services. They cover scenario targeting carrier data rates larger than 10 Gb/s. As new use cases for fully immersive AR and VR experience are introduced in the market, service providers will need to address bandwidth limitations, reduce E2E network latency and improve the overall QoS/QoE for the streaming media services.

The future of multimedia communications depends on 5G. The recently established working groups will consider the LTE of 5G and develop future use cases. The new requirements will influence the main technology trends in its evolution process.

References

1. P.Popovski et al., 5G wireless network slicing for eMBB, uRLLC, and mMTC: A communication-theoretic view, *IEEE Access*, vol.6, Sept. 2018.
2. Q.C.Li et al., 5G network capacity: Key elements and technologies, *IEEE Vehicular Technology Magazine*, vol.9, no.1, pp.71–78, 2014.
3. S.Henry et al., 5G is real: Evaluating the compliance of the 3GPP 5G new radio system with the ITU IMT-2020 requirements, *IEEE Access*, vol.8, March 2020.
4. M.S.Elbamby et al., Toward low latency and ultra-reliable virtual reality, *IEEE Network*, vol.32, no.2, pp.78–84, 2018.
5. ITU-R M. 2083, IMT Vision – *Framework and overall objectives of the future development of IMT for 2020 and beyond*, Sept. 2015.
6. 3GPP TR 38.913, *Study on scenarios and requirements for Next Generation access technologies*, Release 15, June 2018.

7. 3GPP TR 38.801, *Study on new radio access technology: Radio access architecture and interfaces*, Release 14, Apr. 2019.

8. H.C.Rudolph et al., Security challenges of the 3GPP 5G Service Based Architecture, *IEEE Communications Standards Magazine*, vol.3, no.1, pp.60–65, Mar. 2019.

9. 3GPP TS23.501, *System architecture for the 5G System (5GS)*, Stage 2, Release 16, Sept. 2019.

10. 3GPP TS23.502, *Procedures for the 5G System (5GS)*; Stage 2, Release 16, Sept. 2019.

11. 3GPP TS23.503, *Policy and charging control framework for the 5G System (5GS)*, Stage 2, Release 16, Dec. 2019.

12. F-L.Luo, Signal processing techniques for 5G: An overview, ZTE Communications Special Topic: *5G Wireless - Technology, Standard and Practice*, vol.13, no.1, pp.20–27, 2015.

13. Y.Yuan, and X.Wang, 5G new radio: Physical layer overview, *ZTE Communications Special Topic: 5G New Radio (NR) - Standard and Technology*, vol.15, no.s1, pp.3–10, 2017.

14. 3GPP TS38.300, NR: *Overall description*, Stage 2, Release 15, Apr. 2018.

15. D.Kreutz et al., Software-defined networking: A comprehensive survey, *Proceedings of the IEEE*, vol.103, pp. 14–76, Dec. 2015.

16. R.Mijumbi et al., Network function virtualization: State-of-the-art and research challenges, *IEEE Communications Surveys & Tutorials*, vol.18, pp.236–262, Sept. 2016.

17. 3GPP TS38.141, NR: *Base Station (BS) conformance testing Part 1: Conducted conformance testing*, Release 16, Jan. 2020.

18. 3GPP TS38.141, NR: *User Equipment (UE) conformance specification; Applicability of radio transmission, radio reception and radio resource management test cases*, Release 16, Jan. 2020.

19. 3GPP TS38.141, NR: *User Equipment (UE) conformance specification; Radio Resource Management (RRM)*, Release 16, Jan. 2020.

20. 3GPP TS38.508, 5GS: User Equipment (UE) conformance specification; Part 1: Common test environment, Release 16, Jan. 2020.

21. 3GPP TS38.509, 5GS: *Special conformance testing functions for User Equipment (UE)*, Release 16, Jan. 2020.

22. 3GPP TS38.523, 5GS: *User Equipment (UE) conformance specification; Part 1: Protocol*, Release 16, Jan. 2020.

23. ITU-R M.2410, *Minimum requirements related to technical performance for IMT-2020 radio interface(s)*, Nov. 2017.

24. ITU-R M.2412, *Guidelines for evaluation of radio interface technologies IMT2020*, Oct. 2017.

25. 3GPP TR 38.913, *Study on scenarios and requirements for next generation access technologies*, Release 15, July 2019.

26. 3GPP TR38.915, *Release description*, Release 15, Oct. 2019.

27. 3GPP TR21.916, *Release description*, Release 16, Mar. 2020.

28. F.Hartung et al., MBMS - IP Multicast/Broadcast in 3G networks, *International Journal Digital Multimedia Broadcast*, vol.2009, pp.1–25, 2009.

29. D.Lecompte, and F.Gabin, Evolved Multimedia Broadcast/Multicast Service (eMBMS) in LTE-advanced: Overview and Rel-11 enhancements, *IEEE Communications Magazine*, vol.50, no.11, pp. 68–74, 2012.

30. N.Nguyen et al., Implementation and validation of multimedia broadcast multicast service for LTE/LTE-advanced in open air interface platform, in *Proc. IEEE Conf. Local Computer Networks Workshops (LCN Workshops)*, 2013.

31. H.Chen et al., Pioneering studies on LTE eMBMS: Towards 5G point-to-multipoint transmissions, in *Proc. IEEE Sensor Array and Multichannel Signal Processing Workshop (SAM)*, pp.1–5, 2018.

32. W.Guo et al., Roads to multimedia broadcast multicast services in 5G new radio, in *Proc. IEEE International Symposium on Broadband Multimedia Systems and Broadcasting (BMSB)*, pp.1–5, 2018

33. M.Säily et al., Radio access networks: Enabling efficient point-to-multipoint transmissions, *IEEE Vehicular Technology Magazine*, pp.29–37, 2019.

34. J.Lee et al., LTE-advanced in 3GPP Rel-13/14: An evolution toward 5G, *IEEE Communication Magazine*, vol.54, no.3, pp.36–42, 2016.

35. 3GPP TS25.992, *MBMS UTRAN/GERAN requirements*, Oct. 2003.
36. 3GPP TS41.101, *Technical specifications and Technical reports for a GERAN-based 3GPP system*, Dec. 2004
37. 3GPP TR25.905, *Improvement of the MBMS in UTRAN*, Dec. 2007.
38. 3GPP TS25.992, *Multimedia Broadcast/Multicast Service (MBMS); Requirements*, Release 6, Dec. 2009.
39. 3GPP TS25.992, *Multimedia Broadcast/Multicast Service (MBMS)*, Release 10, Mar. 2011.
40. 3GPP TS25.992, *Multimedia Broadcast/Multicast Service (MBMS)*, Release 11, Sept. 2012.
41. 3GPP TS25.992, *Multimedia Broadcast/Multicast Service (MBMS)*, Release 12, Oct. 2014.
42. 3GPP TS25.992, *Multimedia Broadcast/Multicast Service (MBMS)*, Release 13, Dec. 2015.
43. 3GPP TS25.992, *Multimedia Broadcast/Multicast Service (MBMS)*, Release 15, July 2018.
44. 3GPP TS26.346, *Multimedia Broadcast/Multicast Service (MBMS), Protocols and codecs*, Release 16, Mar. 2020.
45. 3GPP TR22.827, *Study on Audio-Visual Service Production (AVPROD)*, Release 17, Jan. 2020.
46. 3GPP TS22.263, *Service requirements for video, imaging and audio for professional applications (VIAPA)*, Release 17, Dec. 2019.
47. 3GPP TS26.116, *Television (TV) over 3GPP services - Video profiles*, Nov. 2017.
48. 3GPP TR26.906, *Evaluation of HEVC for 3GPP services*, Mar. 2017.
49. R.T.Azuma, A survey of augmented reality, *Presence: Teleoperators and Virtual Environments*, vol.6, no.4, pp. 355–385, 1997.
50. A.A.Barakabitze et al., QoE management of multimedia streaming services in future networks: A tutorial and survey, *IEEE Communications Surveys & Tutorials*, vol.22, no.1, pp.1–43, 2020.
51. .Duan et al., Perceptual quality assessment of omnidirectional images: Subjective experiment and objective model evaluation, *ZTE Communications Special Topic: Quality of Experience for Emerging Video Communications*, vol.17, no.1, pp.38–47, 2019.
52. J.van der Hooft et al., Towards 6DoF Virtual Reality video streaming: Status and challenges, *IEEE COMSOC MMTC Communications – Frontiers, Special Issue: Future Network Architecture, Technologies, and Services*, vol.14, no.5, pp.30–38, 2019.
53. U.Yiling et al., Introduction to point cloud compression, *ZTE Communications Special Topic: Next Generation Mobile Video Networking*, vol.16, no.3, pp.3–8, 2018.
54. 3GPP TS26.118, *Virtual Reality (VR) profiles for streaming applications*, Mar. 2020.
55. 3GPP TS26.259, *Subjective test methodologies for the evaluation of immersive audio systems*, Oct. 2018.
56. 3GPP TS26.260, *Objective test methodologies for the evaluation of immersive audio systems*, Mar. 2020.
57. 3GPP TS26.818, *Virtual Reality (VR) streaming audio; Characterization test results*, Oct. 2018.
58. 3GPPTS26.114 IP, *Multimedia Subsystem (IMS); Multimedia telephony; Media handling and interaction*, Mar. 2020.
59. 3GPP TS26.223, *Telepresence using the IP Multimedia Subsystem (IMS); Media handling and interaction*, Mar.2020.
60. 3GPP TR26.919, *Study on media handling aspects of conversational services in 5G systems*, Oct. 2019.
61. 3GPPTS26.114 IP, *Multimedia Subsystem (IMS); Multimedia telephony; Media handling and interaction*, Mar. 2020.
62. 3GPP TS26.223, *Telepresence using the IP Multimedia Subsystem (IMS); Media handling and interaction*, Mar. 2020.
63. 3GPP TR26.891, *5G enhanced mobile broadband; Media distribution*, Dec. 2018.
64. 3GPPTS26.247, *Transparent end-to-end Packet-switched Streaming Service (PSS); Progressive Download and Dynamic Adaptive Streaming over HTTP (3GP-DASH)*, Dec. 2019.
65. 3GPP TS26.501, *5G Media Streaming (5GMS); General description and architecture*, Mar. 2020.
66. 3GPPTS26.238, *Uplink streaming*, Mar. 2020.
67. 3GPPTR22.891, Study on new services and markets technology enablers, Oct. 2016.
68. 3GPPTS22.261, *Service requirements for the 5G system*, Mar. 2020.
69. 3GPP TS26.918, *Virtual reality media services over 3GPP*, Mar. 2018.
70. 3GPP TR26.928, *Extended Reality (XR) in 5G*, Mar. 2020.
71. 3GPPTR26.929, *QoE parameters and metrics relevant to the Virtual Reality (VR) user experience*, Oct.2019.

72. T.X.Tran et al., Collaborative mobile edge computing in 5G networks: New paradigms, scenarios, and challenges, *IEEE Communications Magazine*, vol.55, pp.54–61, Apr. 2017.

73. A.Ghosh et al., 5G evolution: A view on 5G cellular technology beyond 3GPP Release 15, *IEEE Access*, vol.7, pp.127639–127651, Sept. 2019.

74. K.David, and H.Berndt, 6G vision and requirements: Is there any need for Beyond 5G? *IEEE Vehicular Technology Magazine*, vol.13, no.3, pp.72–80, Sept. 2018.

75. P.Yang et al., 6G wireless communications: Vision and potential techniques, *IEEE Network*, vol.33, no.4, pp.70–75, July/Aug. 2019.

76. B.Zong et al., 6G technologies: Key drivers, core requirements, system architectures, and enabling technologies, *IEEE Vehicular Technology Magazine*, vol.14, no.3, pp.18–27, Sept. 2019.

77. Z.Zhan et al., 6G wireless networks: Vision, requirements, architecture, and key technologies, *IEEE Vehicular Technology Magazine*, vol.14, no.3, pp.28–41, Sept. 2019

8

5G Network Slicing: Principles, Architectures, and Challenges

Bojan Bakmaz and Miodrag Bakmaz
University of Belgrade

CONTENTS

8.1 Introduction

Fifth-generation (5G) mobile networks are expected to create a multitenant ecosystem with extremely increased performance for specific types of services and dedicated use cases in order to simultaneously satisfy various users' demands. These requirements stem from a diversity of novel application areas such as Industry 4.0, intelligent transport systems, telemedicine, smart grid and so on. In order to cope with the aforementioned requirements, the concept of network slicing emerged as one of the key prospective techniques that support different independent services through a common infrastructure. In this sense, service necessity, isolation and support on a common physical level are assured. It means that on-demand services can be fully supported, while network resources are efficiently allocated, managed and controlled, according to the users' requirements. Driven by the new requirements, network slicing can reshape and reorganize the network resources optimally for different services, therefore providing more flexibility to apply new techniques into vertical applications.

Although the network slicing concept is still immature, the potential enabling technologies, such as software-defined networking (SDN) and network functions virtualization

(NFV), have many feasible researches with practical solutions (Ordonez-Lucena et al., 2017). Based on SDN and NFV, many user-centric slicing strategies were proposed. By means of SDN and NFV, operators can provide a high degree of flexibility and programmability, allowing legacy functions to be partitioned or migrated in data centers, advancing virtual architectures (Zhang et al., 2017). Currently, network slicing is in the main focus of many standardization bodies, e.g., International Telecommunication Union (ITU), 3rd Generation Partnership Project (3GPP), European Telecommunications Standards Institute (ETSI) and so on. At the moment, efforts are toward developing 5G mobile systems that are in a position to deploy slicing of different structures and sizes (Katsalis et al., 2017).

The remainder of this chapter is organized as follows. The next section presents service requirements in 5G mobile systems. Following that, a brief overview of network slicing origination and main principles is provided. Algorithmic aspects of network slicing embedding are presented, too. As enablers of network slicing, different virtualization technologies are introduced. Then, specific slicing approaches for radio access domain and core domain are analyzed. The main objective is to analyze the current maturity of prospective solutions according to the architectural segments they target and to identify remaining gaps. Finally, some open research challenges are elaborated, and relevant conclusions are provided.

8.2 Diversity of Service Requirements

5G mobile networks can drastically change the architecture and nature of communications. Many use cases are emerging with diverse requirements in terms of throughput, latency, connection density, mobility, reliability, spectral and energy efficiency. Based on the corresponding key performance indicators (KPIs), these use cases may be broadly categorized in the three generic services (ITU-R, 2015), i.e., enhanced mobile broadband (eMBB) communications, massive machine-type communications (mMTCs), also referred to as massive Internet of Things (mIoT) and ultrareliable low-latency communications (uRLLCs), also referred to as mission-critical machine-type communications (cMTCs). Here, KPIs can be treated as technical requirements for 5G services, as presented in Table 8.1. Recently, ultrahigh-speed low-latency communications (uHSLLCs) are envisioned in the forthcoming 5G phase II, as a combination of eMBB and uRLLC (Frascolla et al., 2017).

TABLE 8.1

Typical Requirements for 5G Services

	eMBB	mMTC	uRLLC	uHSLLC
Throughput	High	Low	Low	High
Latency	Medium	Medium	Low	Low
Reliability	Medium	Low	High	High
Mobility	High	Low	Medium	High
Connection density	Medium	High	Low	Low
Spectral efficiency	High	Low	Low	Medium
Energy efficiency	High	Medium	Low	Medium

eMBB, enhanced mobile broadband; mMTC, massive machine-type communication; uHSLLC, ultrahigh-speed low-latency communication; uRLLC, ultrareiable low-latency communication.

eMBB communications use stable connections with very high average user throughput, while generated traffic can be considered as a direct extension of the traditional broadband services. It is characterized by large payloads and stationary device activation pattern over an extended time interval. This allows resource scheduling such that no two devices access the same resource simultaneously. On the other hand, an mMTC device is activated periodically and uses a fixed, typically low, transmission rate in the uplink. A huge number of devices can be connected to a given base station, but at a certain time, only a random subset of them becomes active and attempts to send their data. A large number of potentially active devices make it infeasible to allocate a priori resources to individual devices. Instead, it is necessary to provide resources that can be shared through random access. The objective in the design of mMTC is to maximize the supported arrival rate. Finally, uRLLC supports low-latency transmissions of small payloads with very high reliability from a limited set of terminals (much smaller than for mMTC), which are active according to patterns typically specified by stochastic events. In this case, supporting intermittent transmissions requires a combination of scheduling. In order to avoid that too many resources being idle due to the intermittent traffic, a certain amount of predictability in the available resources is needed. Due to the low latency requirements, a uRLLC transmission should be localized in time. Diversity, which is critical to achieve high reliability, can be achieved using multiple frequency or spatial resources (Popovski et al., 2018). Having in mind the fact that these services cannot be simultaneously supported by the common physical infrastructure, agile and programmable network architecture is envisioned. In that way, each service should have a tailored network instance to satisfy its requirements.

The 5G Infrastructure Public Private Partnership (5G PPP), as a joint initiative between the European Commission and industry, brought interest into the requirements of 5G through the *Mobile and wireless communications Enablers for Twenty-twenty (2020) Information Society* (METIS) project. They are studying the requirements and architectures that will enable network slicing in their first 5G outputs. The network capabilities such as extremely fast (data rates of 10 Gb/s), great service in a crowd (traffic volume of 9 Gb/h), super real-time and reliable connection (latency less than 5 ms and loss probability in order of 10^{-5}) as well as ubiquitous thing communicating (3×10^5 devices/cell) are supporting rapid time-to-market for brand new services and reducing management costs.

3GPP SA WG1 has completed its first technical specification on *service requirements for the 5G system* (3GPP TS 22.261, 2018). This specification contains both requirements on underlying capabilities and performance targets. 5G capabilities related to the scalability and ubiquitous support for a plethora of heterogeneous services and vertical markets can be characterized by

- resource efficiency for existing and new services ranging from low data rate servicing (e.g., mMTC) to high bitrate multimedia services (e.g., virtual and augmented reality);
- capability to allow third-party providers to manage network slices and deploy applications in the operator's hosting environment;
- connectivity from remote devices via relay and seamless service connectivity between indirect and direct connections.

It should be noted that this specification defines performance targets for different scenarios when considering service areas (urban and rural wide area, indoor hotspots, dense urban area, etc.) and applications (crowd broadband access, tactile interaction, transport systems, etc.). For example, user experienced data rates vary from 1 Gb/s in downlink for indoor hot spots to 15 Mb/s in uplink for airplanes connectivity. Also, latency targets are as low as 0.5 ms for tactile interaction, whereas capacity targets can be as high as 15 Tb/s/km^2 with 250,000 users/km^2 for indoor hot spot such as office environments. These service requirements are used as starting guidelines for the others related 3GPP WGs dealing with upcoming standards.

8.3 Origination and Principles of Network Slicing

The period 1960s, when the IBM operating system CP-40 with implemented resource virtualization was developed, can be considered as a pioneering step to the network slicing concept (Goldberg, 1974). Virtualization was widely adopted through data centers in the 1970s, while a decade later, it was applied into the computer networks. In the late 1980s, overlay networks of nodes connected over logical links and forming a virtual environment over physical infrastructure are introduced. They can be seen as an early form of network slicing, combining heterogeneous resources over various administrative domains. During the past decade, several solutions have been proposed in the context of distributed service architectures, such as content delivery networks, large-scale distributed testbed platforms and distributed cloud computing systems (Esposito et al., 2013; Berman et al., 2014). The main idea was to promote research on clean state network while addressing consolidate resources and mobile environment. Network slicing has already been considered in current Long Term Evolution (LTE) networks (Katsalis et al., 2017).

Next-generation mobile network alliance has recognized network slicing as a concept for running multiple independent logical networks on a common physical infrastructure (NGMN Alliance, 2016). Each slice represents an independent virtualized end-to-end (E2E) network that is implemented on top of a physical infrastructure allowing operators to run deployments based on different architectures in parallel. A network slice as a logical E2E construct is self-constrained, having customized functions also including a phase in the user equipment (UE) and using network function (NF) chains for service distribution (Nikaein et al., 2015). Besides customized capabilities required for corresponding services, adaptability to changing requirements is needed, too (Zhang, 2019). Key aspects necessary to realize the network slicing are resources, virtualization, orchestration and isolation.

Resource is a manageable unit defined by a set of attributes or capabilities that can be used for service distribution. In this sense, there are two types of resources to be considered, i.e., NFs and infrastructure resources. NFs as functional blocks provide network capabilities in order to support and realize the particular service to use case demands. They can be physical and/or virtualized. As for infrastructure resources, they include computing hardware, storage capacity and radio interface. Virtualization enables resource sharing among slices. The allocation of resources to each slice can be either statical or dynamical. With static slicing, a fixed portion of the physical resources is assigned to each slice during the entire service time. This amount of resources usually corresponds to the peak service requirements. Despite its simplicity, static slicing does not provide efficient utilization of physical resources. In the case of time-varying service requirements, a static slicing approach may result in the resources overprovisioning. Therefore, a more promising solution is dynamic slicing proposed by Raza et al. (2018).

Orchestration is required to separate processes for creating, managing and distribution of services. Network slicing orchestration cannot be performed by a single centralized entity, not only because of the complexity and broad scope of tasks but also because of its necessity to preserve management independence. Each slice needs to be managed independently as a separate network. This is one of the major requirements for isolation, which needs to be satisfied in order to operate parallel slices simultaneously on a common shared underlying substrate. Isolation is an E2E issue, regardless of the congestion and performance levels of offered slices. Security and reliability in terms of isolation imply that attacks, overload or failures occurring in one slice do not affect the operation of other slices in the system. Thus, each slice needs to involve independent security functions (Arfaoui et al., 2018). The open interfaces that support the programmability of the network introduce security treats to softwarized architecture. The number of virtual network functions (VNFs) that slices have in common or share is in direct relation to the number of security vulnerabilities between them. In the case of network sharing, when slices of different tenants are used, the security vulnerabilities are even more evident. The fact that each of the operating tenants can share resources alongside their isolated slices with different security parameters exposes their individual slices to the intratenant security threats. Thus, multi-level security including policies and mechanisms for software integrity are needed.

In order to enable on-demand mobility management in 5G systems, a mobility-driven network slicing (MDNS) is proposed by Hucheng et al. (2017). MDNS takes individual mobility requirements into account while customizing networks for different services. Within this framework, the actual level of required mobility support is determined by a mobility description system, and network slice profiles with the corresponding mobility management schemes are defined by a network slice description function. By instantiating the network slices, each mobile terminal can select the network slice with the most appropriate mobility management scheme.

5G slice is a bundle of network services functions, applications, resources and equipment. The perceived advantages are significant, particularly useful in the design of next-generation wireless networks, such as slice isolation, simplified service chains, flexible VNF settings and transparent management. The entire isolation allows the simpler and efficient design of each slice and as a consequence overall system, too. The goal is to meet requirements offered by the slice tenant. Simplified service chains mean that in network slicing, each service can rely on a different subset of functions. NFV introduces an additional degree of freedom regarding the placement of these functions in the network. Here, functional components can be deployed in servers or cloud infrastructure instead of network hardware. Intelligent placement can improve network performance and reduce operating costs. Finally, network slicing provides an abstraction of the physical resources and makes slice management transparent to the tenant.

8.4 Generic Network Slicing Architecture

Service providers, physical network operators and virtual network operators can be considered as major players in the life cycle of slices. The role of physical network operator is to provide resources and to ensure orchestration of different slices. In this case, the operator is fully aware of prior knowledge about the service demands and cost/revenue models of every slice (Han et al., 2018). It is able to scale the slices in real time according to their utilization, which can be flexibly defined by the quality of service (QoS) level or the revenue rate. On the other hand, the role of tenants (i.e., service providers and virtual network operators) is to place slice requests and then manage the provided slice. A generic network slicing architecture for mobile systems is presented in Figure 8.1.

New slice template is determined by a tenant according to the functionality and resource requirements. It is obvious that slice templates should be available for the most common types of services. However, in order to be able to support novel services, a more flexible approach is needed. In the specific case, an enterprise (tenant) can reserve resources arbitrarily, similar the way that resources are reserved in the cloud and install on top only the necessary VNFs. After receiving a slice request, the network operator faces the problem of efficient embedding a concrete slice into the physical network. This step requires decisions on placing and interconnecting several VNFs and can be formulated as a constrained optimization problem. Based on the solution to this problem, the network operator establishes the slice using softwarization and virtualization technologies. Once the slice has been configured, the tenant can manage its resources; in the same manner, it would manage a dedicated physical network, by a disposed set of control functions exposed by network operator. In addition, the tenant can request readjustment of reserved resources, periodically. The complexity of tenant and operator interconnections during the lifetime of a slice is a subject of ongoing development and standardization process (ETSI, 2018a).

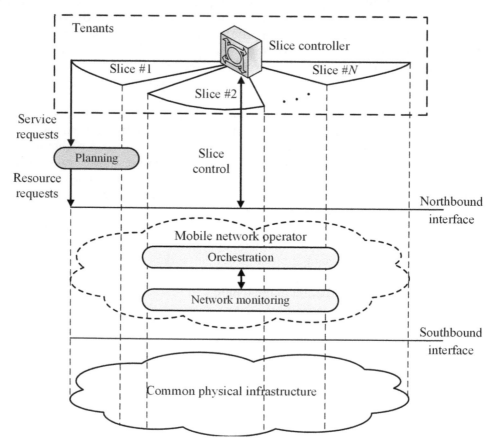

FIGURE 8.1
Generic network slicing architecture.

Even with the same service type, elastic slices can be defined to guarantee an average QoS level for a lower payment, whereas inelastic slices provide guaranteed minimal QoS level for a higher payment. This approach, usually known as slice as a service (SlaaS) introduced by Sciancalepore et al. (2017), improves the sharing efficiency and the resource utilization rate. However, as such slices are operated by tenants, and their scales shall be formulated and protected by agreements, a new agreement between the network operator and tenant may, therefore, be essential to flexibly rescale or terminate a slice at arbitrary time, which leads to higher operational costs. As an efficient alternative, the network operator can offer resources to implement slices of different types and optimize its resource allocation by choosing if to accept or reject every slice creation request from a tenant. On the other hand, as every slice is logically isolated from the others, i.e., a tenant has no access to slices operated by other tenants, but only the administration over its own resources by requesting new slices or terminating active slices of its own. In this case, neither the operator nor the tenants can jointly optimize all slices in a fully dynamic approach, which disables most classical techniques of resource allocation and initiate new challenges in network resource management.

8.5 Algorithmic Formulation of Network Slicing

A virtual link between virtual nodes can be realized as a multihop physical path with reserved resources on all physical links constituting the path. Virtual nodes implement some specific network functionalities as physical nodes (e.g., gateway, firewall, etc.). Network slicing can be seen as an optimization problem of implementing NFs over a set of candidate locations and corresponding interconnections (Vassilaras et al., 2017). In this case, optimal orchestration of network slicing at the operator level can be formulated as a special instance of virtual network embedding (VNE) problem (Fischer et al., 2013). Physical network can be presented as an undirected graph $G=(V, E, \mathbf{R}, \mathbf{c})$ with $v \in V$ nodes and $e \in E$ links, characterized by available resources R_v and R_e, respectively. Each node and link are also associated with a cost, c_v and c_e, respectively. Here, costs can be reflected as real operating costs or performances. The virtual network $H=(N, L, \mathbf{r}, U)$ has resource requirements r_n and r_l for each virtual node $n \in N$ and virtual link $l \in L$, respectively.

Each virtual node n can be embedded in exactly one physical node v from a set of physical nodes U, $U \subset V$. The usage cost is a linear function of the used resources, i.e., the cost of a virtual node n with resources demand r_n using physical node v with cost c_v is $c_v r_n$. A feasible embedding can be achieved just in the case where all node and link resource constraints are satisfied:

$$\sum_{n \in U_n} r_n^v < R_v \quad \text{and} \quad \sum_l r_l^e < R_e, \quad e = (u, v), \quad l = (m, n). \tag{8.1}$$

From the perspective of computational complexity, VNE is an integer linear programming (ILP) problem, which is nondeterministic polynomial time (NP) hard, as it can be reduced to the multiway separator problem (Yang and Guo, 2016). VNE algorithms are mostly heuristic or metaheuristic in the open literature, because they can offer suboptimal solutions with additional challenges of E2E constraints, heterogeneous requirements, nonlinear resource utilization and slice fairness (Bojkovic and Milovanovic, 2017).

Network slicing problem can be decomposed into a node embedding and an integral minimum cost problem with multiple flow demands for link embedding. Slice embedding can be equivalently studied as the embedding of an appropriate virtual network, reflecting the required slice components. In this context, a slice can be defined as $H'=(N, L, r, U)$, where the virtual nodes N represent VNFs, while L are links with corresponding resource requirements r. Connectivity of H' describes how the different VNFs are connected. In this case, the location constraint sets Un can be used to capture both the capabilities of physical nodes to run a specific VNF and the location requirements of the applications/users. The network slicing problem is to find the feasible slice embedding with the least cost. By appropriately selecting the embedding sets Un and virtual links, costs, resources and demands, any network slicing problem can be formulated as an extended VNE problem.

As a representative solution, network slice embedding model under the reliability constraints is proposed by Tang et al. (2018). The model jointly maximizes the number of network slice requests and minimizes the failure rate of network slicing. Furthermore, the trade-off between system stability and network slicing reliability is investigated. The slice embedding problem is observed as a mixed ILP problem and the heuristic queue-aware algorithm. Here, Lyapunov optimization (Neely, 2010) is applied for resource allocation while ensuring the stability of the queue. An online mechanism based on time window for processing incoming network slicing requests is introduced. Both theoretical and simulation analyses results demonstrate that the proposed algorithm can improve the throughput while guaranteeing the network reliability and stability.

8.6 Key Enabling Technologies

5G will guarantee levels of network performance with respect to the QoS metrics such as throughput, latency, availability, reliability and so on. For realizing this vision, mobile systems need to evolve into the more virtualized, cloud-like environment, using prospective technologies such as SDN and NFV. The growing interest in these technologies is motivated by the novelty of the overall context, especially having in mind their sustainability and high performance. Together with cloud, edge and fog computing, these technologies can be seen as facets of a broad innovation wave, called softwarization (Manzalini and Crespi, 2016).

8.6.1 Software-Defined Networking

SDN is essentially a centralized paradigm, in which the network intelligence (i.e., the control function or the control plane) is logically consolidated at one or a set of control entities (i.e., SDN controllers), while the data plane is simplified and abstracted for applications and services. SDN was originally designed for flexible and dynamic control of data path functions such as traffic routing and management. However, characteristics such as flexibility, scalability and robustness become a significant stimulus for SDN implementation in the slicing concept. A SDN controller alleviates third parties over the virtualizer, and through the means of an agent, it provides multitenancy support. Each tenant is assigned a policy that determines its programming capabilities for the underlying data layer using the data control plane function.The SDN-based network architecture provided by the Open Networking Foundation (ONF TR-526, 2016) features a unified plane, where hierarchical

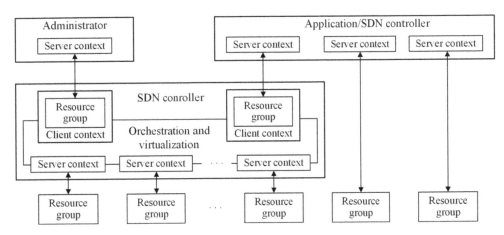

FIGURE 8.2
SDN architecture provided by the Open Networking Foundation. SDN, software-defined networking.

controllers are used to achieve differentiated services in user access layers. The major SDN architectural components are resources and controllers (Figure 8.2).

By applying the SDN architecture, the client context provides the complete abstract set of resources (resource group) and supporting control logic that constitutes a slice, including the complete collection of related client service attributes. Resource group includes infrastructure resources and NFs, as well as network services. SDN controller mediates between clients and resources, acting simultaneously as server and client via corresponding contexts. Server context represents all the information the controller needs to interact with a set of resources, included in a resource group. On the other hand, client context provides all the necessary information to the controller in order to support and communicate with a given client. It comprises a resource group and a client support function. This is in accordance with the key principles of network slicing, having in mind a wide range of service demands, which need to be satisfied in a flexible and cost-effective way (Ordonez-Lucena et al., 2017).

SDN controller performs virtualization and orchestration functions in the process of transforming the resource groups accessed through server contexts to those defined in separate client contexts. Through virtualization, the SDN controller carries out abstraction and sharing of the underlying resources. In that way, each client context provides a specific resource group that can be used by the client associated with that context to realize desired service. When performing orchestration, the SDN controller optimally distributes the selected resources to such separate resource groups. The interplay of both controller functions enables the fulfillment of the diverse demands from all clients while providing isolation among them. The SDN architecture also includes an administrator, with instantiating and configuring functions, including the creation of both server and client contexts, as well as their joint policies.

8.6.2 Network Function Virtualization

It is obvious that SDN architecture lacks capabilities that are crucial for efficient management of network slices life cycle and corresponding structural resources. Regarding this fact, the NFV approach is ideal to play this role, because it can manage the infrastructure resources and orchestrates the allocation of such resources. NFV can be seen as software abstraction of

NFs in next-generation communications systems, which otherwise require dedicated hardware concerning traditional infrastructure. Virtualization provides an important foundation for network slicing by enabling flexible slice creation on shared physical resources.

With NFV, services are described as a forwarding graph (Abdelwahab et al., 2016) of connected NFs. It defines the sequence of NFs that serve different E2E traffic flows. Generally, virtualization can be conducted on mobility management entities, serving gateways, baseband processing units (e.g., medium access control (MAC), radio link control (RLC), radio resource control (RRC), etc.), switching function, traffic balancing, etc.

To achieve the full potential from the management and orchestration functionalities, appropriate cooperation between SDN and NFV is required. It is important to note that these two concepts are not dependent on each other, although they are mutually beneficial. It means that the NFs can be deployed without SDN and vice versa. However, the integration of SDN and NFV architectures into a common framework is a complex issue (ETSI, 2015). The NFV framework defines the following:

a) VNFs that represent software components of NFs deployed in virtual environments.
b) NFV infrastructure (NFVI) that comprises the logical building blocks, i.e., storage, computing, network and corresponding hardware.
c) Management and orchestration (MANO) systems that are responsible for full automation, management and coordination of VNFs and NFVI.

In the context of network slicing, this framework provides service chaining, resource-oriented embedding, as well as management of VNFs.

8.6.3 Cloudification

Cloud computing technology allows accessing a set of shared and configurable computing resources (e.g., servers, storage facilities, infrastructure, applications) offered as services. It can be seen as a solution for serious challenges such as high availability, load balancing and high performances. The long-term solution for growing traffic demands on the current cellular architecture is the possibility of excluding computing beyond data centers toward mobile end user, providing E2E connectivity as a cloud service. A set of techniques such as mobile node function virtualization and caching are also introduced for the on-demand provision of a decentralized and elastic network as a cloud service over a distributed network of data servers. Operators can enter the cloud computing market and create new value-added services and experiences by integrating industry content and applications.

Cloud computing makes user obtain much more real-time applications to utilize 5G network efficiently. This can be realized through application, platform, as well as the infrastructure segment. Application is based on demand software services. They vary in their pricing and distribution schemes. The platform segment refers to products that are used to improve the future Internet. Finally, the infrastructure can be seen as the backbone of the entire concept.

For enabling a network slicing process, edge computing together with cloud computing offers single or multiple platforms. In addition, edge computing enables diverse applications, data management and service acquisition in close proximity to the end users. In that way, a kind of edge-centric networking will be allowed, while data proximity, ultralow latency, high data rates and control are permitted (Lopez et al., 2015; Taleb et al., 2017).

Among numerous edge computing realization, one of the most popular is multiaccess edge computing (MEC) originally focusing on radio access network (RAN) (ETSI, 2016) and later on fixed access, too (ETSI, 2018b). MEC is advantageous to the RAN, especially with respect to latency and bandwidth utilization, its enabling technologies and orchestration functions. Moreover, MEC platform and a 5G system can collaboratively interact in traffic routing and policy control–related operations. By offloading computation-intensive tasks to the servers in close proximity to users, MEC can not only reduce the traffic bottlenecks in the core network (CN) and backhaul but also reduce both the processing delay and the energy consumption at UEs (Guo et al., 2018). Also, it brings new business-related opportunities allowing operators to monetize combined cloud and network resources as well as particular services to tenants.

The second preferred realization is fog computing. From a systematic perspective, fog environment provides a distributed computing system with a hierarchical topology. Fog environment is characterized by stringent latency requirements and reduced power consumption of end devices, while providing real-time data processing and control with localized computing resources. Available resources are typically much more heterogeneous, and the criticality of some applications requires special treatment. The complementarity between fog and cloud has traditionally been seen as a mandatory feature. However, Lingen et al. (2017) advocate for a different approach. Instead of specifying an architecture where fog and cloud are complementary by design, the focus is on a service management architecture that fuses fog and cloud. With this solution, an infrastructure composed of network, cloud and fog nodes can be provided as a promising paradigm to service administrators as a unified resource pool. Administrators can then define where to allocate resources according to the service requirements. The requirements to host and manage NFV, MEC and fog computing services are doubtless similar. It is only a matter of time until fog computing becomes part of the convergence that is already existing between NFV and MEC, driven by operators, enterprises and third-party investments. The goal is to conceal underlying complexity and to promote service management into simple and intuitive operations. Overlapping needs and challenges, when considering MEC and fog integration, can be identified as virtualization, multitenancy, management of network and services, billing and pricing, security, etc.

8.7 End-to-End Network Slicing

Since 5G systems provide ubiquitous connectivity for everything, both the RAN and CN domains need evolution to support E2E network slicing. In ultradense heterogeneous networks, the cooperation of multiple RANs should be considered to provide seamless mobility and high throughput. Although service providers and operators have started to pave the way for slicing solutions, the management remains at an early stage in terms of its development. Many technologies are introduced to create, activate, maintain and deactivate network slice at the service level.

The open research challenge refers to the cooperation with traditional and emerging technologies such as broadband transmission, mobile cloud engineering, SDN and NFV. The cloud architecture of RANs and CN have advantages of physical resource pooling, distribution of software architectures and centralization of management. It should be noted that there is still a problem with the integration of network slicing with centralized RAN, SDN and NFV to provide point-to-point connection between physical radio equipment and controller.

8.7.1 Network Slicing in Radio Access Network Domain

One of the main challenges in implementing slicing in the RAN domain consists of designing and managing several virtual instances on the same shared infrastructure in an efficient manner, while guaranteeing the required QoS for each of them. In order to manage limited frequency spectrum resources, the following characteristics are required: dynamic resource management, resource isolation and sharing, as well as functional requirements. Dynamic resource management implies flexible and programmable procedures, enabling the advantages of application programming interfaces (APIs). In this case, resource sharing is provided with the help of KPIs for each slice. For example, eMBBC slice seeks high throughput, while on the other hand, uRLLC slice needs very low latency (Bojkovic et al., 2017; Sachs et al., 2018).

Two slices are isolated as long as the actions performed on one slice do not result in performance degradation in the other slice. This concept is not clear today when it comes to resource allocation, while it is easy to understand in other domains, e.g., security. It is common understanding to consider that RAN domain is characterized by the fact that slices are in a position to provide self-contained networks together with the corresponding degree of isolation. Spectrum isolation leads to be a bottleneck because of limitations in multiplexing gains. The goal of functional requirements is that each network slice needs a different control/user plane functional split and corresponding VNF placement to provide optimal performance. The interface between network slicing and RAN is shown in Figure 8.3.

The network slice selection function introduced by Rost et al. (2017) is significant for selecting the corresponding slice per user. As for RAN–CN interface, it is carried out in a way that the control and user plane traffic are routed to functional entities in the CN slice. User plane anchor (UP-anchor) is responsible for traffic distribution according to the slice policy and for encryption. The radio flow in slice #1 is configured with two connections, whereas slice #2 is configured with only one connection according to the provided policy configuration.

Radio resource management and control in the base station is responsible for configuring the RAN protocol stack and QoS according to the slice requirements. For example, in eMBB slice with high capacity requirements, radio bearers are configured to support multiconnectivity, similar to the dual connectivity in LTE systems. On the other hand, in uRLLC slices, lower frame error rates, as well as multipoint diversity techniques, need to be utilized. Also, flexible architecture incorporating mmWave (millimeter wave) support is required to meet different slice requirements. This support is essential for multiconnectivity realization covering both eMBB and mMTC slices.

Depending on the way of spectrum sharing and slices multiplexing, the degree of slice isolation from a performance perspective varies from a slice corresponding to a standalone network (with its specific spectrum and infrastructure) to the CN-limited slices (resource allocation in the RAN is slice independent). Intermediate solutions are also possible as proposed by Elayoubi et al. (2019). Here, physical and link layers are common to all slices, except those on dedicated or unlicensed spectrum. On the other hand, the higher layers are slice specific, i.e., specific NFs (selection, configuration and chaining) are to be performed for each slice. This chaining aims to achieve specific requirements, e.g., some processing functions for latency reduction.

Tiling scheme is an essential enabler for slicing in the RAN domain, while scheduling is a practical implementation of the flexible numerology concept in 5G systems (Zaidi et al., 2016). This offers the opportunity to provide heterogeneous services using different

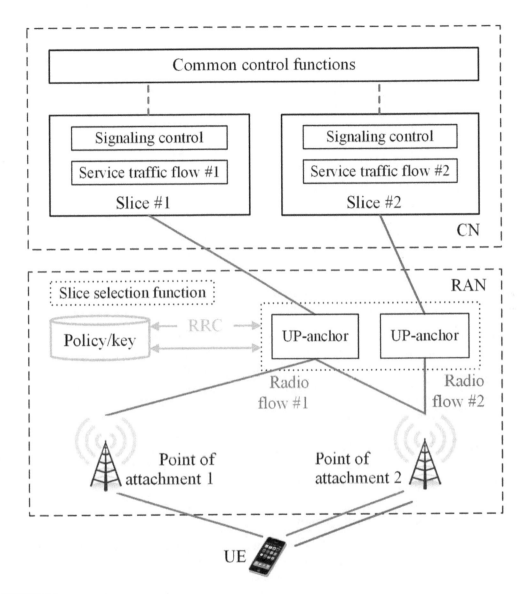

FIGURE 8.3
Interface between network slicing and RAN. RAN, radio access network; UE, user equipment.

subcarrier spacing (SCS) and/or transmission time interval (TTI). In principle, resources with the same numerology are grouped within a tile (resource block group) in order to reduce scheduling overhead. An illustrative example of the tiling concept is presented in Figure 8.4.

eMBB tiles are characterized by short TTI size to quickly overcome the slow start phase; then a longer TTI (1 ms) can be used to minimize control overhead. mMTC tiles use the combination of a lower SCS (15 kHz) and a longer TTI to increase energy efficiency and coverage. On the other hand, uRLLC tiles use short TTIs (0.25 ms) to meet latency requirements. Larger SCS (120 kHz) can also be useful for some uRLLC use cases. The scheduler has the role of allocating resources to comply with the corresponding QoS requirements for the different slices. Its complexity can be simplified by dynamically determining the

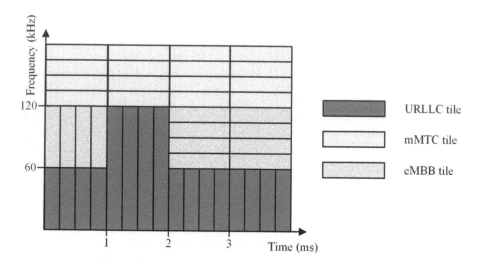

FIGURE 8.4

Example of the tiling concept. eMBB,enhanced mobile broadband; mMTC,massive machine-type communication; uRLLC, ultrareliable low-latency communication.

tile composition and by mapping the slices to these tiles. Multiplexing in the time domain is simpler as 3GPP imposes symbol alignment between tiles, ensuring orthogonality. As for multiplexing in the frequency domain, 3GPP advocates the insertion of guard bands between adjacent tiles with different SCS in order to guarantee orthogonality.

The coexistence of a large number of slices poses many challenges related to resource allocation to slices and flows, considering not only radio resources but also processing ones (for VNF and MEC). One of the major open research problems consists of adapting existing resource allocation schemes, designing and developing new ones and fitting them into the new context of RAN slicing. These schemes have to jointly ensure QoS for individual slices, fairness between slices and overall resource efficiency.

8.7.2 Slicing in Core Network Domain

Traditionally, the core domain has been designed as a single network architecture serving multiple purposes, addressing a range of requirements and supporting backward compatibility and interoperability. During the past decade, the evolution of mobile CN was evident. The beginning was with LTE and a full IP CN, then passing through softwarization and virtualization of the CN elements and ending by network slicing. Today, SDN and NFV have become the key enablers in order to obtain more dynamic evolved packet core (EPC) networks. It should be noted that the main EPC entities have been divided into granular NFs. Moreover, 3GPP has completely reshaped the CN with modular architecture and network slicing concept for many types of different services.

Entities such as mobility management entity (MME), home subscriber server (HSS), serving/packet gateway (S/P-GW) and policy and charging rules function (PCRF) can now be deployed on sophisticated virtual platforms, thanks to the progressive standardization activities in the area of NFV. The fact that these building block entities can be deployed as virtual instances brings more flexibility, elasticity and QoS assurance to the service provisioning. The flexibility and elasticity in service provisioning imply that network operators can now deploy multiple instances of the EPC, all at the same time, to serve different categories of users based on their service requirements. Moreover, while some services may need all the components that constitute the EPC, others do not, e.g., mMTC with limited mobility does not need mobility management. Therefore, the notion of CN slicing is centered around the possibility to deploy multiple instances of virtual EPC (vEPC) running in parallel to fulfill different service demands, e.g., uRLLC services can require a distributed vEPC closer to the end users.

From the introduction of network slicing on the mobile network layer, new requirements as the result of autonomous and dynamic network configuration are raised for the transport layer. As a consequence, there is a need for a new interface between the RANs and CNs (3GPP TR 28.801, 2018). The purpose of introducing such an interface is to connect the management system of the RAN with the transport network controller. One of the main objectives is to provide and facilitate the mapping of RAN slice to CN resources (Afolabi et al., 2018).

In fact, vEPC can be orchestrated and managed over cloud platforms. Here, different orchestration schemes can be utilized offering efficient management and operation of the EPC entities. Control plane and user plane entities can be provided as a service, bringing a new dimension to the operating models of the vertical markets. The EPC as a service (EPCaaS) model proposed by Taleb et al. (2015) leverages the cost efficiency offered by deploying vEPCs on the cloud to introduce two major virtualization approaches and different architectural implementation scenarios. First one is a full virtualization approach where both the control and user plane entities are virtualized. For the second approach, only the control plane entities are virtualized, while the data plane components are deployed on proprietary hardware in order to ensure high throughputs and to enable the implementation of traffic policies. The suggested implementation scenarios include 1:1 mapping (where each functional component runs on a virtual machine), 1:N mapping (where each functional component runs on multiple virtual machines, N:1 mapping (where all functional components run on virtual machine) and N:2 mapping (where the control plane and data plane components run on a virtual machine each).

Next-generation CN (NGCN) of 5G systems architecture, which is based on softwarization and virtualization concepts, not only has advantages such as flexibility and backward compatibility but also overcomes drawbacks of the vEPC architecture with the introduction of SDN. SDN controller, which can be either virtualized or nonvirtualized, is in charge of interpreting the signaling messages received from the control plane and responsible for installing the forwarding rules into the user plane via an API. Moreover, the control and user planes are completely separated, and they can get scaled independently in a cost-effective manner. SDN brings flexibility of flow distribution and thus provides improved mobility management. Control and user plane functions can be flexibly placed around the network, for example, closer to the edge or users, thus reducing the network latency. This encourages the development and implementation of MEC and its use cases including traffic offloading or local breakout, content distribution and service caching and so on (Nguyen et al., 2017). However, introducing a new SDN controller and its interfaces causes additional latency, while the scalability of the SDN controller is also a major problem. It could be overcome by using multiple controllers or through a hierarchical design of controllers, but in the context of the mobile network, these solutions are still indeterminate.

Similar to the traditional 3GPP architecture, the NGCN architecture connects the different 5G core NFs together and with UEs as well as the RAN via reference interfaces, as shown in Figure 8.5. However, defining the CN in this way, complexity is introduced to add new network elements/instances, as it requires the operator to reconfigure multiple E2E interfaces. The access control and session management are separated in NGCN to

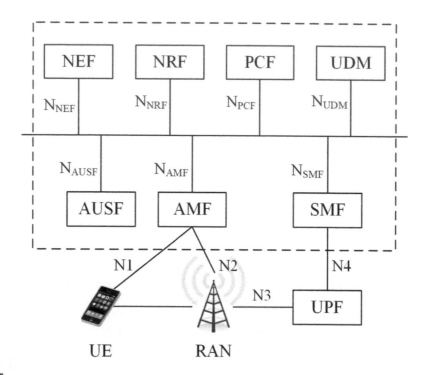

FIGURE 8.5
NGCN architecture. AMF, access and mobility management function; AUSF, authentication server function; NEF, network exposure function; NGCN, next-generation CN; NRF, network function repository function; PCF, policy control function; RAN, radio access network; UDM, unified data management; UE, user equipment; UPF, user plane function.

better support fixed access and ensure scalability and flexibility. The most important NFs defined in NGCN are as follows:

- Access and mobility management function (AMF), besides handling access control and mobility, also integrates network slice selection functionality as part of its basic set of functions.
- Session management function (SMF) handles user sessions according to the network policy.
- Policy control function (PCF) integrates a policy framework for network slicing.
- User plane function (UPF) can be deployed based on the service type, in several configurations and locations.
- Unified data management (UDM) is with similar functionality as HSS, i.e., integration of subscriber information for both fixed and mobile accesses.
- NF repository function (NRF) provides registration and discovery functionality allowing NFs to discover each other and communicate via APIs.
- Network exposure function (NEF) defines an API layer related to the 5G CN, which can effectively control network slices.

One of the advantages of having fine granularity is the possibility to share some NFs among slices while

- reducing management complexity of network slices by sharing the authentication server function (AUSF), UDM and mobility management procedures (i.e., AMF);
- reducing the signaling with a higher number of shared control plane NFs;
- simpler managing common hardware, in the case of NFs which cannot be deployed in a software environment.

Accordingly, it is important to identify common NFs that need to be shared by several E2E network slices.

8.8 Concluding Remarks

Although network slicing is currently undergoing a standardization phase, its realization is full of different challenges. This chapter focuses on some relevant problems concerning principles and enabling technologies of slicing in 5G networks. Strong potential of network slicing was revealed for addressing the diverse requirements of future services. Having in mind the current maturity state, the role of network slicing in the development of the 5G architecture is provided. At the same time, the enablers of network slicing such as the NFV, SDN and cloud technologies are introduced considering various scenarios, resource allocation principles for QoS requirements and interface constraints. Significant efforts are still needed to develop optimal architectural configuration in order to realize a cooperative environment between enablers technologies and take benefits from their advantages. With regard to E2E architecture, this chapter surveys feasibility of network slicing implementation in RAN and CN domains with the objective of identifying the plausible options when responding to the needs of verticals.

References

3GPP TS 22.261 V.16.6.0 (2018). Service requirements for the 5G system.

3GPP TR 28.801 V.15.1.0 (2018). Study on management and orchestration of network slicing for next generation network.

Abdelwahab, S., et al. (2016). Network function virtualization in 5G. *IEEE Communications Magazine*, 54(4), 84–91.

Afolabi, I., et al. (2018) Network slicing & softwarization: A survey on principles, enabling technologies & solutions. *IEEE Communications Surveys & Tutorials*, 20(2), 2429–2453.

Arfaoui, G., et al. (2018). A security architecture for 5G networks. *IEEE Access*, 6, 22466–22479.

Berman, M., et al. (2014). GENI: A federated testbed for innovative network experiments. *Computer Networks*, 61, 5–23.

Bojkovic, Z., Bakmaz, B. and Bakmaz, M. (2017). Vision and enabling technologies of tactile Internet realization. *Proceedings of 13th International Conference on Advanced Technologies, Systems and Services in Telecommunications TELSIKS* (pp. 113–118). Nis, Serbia: IEEE.

Bojkovic, Z. and Milovanovic, D. (2017). Optimal slicing in wireless 5G networks. In *Book of Abstracts 5th Conference on Information Theory and Computer Systems TINKOS* (pp. 16–17). Belgrade, Serbia: MISASA.

Elayoubi, S. E., et al. (2019). 5G RAN slicing for verticals: Enablers and challenges. *IEEE Communications Magazine*, 57(1), 28–34.

Esposito, F., Matta, I. and Ishakian, V. (2013). Slice embedding solutions for distributed service architectures. *ACM Computing Surveys*, 46(1), 6:1–6:29.

ETSI GS NFV-EVE 005 (2015). Network Functions Virtualization (NFV); Ecosystem; Report on SDN Usage in NFV Architectural Framework, v. 1.1.1.

ETSI GS MEC 003 (2016). Mobile Edge Computing (MEC); Framework and Reference Architecture, v. 1.1.1.

ETSI GR NGP 011 (2018a). Next Generation Protocols (NGP); E2E Network Slicing Reference Framework and Information Model, V1.1.1.

ETSI GR MEC 024 (2018b). Multi-access Edge Computing (MEC); MEC Support for Network Slicing, v 2.0.5.

Fischer, A., et al. (2013). Virtual network embedding: A survey. *IEEE Communications Surveys & Tutorials*, 15(4), 1888–1906.

Foukas, X., et al. (2017). Network slicing in 5G: Survey and challenges. *IEEE Communications Magazine*, 55(5), 94–100.

Frascolla, V., et al. (2017). 5G-MiEdge: Design, standardization and deployment of 5G phase II technologies: MEC and mmWaves joint development for Tokyo 2020 Olympic Games. *Proceedings of the Conference on Standards for Communications and Networking CSCN* (pp. 54–59). Helsinki, Finland: IEEE.

Goldberg, R. (1974). Survey of virtual machine research. *IEEE Computer*, 7(6), 34–45.

Han, B., Lianghai, J. and Schotten, H. D. (2018). Slice as an evolutionary service: Genetic optimization for inter-slice resource management in 5G networks. *IEEE Access*, 6, 33137–33147.

Guo, H., Liu, J. and Zhang, J. (2018). Computation offloading for multi-access mobile edge computing in ultra-dense networks. *IEEE Communications Magazine*, 56(8), 14–19.

Hucheng, W., et al. (2017). Mobility driven network slicing: An enabler of on demand mobility management for 5G. *The Journal of China Universities of Posts and Telecommunications*, 24(4), 16–26.

ITU-R Rec. M.2083-0 (2015). IMT Vision – Framework and overall objectives of the future development of IMT for 2020 and beyond.

Katsalis, K., et al. (2017). Network slices toward 5G communications: Slicing the LTE network. *IEEE Communications Magazine*, 55(8), 146–154.

Lingen, F. van, et al. (2017). The unavoidable convergence of NFV, 5G, and fog: A model-driven approach to bridge cloud and edge. *IEEE Communications Magazine*, 55(8), 28–35.

Lopez, G. P., et al. (2015). Edge-centric computing: Vision and challenges. *ACM SIGCOMM Computer Communication Review*, 45(5), 37–42.

Manzalini, A. and Crespi, N. (2016). An edge operating system enabling anything-as-a-service. *IEEE Communications Magazine*, 54(3), 62–67.

NGMN Alliance (2016). Description of network slicing concept. Retrieved from http:// www.ngmn. org

Neely, M. J. (2010). *Stochastic Network Optimization with Application to Communication and Queuing Systems*. San Rafael, CA: Morgan & Claypool Publishers.

Nguyen, V., et al. (2017). SDN/NFV-based mobile packet core network Architectures: A survey. *IEEE Communications Surveys & Tutorials*, 19(3), 1567–1602.

Nikaein, N., et al. (2015). Network store: Exploring slicing in future 5G networks. *Proceedings of 10th International Workshop on Mobility in the Evolving Internet Architecture* (pp. 8–13). Paris, France: ACM.

ONF TR-526 (2016), Applying SDN architecture to 5G slicing.

Ordonez-Lucena, J., et al. (2017). Network slicing for 5G with SDN/NFV: Concepts, architectures, and challenges. *IEEE Communications Magazine*, 55(5), 80–87.

Popovski, P., et al. (2018). 5G wireless network slicing for eMBB, URLLC, and mMTC: A communication-theoretic view. *IEEE Access*, 6, 55765–55779.

Raza, M. R., et al. (2018). Dynamic slicing approach for multi-tenant 5G transport networks. *IEEE/OSA Journal of Optical Communications and Networking*, 10(1), A77–A90.

Rost, P., et al. (2017). Network slicing to enable scalability and flexibility in 5G mobile networks. *IEEE Communications Magazine*, 55(5), 72–79.

Sachs, J., et al. (2018). 5G radio network design for ultra-reliable low-latency communication. *IEEE Network*, 32(2), 24–31.

Sciancalepore, V., Cirillo, F. and Costa, X. P. (2017). Slice as a Service (SlaaS) optimal IoT slice resources orchestration. *Proceedings of Global Communications Conference GLOBECOM* (pp. 1–7). Singapore: IEEE.

Taleb, T., et al. (2015). EASE: EPC as a service to ease mobile core network deployment over cloud. *IEEE Network*, 29(2), 78–88.

Taleb, T., et al. (2017). On multi-access edge computing: A survey of the emerging 5G network edge cloud architecture and orchestration. *IEEE Communications Surveys & Tutorials*, 19(3), 1657–1681.

Tang, L., et al. (2018). Queue-aware reliable embedding algorithm for 5G network slicing. *Computer Networks*, 146,138–150.

Vassilaras, S., et al. (2017). The algorithmic aspects of network slicing. *IEEE Communications Magazine*, 55(8), 112–119.

Yang, Z. and Guo, Y. (2016). An exact virtual network embedding algorithm based on integer linear programming for virtual network request with location constraint. *China Communications*, 13(8), 177–183.

Zaidi, A. A., et al., (2016). Waveform and numerology to support 5G services and requirements. *IEEE Communications Magazine*, 54(11), 90–98.

Zhang, H., et al. (2017). Network slicing based 5G and future mobile networks: Mobility, resource management, and challenges. *IEEE Communications Magazine*, 55(8), 138–145.

Zhang, S. (2019). An overview of network slicing for 5G. *IEEE Wireless Communications*, 26(3), 111–117.

9

Integrated Satellite Communications: Remote Sensing in Natural Hazards

Jahangir Khan

Sarhad University of Science and Information Technology
China Agricultural University

CONTENTS

9.1 Introduction

Satellite communication complements data products with certain capacity, from acquiring to broadcast, in order to disseminate the information's using fixed and wireless communication expansions. This expresses the integration of satellites using different techniques, platform, algorithms and methods for the next-generation trends in data development for global research and their manifestations. In present practices, the acquisition of information and the expression of remote sensing are usually referred to make use of satellite sensor technologies to discover and sort out the information on the earth surface, based on disseminating signals, and share the data products in real and near-real-time systems. In this watch, National Aeronautics and Space Administration's (NASA's) Earth System Science conducts and sponsors research, collects new observations, develops technologies and extends the technology education to learners of all ages, as a key contributor in the field of earth and space sciences. The NASA's Earth Science Division is an integrated, multisatellite, long-term program design for global observations and monitoring of the global land surface, biosphere, atmosphere, glaciers and oceans. Since the late 1980s, NASA's Earth Observing System (EOS) Program is designed to develop a scientific understanding of Earth's system in response to climatological and hydrological changes and to improve prediction of climate, weather and natural hazards. This coordinated approach enables an improved understanding of the Earth as an integrated system.

The rapid expansion of satellite remote sensing since decades as a counterpart of the 3S technology is tuned as integration of global positioning system (GPS), remote sensing (RS) and geographic information system (GIS) technologies. To explore the potential of 3S (GIS, RS and GPS) technology in different areas is a profound evolution in the field of science and technology. The 3S technology integration in particular presented as GIS is the core, whereas GIS is based on the RS and GPS. The relationship between these three is as a brain and two eyes. GIS plays a key role to manage and analyze the data, whereas RS and GPS collect data from the sensors (Cheng, 2011).

To date, however, there is no definitively processed long-term land data records generated, and those that are publically available are subject to scientific debate. The understanding of this study is to generate and distribute the global decadal scale coarse spatial resolution land products derived from moderate-resolution imaging spectroradiometer (MODIS) and tropical rainfall measuring mission (TRMM) as well as TRMM multisatellite precipitation analysis (TMPA) satellites with high-quality resolution to monitor the land cover changes vigorously. The TRMM-based TMPA estimate in real-time to apply new concepts in merging quasi-global precipitation estimates and to take advantage of the increasing availability of input data sets in near-real-time. The overall system is referred to as the real-time TRMM multisatellite precipitation analysis (TMPA-RT). This dataset is the output from the TMPA 3B40RT, 3B41RT and 3B42RT and provides precipitation estimates in the TRMM regions that have the (nearly zero) bias of the TRMM (satellite 3B43) combined instrument precipitation estimation and dense sampling of high-quality microwave data with fill-in using microwave-calibrated infrared estimates with granule size of 3 hours, located at http://pps.gsfc.nasa.gov/ (Wu et al., 2012 and 2014).

This study utilizes the MODIS temporal and spatial resolution at 1 km pixel area (i.e., NDVI and LST data products) and the TRMM and TMPA temporal resolution (3 hourly) at relatively high spatial (~1 km) and flood stream flow at high spatial (~12 km) resolution during 2010. Due to the weather extremes and flood inception, Pakistan has faced the worst flood in 2010 and affected 24 million peoples in total and damaged more than 2 million hectare crops and 10 billion (US$) economic losses. Therefore, this is must to explore the region with associated events, using vegetation temperature condition index (VTCI) drought monitoring approach in order to settle on the dry and wet conditions as well as global flood monitoring system (GFMS) model for the flood detection/intensity and stream flow during the year of 2010 over Punjab, Pakistan.

9.2 Information and Communication Technologies

Besides, communication technologies reduce the gaps and allow information to share in all fields of sciences, which was not possible several decades ago. Fast communication and transformation of the information in gigabytes and terabytes in a second through physical or wireless medium is a good sign in the era of technological advancements. Information and communication technology (ICT) in a unified way of the integration of telecommunications, computers as well as necessary enterprise software, medium, storage and multimedia systems, which enable users to access, store, transmit and manipulate information. Evolution of the fifth-generation (5G) and post technology will support the combination of broadband and cellular networks to facilitate the horizontal and vertical hands-off, in centric and uncentric environment, and to interconnect the entire world without limits (Khan et al., 2011a, b; Soret et al., 2019). Based on ICT facts and figures, the technological advancement and implementation, one third of the world's population is online, and based on these statistics in the world, these facts increase rapidly up to 10 billions, which is a positive sign by implementations of the broadband and cellular technologies, which also indicate a direct impact on the satellite(s) communication technologies. ICT sector has been confronted with the convergence phenomenon for the sophistication of the technology and is often used as an extended synonym for information technology (IT). Due to which digital divide takes place in the technological advancement especially in ICT, in terms of ICT impacts and with implementation of ICT for the acquisitions of data products using the 5G technology in a real-time and near-real-time illustration for the study of cross- and interdiscipline areas of science and technology. The current study will lead the digital divide in the perspective of regional development and implementation of ICT in the area of natural hazard to acquire the real-time and near-real-time imagery and data products.

9.2.1 5G Systems

The 5G network will lead to a huge shift toward a landscape dominated by wireless connectivity. In order to support this shift, architectural changes exist. For example, network function virtualization and software-defined networking, both at the core and edge of the network, are applied to provide increased computing power scalability, reduced operation costs and business models. Service providers need to provide seamless connectivity between terrestrial and satellite. As for traffic, it will be directed to the best transport options, which are available according to bandwidth, latency network conditions and other application-specific requirements. The networking between terrestrial and satellite is becoming known in the 3rd Generation Partnership Project (3GPP) standards. Quality of service and operational expense are key as the landscape becomes more competitive. New opportunities for extending satellite services in urban and rural areas emerge multicast and network offload schemes, area and maritime mobility (seamless connectivity plan, emergency services, broadcast). Satellite industry needs to participate in various committees including 3GPP to ensure that satellite systems are integrated as part of 5G and to support highly available and reliable connectivity using satellite for use cases such as ubiquitous coverage, disaster relief, public safety requirements, emergency response, remote sensor connectivity, broadcast service and so on. Satellite operators and industry observers are optimistic that space-based communications will play a significant role in connecting 5G devices. Satellite has been included into 5G standard and now can become an integral part of 5G. One of the main goals of 5G is to reach ultralow latency along with highest efficiency. This requires a greater separation between the control plane and user plane and for the radio access network functionality to be split. Also, this is specified in 3GPP, Release16, with a study to identify use cases for the provision of services with satellite integrated in the 5G system (Soret et al., 2019).

9.2.2 Satellite Integration

Satellite will continue to be the most effective means for reaching areas beyond terrestrial coverage. The main characteristics are capacity, resilience, reliability and latency. Many services are effectively provided by satellites (broadcast, multicast). Satellite industry needs to participate in various committees, including in 3GPP and ITU-T to ensure that satellite systems are integrated as an intrinsic part of the space, e.g.: (1) to support highly available and reliable connectivity using satellite for use cases such as ubiquitous coverage, disaster relief, public safety requirements, emergency response, remote sensor connectivity, broadcast service and so on, (2) to support an air interface with one latency of up to 275 ms when satellite components are involved and (3) to support seamless mobility between terrestrial and satellite-based networks with widely varying latency (Sui and Goodchild, 2011). The 5G use cases can be efficiently supported by satellite, i.e., service categories such as multimedia delivery, broadband, machine-type communication and critical communication (disaster management, air traffic management, reliable communication and vehicular communication). There are many satellite roles in 5G vision. Satellites will integrate with other networks rather than to be a standalone network to provide 5G, and it is this integration that forms the core of vision. Satellite systems are fundamental components to deliver reliably 5G services in all regions of the world, all the time and at acceptable cost. Recognizing their inherent characteristics, the satellite component will contribute to augment the 5G service capability and address some of the major challenges in relation to the support of multimedia traffic growth, ubiquitous coverage, machine-to-machine (M2M) communications and critical telecom missions, while optimizing the value for money to the end users. Satellites can be part of a hybrid network configuration and consists a mix of broadband infrastructures and broadband infrastructures managed in such a way that it brings seamlessly and immediately converged services to all end users.

5G wireless will support a heterogeneous set of integrated air interfaces: from evolution of current access schemes to new technologies. To achieve the expected capacity, coverage, reliability, latency and improvements in energy consumption, the 5G architecture is expected to run over a converged optical–wireless–satellite infrastructure for network access, with the possibility of transmitting digital and modulated signals over the physical connections.

The integration of satellites data products, M2M connectivity, GPS that supports RS infrastructure, with high-resolution radiometers and moderate-resolution imaging spectrometers enables to study the earth terrestrial and polar regions in a variety of sharing and probing the information. In the provisions of ICT, the traditional radio frequency (RF) and free space optical (FSO) are well thought-out for intersatellite and ground-to-satellite links (GSLs). The FSO links have very narrow beamwidth with increased transmission range as well as are highly susceptible to atmospheric effects and pointing errors and offer high transmission rates and produce less interference. FSO has been demonstrated in ground-to-satellite communication in numerous scientific missions. The planned commercial LEO constellations, such as SpaceX, Telesat and LeoSat, deploy laser communication equipment for high-throughput FSO intersatellite links (ISLs). However, RF links are decisive as fallback solution, if FSO communication is infeasible (bad weather) in hybrid RF-FSO systems, to enable the integration into RF-based systems (e.g., ground-to-satellite communication in 5G NR will either take place completely in the S band around 2 GHz or in the Ka band, where the downlink operates at 20 GHz and the uplink at 30 GHz). There are three types of data traffic in a LEO constellation: (1) user data, (2) control data, (3) telemetry and (4) telecommand (TMTC) data. Next in wireless networks, user data and control data are typically separated to facilitate a more efficient management of the wireless resources. For machine-type communications, the size of the control information is similar to the data size, and a massive number of devices need to be handled in stringent latency constraints. These requirements are in stark contrast to classical mobile broadband communications. The physical links in satellite communication are broadly classified in

GSLs and ISLs. Ground refers to any transceiver located on Earth, which can be either GS, a UE or a gateway (Soret et al., 2019).

The successor of satellites, such as M2M connectivity and GPS that supports RS infrastructure, leads to the EOS' series of satellites, science components, and a data system called Earth Observing System Data and Information System (EOSDIS) for the interdisciplinary studies of earth system science and associated services. The employed satellite called MODIS is the more accurate and versatile instrument and is more sensitive to changes in vegetation dynamics as well as monitors the land-cover changes. The continuation of MODIS-Terra and MODIS-Aqua satellites is certain over time with successor satellite and sensor systems previously planned and was assured until 2018 with National Polar-Orbiting Operational Environmental Satellite System (NPOESS) satellites series to make possible the potential of multi-data comparison. In addition to MODIS, the TRMM and TMPA satellites with high-quality spatial and temporal resolution's observations enable to conclude the land cover changes vigorously (Friedl et al., 2002). The GFMS model was evaluated in terms of performance and accuracy for flood detection against flood event, which utilized the real-time integrated satellite TRMM and TMPA data products obtained from the NASA Goddard TRMM/GPM Precipitation Processing System (PPS) (Wu et al., 2012 and 2014). The TRMM (NASA and the Japan Aerospace Exploration [JAXA] Agency joint mission) contributes 17-years (November 1997–April 2015) datasets of global precipitation and lightning for measuring the rainfall events, flood, drought and weather forecasting, located at https://trmm.gsfc.nasa.gov/.

Recently, satellites are by now being installed to acclimatize the weather extremes and impacts of natural hazards to share information about the climates to disaster respondents. The satellite integration would then be able to deliver early warnings to disaster-stricken areas and vulnerable groups in a real-time and near-real-time basis of the climate extreme's crossways expansion to show the spatial extent of severity and timescale with related impacts unforced natural variations. And also to facilitate and share the medium and data products in purpose, go through a series of chains from probing to disseminate the information.

9.3 Remote Sensing Services

Satellite remote sensing is a technology for condensed acquisition of spatial data and extraction of some information. It involves the use of both systems and algorithms in order to record information about the object from a remote location. The principles of remote sensing result from the characteristics and interactions of electromagnetic radiation (EMR) from source to sensors. They are based on (1) the source of energy and the type as well as amount of energy it provides, (2) the absorption and scattering effects of the atmosphere on EMR, (3) the mechanisms of EMR interaction with Earth surface and finally, (4) the nature of sensor response, determined by the type of source. The reason is to address time-critical applications (Sui and Goodchild, 2011).

The trend concerning open earth research observation data has received an interest in satellite-based monitoring and mapping of the earth surface. The observation of the natural surface monitoring, natural and man-induced phenomena goes back to the early 1970s with LANDSAT program (Harris and Baumann, 2015). The incorporation of social media information and remote sensing-based spatial technologies leads to joint exploitation of both kinds of data, i.e., two sources of data (Murakami et al., 2016). Since decades, there have been attempts toward the application of remote sensing data for land cover changes and environmental monitoring (Honicky et al., 2008).

The real-time software telemetry processing system (RTSTPS) is baseline technology for NASA's standard protocol processing and level-0 data production elements along with processing support of Earth remote sensing mission and is responsible for space data system protocol and level-0 processing system data gateway as well as processing for legacy and upcoming gateway. RTSTPS ingests satellite data after processing and to provide level-0 data distribution to end users. RTSTPS processing has been verified with live Terra and Aqua MODIS data. Concerns about the ability of commercial software to process its high data volumes led the Goddard Earth Sciences Distributed Active Achieve Centre (GES DAAC) to develop contingency science processing system. Nevertheless, commercial software option was used when data was first flowing from GES DAAC from MODIS instruments aboard for a system processing of MODIS data received by direct broadcast stations. NASA's MODIS data are updated and are free of charge and have improved spatial and spectral resolutions, better sub-pixel geometric registration with superior calibration, cloud screening and atmospheric correction at level-3 (Friedl et al., 2002; Wu et al., 2012 and 2014). The general trend results in speeding up consistently the development of new algorithms and techniques but is still power less against some spots of space home earth observation with regional applications for dry and wet conditions and of high importance to explore the technologies integration in cross disciplines.

9.3.1 Study Site

The geographic location of Pakistan is approximately 23°N–37°N and 60°E–79°E, in the western zone of south Asia. The Pakistan's proposed study site area of the Punjab province is 205,344 km^2 with altitude of 100–260 m above sea level, approximately. The region's climate is dry, with an annual precipitation gradient from 100 mm in the south to 600 mm in the northwest, which reaches up to 1000 mm along the north-eastern border and, during the Rabi season, typically ranges from 200to 500 mm and can reach up to 800 mm under heavy rainfall and beyond. In Pakistan, drought and flood are associated with rains coming from southwest monsoons, which appear to be related to La-Nina and El-Nino events. ENSO has a direct impact on drought and flood in the region, with a negative impact of rainfall in the summer on irrigated agricultural lands of Pakistan (Khan et al., 2018). The current study focused on the region defined as (23.57°N–37.07°N, 60.46°E–77.08°E) for Pakistan and (27.41°N–34.02°N, 69.15°E–75.23°E) for this calculation of Punjab. In this research study, a 633×754 km^2 area of Punjab land was selected to assess the VTCI dry and wet conditions as well flood events over the region as shown in Figure 9.1.

9.3.2 Climate Extremes

The global warming causes the global climate changes, and Pakistan is listed in number 8 of the most affected ten countries from 1995–2014 in the world (Thenkabail and Gamage, 2004; Dai, 2011, 2013; Kreft et al., 2015; Dow and Downing 2016; Glanemann et al., 2017). By large, Eckstein et al. (2019) listed Pakistan on number 5 with impacts of weather-related

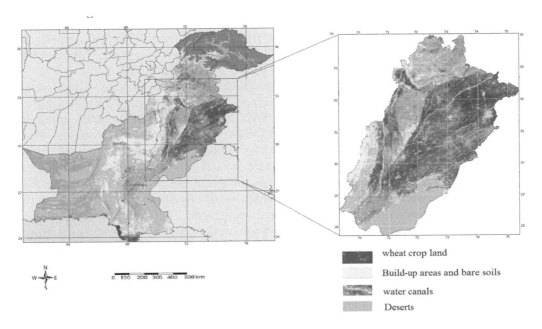

FIGURE 9.1
Geographic position of Punjab (Pakistan) with sinusoidal grid mapped plate's h23/v05, h23/v06, h24/v05 and h24/v06 with a spatial resolution of 1 km square defined and coordinated in (27.41–34.02°N, 69.15–75.23°E).

loss events in the form of storms, floods, heat waves and so on. Since 1900, the natural disasters in response to drought and flood are the most common and frequently occurring hazards due to environmental, climatic and ecosystem anthropogenic factors in the region like Pakistan. Droughts and floods are the world's costliest natural disasters due to extreme climate events under global warming. The employed satellites data products in this study explore the natural hazards, such as droughts and floods. Drought is a complex phenomenon of environmental processes and considered as the stochastic nature of hazards with a prolonged scarcity and dearth character that results from the meager annual precipitation and superficial water contents with impacts on the agricultural, meteorological, ecological and socioeconomic hooked on with potentially significant consequences. Pakistan lies in the south Asian region, where the reliance on ground data is not sufficient for drought monitoring and flood events. This is due to the lack of reliable and complete data and proper information network systems and the lower availability of weather stations in the region. This is of high importance to use the satellite remote sensing data products in the region to determine the weather extremes, as key indicator of drought and flood.

9.3.3 Processing and Modeling of NDVI and LST Data Using VTCI Approach

In this research work, the MODIS-Aqua data products were used to expose the region. The 8-day LST (MYD11A2) data products with known emissivity in bands 31 and 32 and NDVI (MYD13A2) data products acquired over 16-day interval in bands 1 and 2, to effectively distinguish the states and processes of vegetated surfaces with approximately 1 km resolution, overlap the h23/v05, h23/v06, h24/v05 and h24/v06 tiles in the sinusoidal projection during the period of 2010, located at https://lpdaac.usgs.gov/. The latitude and longitude of the study area are placed to 30°N and 75°E with Lambert Azimuthal projection for the horizontal tiles 23 and 24 along with vertical tiles 5 and 6, located at https://modis.gsfc. nasa.gov/related/organigram.php. The MODIS VTCI imagery was utilized for the VTCI time series values at the five weather stations over the region of Punjab.

In addition, this study utilized the real-time TRMM rainfall-incorporated data product and TMPA precipitation data product obtained from the NASA Goddard TRMM/GPM precipitation processing system (PPS) (Wu et al., 2012). The defined TRMM/TMPA anomaly is considered for the 1-year (2010) flood in a retrospective manner against the region to expose the runoff streamflow and flood detection/intensity observations of flood events through north to south, over the plain of Punjab, Pakistan, located at http://flood.umd.edu/.

The VTCI approach was adopted for the integration of the composite 16-day LST and NDVI data products in response to VTCI, which is an effective approach to extract the soil moisture, dry and wet conditions information vigorously. For the background study, the VTCI approach can be computed for the warm and cold edges as well as for drought conditions as presented in Equations (9.1) and (9.2) (Tian et al., 2016; Khan et al., 2018; Khan and Wang, 2018a, b).

$$\mathrm{VTCI} = \frac{\mathrm{LST}_{\mathrm{NDVI}i,\max} - \mathrm{LST}_{\mathrm{NDVI}i}}{\mathrm{LST}_{\mathrm{NDVI}i,\max} \quad \mathrm{LST}_{\mathrm{NDVI}i,\min}} \tag{9.1}$$

$$\mathrm{LST}_{\mathrm{NDVI}i,\max} = a + b\mathrm{NDVI}_i$$
$$\mathrm{LST}_{\mathrm{NDVI}i,\min} = a' + b'\mathrm{NDVI}_i \tag{9.2}$$

The plots reside in LST-NDVI space are the most appropriate method for the determination of warm and cold edges in the expression of the soil moisture. The LST-NDVI scatter plots space called the warm edges ($\mathrm{LST}_{\max}$) and cold edges ($\mathrm{LST}_{\min}$) against the VTCI drought monitoring approach. The coefficients are estimated from the scattered plots of the LST and NDVI in the area and yield the VTCI imagery from D 009 to D 361 in the year of 2010. Subsequently, the time series VTCI imagery pixel area values at five weather stations were considered for this calculation. And the TRMM/TMPA data products, spatial and temporal observations, define the flood events over the region of Punjab (Pakistan).

9.3.4 MODIS VTCI Time Series Dry and Wet Conditions

The determination of dry and wet conditions in the plain by using the VTCI approach with LST-NDVI feature plots represents the single-year warm and cold edges in response to soil moisture as well as VTCI imagery during the year of 2010 based on the VTCI multiyear approach conception (Tian et al., 2016; Khan et al., 2018). The single-year 16-day composite LST and NDVI products were utilized to identify the warm and cold edges from D 009 to D 361 during 2010 in response to soil moisture and VTCI imagery. The current study utilizes the VTCI imagery time series values at five weather stations. The demonstrated Equations (9.1) and (9.2) of VTCI present the equations for single-year warm and cold edges as well as VTCI imagery from D 009-361 in the year of 2010, whereas D termed as day of year (DOY). Based on multiyear VTCI approach (Khan et al., 2018), the single-year warm and cold edges present the warm edge (upper limits of the scatter plots) observe as $\mathrm{LST}_{\max}$ responds to less soil moisture availability and is under dry conditions, whereas the cold edge (lower limits of the scatter plots) observe as the $\mathrm{LST}_{\min}$ responds to no water scarcity.

Hence, the utilized VTCI imagery's time series values in this calculation vary with the feature space plots and depend on the time and seasons in the study area.

In the computation of edges, an unbalanced temperature increase and decrease from winter into summer and summer into winter seasons were observed for D 009–361 in the year 2010. The determination of dry and wet spells depends on the variations in warm and cold edges. The dry and wet spells dissimilarity in the single year's warm and cold edges exposes the dry and wet conditions and flood events. Based on the weather extremes and flood events in 2010, the temporal changes of the land surface soil moisture conditions were disclosed. Consequently, the retreat states of edges were significant and insignificant for the agricultural practices and represent dry and wet conditions in the region. Subsequently, for the drought and flood categorization, the VTCI imagery's time series values over five weather stations from D 009–361 were presented during 2010.

By and large, the VTCI times series values reveal the dry and wet conditions on the VTCI fractional scale (dry to wet) or (0.0–1.0) in the region. The determined VTCI imagery's time series values have been utilized for dry and wet scale; this reveals the dryness and wetness conditions at five stations. The VTCI values are categorized normal (0.58–0.77) in the blooming stage (early-February) in winter-wheat-crop season in D-041 as the favorable conditions in the green-up stage in the year of 2010. For the categorization of VTCI imagery, values from D 009–361 in the year of 2010 for droughts and the wetness are collected over five metrological stations (located in the south, central and northeast in the plain of Punjab). Moderate and mild drought as well as wetness observed from the D 009–361 over five weather stations varies from (0.28–1.0) and highlights the weather extremes in the form of severe drought to wetness.

Generally, the representation of the dry and wet conditions from D 009–361 in the year of 2010 at five stations with total VTCI time series values occurrences (i.e. $n=115$) observed in the plain from severe drought to wet (0.28–1.0). Based on the drought classification in the given occurrences, it was established that the severe drought ($n=13$), moderate drought ($n=30$), mild drought ($n=31$) and wetness ($n=41$) occurrences' exposed in results of the total 115 VTCI occurred values. In the year of 2010 flood events, the VTCI values in D-201 are "1" at five weather stations, whereas in D-217 over four stations, the values were presented as "1" excluding LHR. These VTCI values show the wetness/flood events (time, duration and intensity). Moreover, the drought severity and wet spells depend on the soil characterization: lower value of the VTCI results in dryness severity, and higher values of VTCI result in wetness. The VTCI values show severe drought to wet conditions in D 009–361 during 2010. It was established that in D–185, heavy rainfall was recorded and, during the D 201–232, the wetness/flood events were noticed over five stations, which shows the flood's time, duration and intensity.

The study showed that the plain faced heavy rainfall in summer season as well as the streamflow of river in the region. In the broad picture, in terms of the winter season, the plain salvages the normal condition for the upcoming winter crop (2010–2011) with somehow significant conditions (Khan and Wang, 2018a, b). This presents the complete structure of drought and flood intensity and lasting of flood events. This also indicates that the farming practices over the plain are irrigated and rain-fed and categorized significant in the northeast as compared with the south of the region and demonstrates the better practices of VTCI monitoring in both rain-fed and irrigated lands with very promising results during the winter-wheat-crop seasons (Rabi season). This shows the winter-wheat-crop (2009–2010) harvested before the inception of the summer rainfall, whereas the inception of flood takes place in Kharif season and the region was exposed to flood events/spells, as the plain faced the worst flood in the history of the past five decades with huge losses.

By large, this is essential for comparing and analyzing the satellite products, in the consideration of VTCI approach for the dry and wet conditions using MODIS LST NDVI products, and is obvious to confirm the VTCI time series values with TRMM/TMPA data products, spatial and temporal observations, for flood events (time, duration and intensity) using GFMS model in Punjab.

9.3.5 TRMM and TRMM Satellites Flood Observations

In Asia, the monsoon rain reaches the foothills of the Tibetan Plateau in early-July to late-July (JD 180–200) and the withdrawal in early-September (around JD–250) to express the monsoon onset and retreats, demonstrated in the Julian Day (JD). In this study, the flash flood occurs due to monsoon heavy rains in 2010 and found the flood events in the JD 200–229 crashing down from the Himalayan mountain ranges (called "The Third Pole" of the world) in the northern part of Pakistan (Zeng and Lu, 2004; Bashir et al., 2017). The spatial and temporal observations of Figures 9.2, 9.4 and 9.6 show the river flow/routing and flood events using MODIS imagery for the description of the plain as well as a presentation for the TRMM/TMPA anomaly's flood events in the region of Punjab. Hence the plain of Punjab lies on Indus Basin which contains five rivers. And the five rivers of Punjab

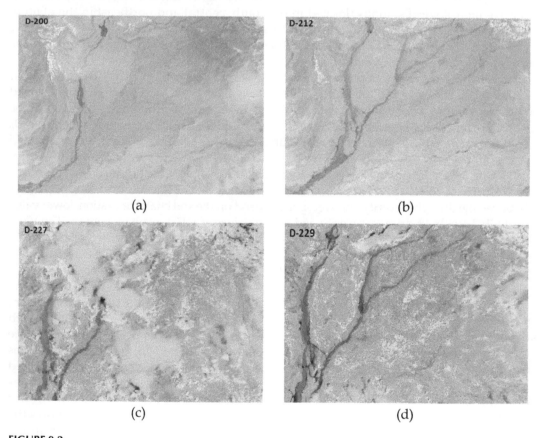

FIGURE 9.2
MODIS imagery of the flash flood from monsoon rains during 2010 in: (a) JD-200, (b) JD-212 and (c) lasting in the JD-227 as well as (d) in the JD-229, deafening down from the Himalayan mountain ranges (Northern Pakistan) and swept over the south plain of Punjab and across the region.

mainly flooded during the summer 2010 (monsoon) due to heavy rainfall and snow melting of Himalayan mountain ranges in the north of Pakistan (Bashir et al., 2017). Through the north of Pakistan, the flood streamflow ends into the Arabian Sea, located in the south of the country.

The defined TRMM/TMPA anomaly is considered for the 1-year (2010) flood in a retrospective manner against the region and five weather stations runoff streamflow, streamflow and flood detection/intensity observations of the flood events through north to south, over the plain of Punjab, Pakistan, located at http://flood.umd.edu/ (Wu et al., 2012, 2014). The TRMM/TMPA flood events (runoff streamflow at ~1 km, streamflow events at ~12 km and flood detection/intensity at ~1 km) are presented in Figures 9.3–9.6. Subsequently, the streamflow above flood threshold swept over to the south of the plain through the Indus Basin of Punjab's Plain and at five weather stations. Hence, flooding in the main Indus River of the plain passes through the northeast and northwest as well as from southwest (contiguous area of Indian-Punjab) and lasting in the south of the plain (i.e., MLN 1 and MLN 2), which shows the runoff streamflow above the threshold at 1 km resolution. The flood events are presented at FSD 1 and FSD 2 located at Faisalabad (center of the plain) and LHR at Lahore (northeast of plain) with high runoff streamflow (Figure 9.3a), whereas, MLN 1 and MLN 2 are located in Multan (south of the plain) with low runoff streamflow (Figure 9.3b). This describes the flood time, duration and intensity at five weather stations. In contrast, the flood streamflow at 12 km resolution in the region is presented in Figure 9.4 in the time of JD 201–281 in 2010.

It was distinguished that the streamflow lasting at 12 km resolution is due to October 08, 2010, presented in Figure 9.4, whereas, at 1 km resolution, the flood detection/intensity is due to September 08, 2010, presented in Figure 9.6, respectively. This shows dry and wet conditions for the winter-wheat-crop farming subsequent to the lasting of flood spells. Figure 9.4 shows maximum water flow above the threshold and ends in the south of the region and defines the plain to normal condition and water flow in the Indus Basin and also categorized significant and insignificant for the crop seasons. Moreover, Figure 9.4b–i shows the flood streamflow events occurred on July 20, July 29, Aug 02, Aug 09, Aug 13,

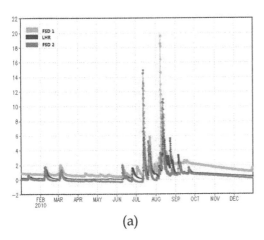
(a)

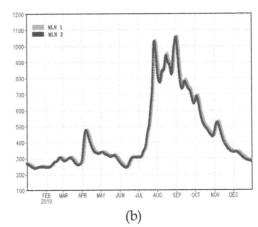
(b)

FIGURE 9.3
The captured runoff streamflow above flood threshold events with temporal (3 hours interval) at relatively high spatial resolution (~1 km) over five weather stations; (a) and (b) highlight the flood runoff streamflow above threshold at time series (21Z01Jan2010 to 21Z31Dec2010), while (a) represents FSD 1, FSD 2 and LHR stations located in center and northeast and (b) represents MLN 1 and MLN 2 stations located in south of the plain, respectively, which shows a maximum water flow above the threshold in the south of the region.

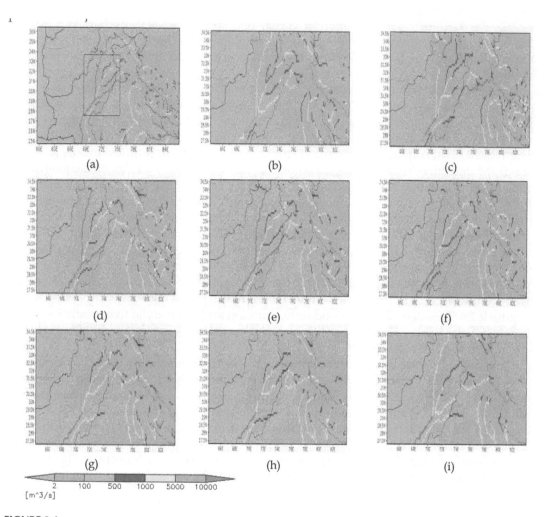

FIGURE 9.4
The streamflow events with temporal 3 hours' interval at ~12 km spatial resolution over the region (b to i) highlight the flood streamflow at time series (21Z20July2010–21Z08Oct2010), which shows a maximum water flow above the threshold and ends in the south of the region and defines the plain to normal condition and water flow in the Indus Basin and also categorized normal for winter-wheat-crop seasons. Whereas (b), (c), (d), (e), (f), (g), (h) and (i) show the flood streamflow events occurred on July 20, July 29, Aug 02, Aug 09, Aug 13, Aug 26, Aug 31 and Oct 08, 2010, respectively, while (a) presents the flood streamflow over the entire region (Pakistan).

Aug 26, Aug 31 and Oct 08, 2010, respectively, while Figure 9.4a presents the flood streamflow over the entire region and especially the demographic area of Pakistan.

Subsequently, Figures 9.5 and 9.6 show the flood detection and intensity (depth above threshold) with ~1 km pixel area over the selected region and demonstrate that the Indus River and four shallow rivers contribute in the appearance of flood events. Figure 9.5a plots present the intensity at the JD 201–232 with wet conditions over LHR, FSD1 and FSD2, whereas Figure 9.5b shows elevated flood intensity due to water flow at MLN 1 and MLN 2 located in the south of the plain; therefore, the two stations present the higher severity of flood and show the normal conditions in first 10 days of September 2010. Figure 9.6a demonstrates the complete establishment of flood events and spells over the region/country (Pakistan). That is why, Figure 9.6b presents the opening spell of the flood, whereas

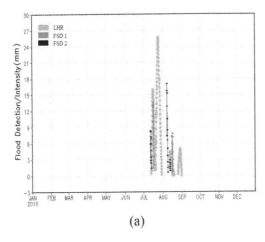 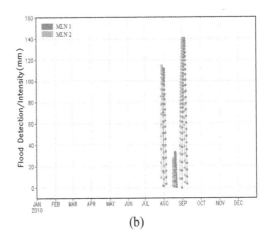

(a)   (b)

FIGURE 9.5

Flood detection and intensity above flood threshold at 3 hours' time interval (01Jan2010 to 31Dec 2010) with water depth in mm at 1 km pixel area over five stations, (a) represents FSD 1, FSD 2 and LHR stations located in center and northeast and (b) represents MLN 1 and MLN 2 stations located in south of the plain, respectively, which illustrate a maximum flood intensity above the threshold in the south of the region.

(Figure 9.6c–i) demonstrates the flood spells with high intensity to the normality of flood events in the plain of Punjab.

Next, Figure 9.6d–f shows elevated intensity of flood events from north to south of the plain, which shows flood intensity and high peak termed as above the threshold in the south of the region in comparison with north and central area of the plain followed by Figure 9.6g. Figure 9.6h and i shows the flood events with its normality. This presents the plain provisions to wetlands due to extreme moisture contents at the JD 201–232 in the year 2010.

Next, the spatial and temporal observations attribute the TRMM and TMPA satellites in significant agreement with the MODIS satellite's VTCI time series values at five weather stations. The TRMM and TMPA signify the observed events in the inundation plot time series (21Z01Jan2010 to 21Z31Dec2010) at each five stations/region for flood runoff streamflow and streamflow at 1 and 12 km resolutions and flood detection/intensity at 1 km resolutions, respectively. This shows the flood events in the year of 2010 with demonstrated significant spatial and temporal observations over the chosen region. The TRMM/TMPA data anomalies make obvious the flood detection and intensity in the JD 201–232 with collateral information with high accuracy at five weather stations and over the region.

Finally, in the year 2010, flood events demonstrate that the flood time, duration and intensity utilizing the MODIS VTCI imagery (VTCI time series values) during 2010 over the plain of Punjab (Pakistan) are in good agreement with TRMM and TMPA satellites observations. This shows that the real-time TRMM flood events are in considerable agreement with high accuracy in determination of flood events in the region. Significantly, the determined flood detection and intensity as well as streamflow runoff above flood threshold, utilized the Goddard TRMM/GPM data products for flood events detection against flood validate the VTCI monitoring approach for dry and wet conditions. The TRMM and TMPA satellites observations are very promising to MODIS VTCI time series values. Largely, the utilized MODIS, TRMM and TMPA data products illustrate the flood events in the region vigorously and revolve to significant and insignificant drought and flood conditions with all calculations for the year of 2010 crop seasons.

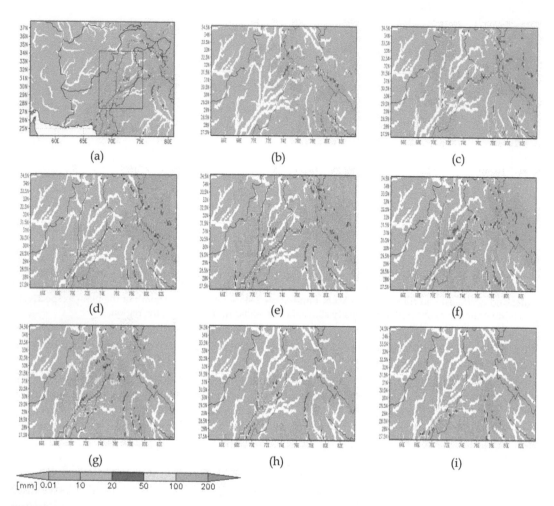

FIGURE 9.6
Flood detection/intensity above flood threshold during (21Z20July2010–21Z06Sep2010) with water depth in mm at 1 km pixel area resolution, and the maximum flood intensity above the threshold in the region and the conditions for the next winter season during 2010–2011 presented in (a) whereas (b), (c), (d), (e), (f), (g), (h) and (i) show the flood streamflow events occurred on July 20, July 29, Aug 02, Aug 09, Aug 13, Aug 26, Aug 31 and Sep 08, 2010, respectively, while (a) presents the flood detection/intensity over Pakistan.

9.4 Conclusions

The satellite integration in connection with 5G network and beyond will lead to a huge shift toward the landscape optimizations by the broadband and wireless connectivity expansions in order to share and broadcast the satellite data products in real-time and near-real-time basis against the climate extremes. The descendant of satellites with associated services in the digital technology innovation is the foremost understandability in the fields of sciences and technology with its applications. The successor of 5G complements the satellite technology from the acquisitions to share the information of MODIS, TRMM and TMPA satellites data products against natural hazards and weather extremes. This learning shows significant capability of the MODIS VTCI imagery (VTCI time series values) for dry and wet conditions (time, duration and intensity) during the year of 2010 using MODIS LST and NDVI data products. This demonstrates the dry/wet conditions and depends on the soil characterization: lower values of the VTCI result in dryness severity, and higher values of the VTCI result in extreme wetness. Also, the employed GFMS model using the TRMM and TMPA satellites data products for the deluge plot time series (flood events and spells) over the region and five weather stations with spatial and temporal observations conclude with: the streamflow events with temporal and spatial resolution over the region highlight the flood streamflow at time series (21Z20July2010 to 21Z08Oct2010), which shows a maximum water flow above the threshold and ends in the south of the region and presents the normality in October 08, 2010, whereas the flood detection/intensity above flood threshold during 21Z20July2010 to 21Z06Sep2010 with water depth in mm at 1 km pixel area resolution with maximum flood intensity above the threshold presents the normality in September 08, 2010, and demonstrates the time, duration and intensity of flood and heavy rainfall in the region. This exhibits the flood streamflow, detection and intensity in very decisive and promising winding up and presents the plain provisions to wetlands due to extreme moisture contents during the summer in the year 2010. Usually, the flood events turn to normality in October 08, 2010, from beginning to end in broad picture over the region. Overall, this indicate the monsoon's heavy rain and flood effects in the region and presents significant outcomes of natural hazards associated with climate extremes during the year of 2010.

By and large, the achievable allusion and advice be payable to climate changes threat toward Asia and specific to the region (e.g. Dai, 2011, 2013; Kreft et al., 2015; Dow and Downing, 2016; Glanemann et al., 2017; Bashir et al., 2017; Eckstein et al., 2019) in the appearance of temperature extremes, glacier melts, flooding and drought conditions. This learning suggests the problem of climate changes with an uphold form of the natural disasters and urges the policy makers to recognize the extreme vulnerability impacts of temperature changes, heat extremes, heavy rainfalls, glacier melts, river/stream flow and agriculture in the region of Punjab, Pakistan.

References

Bashir, F., Zeng, X., Gupta, H., & Hazenberg, P. (2017). A hydrometeorological perspective on the Karakoram anomaly using unique valley-based synoptic weather observations. *Geophysical Research Letters*, 44(20), 10–470

Cheng, W. (2011). The application of——3S(GIS, RS and GPS) technology to the study of China FangZhen period. In *Proceedings of the 2011 Third International Workshop on Education Technology and Computer Science-Volume 01* (pp. 205–208). Washington, DC: IEEE Computer Society.

Dai, A. (2011). Characteristics and trends in various forms of the Palmer Drought Severity Index during 1900–2008. *Journal of Geophysical Research: Atmospheres*, 116(D12), 1–26.

Dai, A. (2013). Increasing drought under global warming in observations and models. *Nature Climate Change*, 3(1), 52–58

Dow, K., & Downing, T. E. (2016). The Atlas of climate change: Mapping the world's greatest challenge. Oakland, CA: University of California Press.

Eckstein, D., Künzel, V., Schäfer, L., & Winges, M. (2019). *Global climate risk index 2020*. Bonn: Germanwatch.

Friedl, M. A., McIver, D. K., Hodges, J. C., Zhang, X. Y., Muchoney, D., Strahler, A.H., Woodcock, C.E., Gopal, S., Schneider, A., Cooper, A., & Baccini, A., (2002). Global land cover mapping from MODIS: Algorithms and early results. *Remote sensing of Environment*, 83(1–2), 287–302

Glanemann, N., Willner, S. N., & Christian, O. (2017). Climate change and trade networks, Asian Development Bank.

Harris, R., & Baumann, I. (2015). Open data policies and satellite Earth observation. *Space Policy*, 32, 44–53.

Honicky, R., Brewer, E. A., Paulos, E., & White, R. (2008). N-smarts: Networked suite of mobile atmospheric real-time sensors. In *Proceedings of the Second ACM SIGCOMM Workshop on Networked Systems for Developing Regions* (pp. 25–30), Seattle, WA.

Khan, J., Abbas, A., & Khan, K. (2011a). Cellular handover approaches in 2.5 G to 5G technology. *International Journal of Computer Applications*, 21(2), 0975–8887.

Khan, J., Bojkovic, Z. S., & Marwat, M. I. K. (2011b). Emerging of mobile ad-hoc networks and new generation technology for best QOS and 5G technology. In *International Conference on Future Generation Communication and Networking* (pp. 198–208), Springer, Berlin, Heidelberg.

Khan, J., Wang, P., Xie, Y., Wang, L., & Li, L. (2018). Mapping MODIS LST NDVI imagery for drought monitoring in Punjab Pakistan. *IEEE Access*, 6, 19898–19911

Khan, J., & Wang, P. (2018a). Remote sensing spatial and temporal earth observations in Pakistan. *Biomedical Journal of Scientific &Technical Research*, 7(3), 5891–5892. DOI:10.26717/bjstr.2018.07.001498

Khan, J., & Wang, P. (2018b). Geospatial Earth observations using vegetation temperature condition index for drought conditions over the cropland of Punjab Pakistan. Research & Reviews: *Journal of Botanical Sciences*, 7(4), 9–10.

Kreft, S., Eckstein, D., Dorsch, L., & Fischer, L. (2015). Global climate risk index 2016: Who suffers most from extreme weather events, weather-related loss events in 2018 and 1995 to 2018. Electronic version, Germanwatch Nord-Süd Initiative.

Murakami, D., Peters, G. W., Yamagata, Y., & Matsui, T. (2016). Participatory sensing data tweets for micro-urban real-time resiliency monitoring and risk management. *IEEE Access*, 4, 347–372

Soret, B., Leyva-Mayorga, I., Röper, M., Wübben, D., Matthiesen, B., Dekorsy, A., & Popovski, P. (2019). LEO small-satellite constellations for 5G and beyond-5G communications. arXiv preprint arXiv:1912.08110.

Sui, D., & Goodchild, M. (2011). The convergence of GIS and social media: Challenges for GIScience. *International Journal of Geographical Information Science*, 25(11), 1737–1748

Tian, M., Wang, P., & Khan, J. (2016). Drought forecasting with vegetation temperature condition index using ARIMA models in the Guanzhong Plain. *Remote Sensing*, 8(9), 690.

Thenkabail, P. S., & Gamage, M. S. D. N. (2004). The use of remote sensing data for drought assessment and monitoring in Southwest Asia (Vol. 85). Iwmi

Wu, H., Adler, R. F., Hong, Y., Tian, Y., & Policelli, F. (2012). Evaluation of global flood detection using satellite-based rainfall and a hydrologic model. *Journal of Hydrometeorology*, 13(4), 1268–1284

Wu, H., Adler, R. F., Tian, Y., Huffman, G. J., Li, H., & Wang, J. (2014). Real-time global flood estimation using satellite-based precipitation and a coupled land surface and routing model. *Water Resources Research*, 50(3), 2693–2717

Zeng, X., & Lu, E. (2004). Globally unified monsoon onset and retreat indexes. *Journal of Climate*, 17(11), 2241–2248

10

Drone Communications in 5G Network Environment

Zoran Milicevic
Serbian Armed Forces

Bojan Bakmaz
University of Belgrade

CONTENTS

10.1 Introduction

Drones imply self-piloted, radio-controlled aerial vehicles that can carry controllers, different sensors, communication interfaces and other types of electronic equipment. They appear in the form of fixed-wing aircraft, unmanned helicopter, multirotor unmanned aerial vehicle and so on. The rapid deployment of low-cost embedded micro-computers, sensors and radio interfaces has led to extensive usage of drones for various applications in both the military and civilian domains. There are some situations and particular environments where modern technology fails in providing required services, such as infrastructure damage or remote underserved areas. In these cases, one or more drones can operate as airborne base stations or flying relays in order to improve communication system capacity and availability. Moreover, cellular-connected drones lead to extended application use cases such as drone-enabled product delivery systems, drone-based real-time multimedia streaming and drone-assisted intelligent transportation systems. It should be emphasized that artificial intelligence–based solutions represent a significant tool in these areas of application. Various types of aerial nodes are of huge interest in improving the performance, agility and flexibility of 5G mobile networks.

Drones are becoming very attractive not only because of fast deployment and low cost but also because of their high maneuverability and ability to hover (Hayat et al., 2016). The size of a drone cells can be adjusted by changing their altitude, transmission power, antenna directivity and other parameters, providing more adaptability to unstable traffic loads and variable user distributions (Ferranti et al., 2018). A novel concept of three-dimensional (3D) cellular networks, which integrate aerial base stations and cellular-connected drone users in beyond 5G environment, is introduced by Mozaffari et al. (2019a).

This chapter surveys the state of the art of drone application in 5G cellular network environment by focusing on communications at the physical layer. Drone-assisted transmission, millimeter wave (mmWave) communication, nonorthogonal multiple access (NOMA), as well as cognitive drone communications, are pointed out. With software-defined networking (SDN), the challenges such as resiliency, reconfigurability, reusability and energy awareness can be easily addressed. The opportunity of employing this paradigm in drone surveillance applications and drone base station (DBS) networks cannot be denied due to the increased number of drones. A trend of drone-based mobile edge computing (MEC) is emphasized together with open issues when combining 5G technologies with drones. These advances are significant from both industry and academia point of view.

10.2 General Structure of a Drone System

Drone systems come in many varieties. The structure of a drone system can provide a deeper understating of each component (Al-Turjman et al., 2020). Widespread commercial drones are equipped with barebone devices necessary for flying. They are capable of recording and streaming videos. In civil applications, such as remote sensing, larger drones with more advanced sensors are required. The most complex drone system can be found in military applications where they use sophisticated components such as stealth technology and hyper sensors.

Generally speaking, a drone system is composed of drone, ground station and communication link as shown in Figure 10.1. The main parts of a drone constitution are master controller, sensors, actuators and power system. The ground station is a part of system with a function to control the drone. Controlling and monitoring drones in continuous fashion are performed through a communication link. The data collected by drone sensors and the operating data are transmitted to the ground station intelligent terminal. There are three functions carried out by ground station: radio control, data processing as well as system testing.

Besides the terrestrial communication link, which transmits telemetry information, video and audio from drones to the ground station (aerial-to-ground, A2G), there exist two more types of links (He et al., 2019). The satellite communication link transmits, for example, positioning and metrological information, while an important segment of the

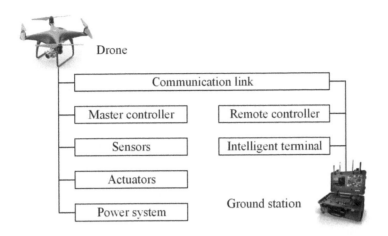

FIGURE 10.1
General structure of a drone system.

drone network belongs to the drone-to-drone (aerial-to-aerial, A2A) communication link. This link is used for data exchange between drones.

10.3 Drones in Mobile Networks

Today, drones are used in many applications, e.g., surveillance, monitoring, transport, logistics and so on. From the wireless networks point of view, a drone is a user equipment (UE) that may also include private networks. Today, drones are using unlicensed links to communicate with the ground control stations. Unfortunately, unlicensed links have limited reliability and range in beyond virtual line of sight (BVLoS) operation. On the other hand, mobile networks are in a position to enable BVLoS operations owing to their wide coverage and reliability. As stated by Bor-Yaliniz et al. (2019), drones in mobile networks can be categorized as mobile-enabled drones (MEDs) and wireless infrastructure drones (WIDs). Here, MED and WID correspond to diverse operational modes rather than different devices.

There are two types of links with a ground control station established by MEDs: (1) the command and control links used for remote piloting, telemetry data, identity, navigation and so on and (2) the type of link established with a ground control station as applicable for delivering sensing data, as well as multimedia content. Different from command and control data, application data are less critical in many cases.

On the other hand, WIDs are classified taking into account their functionalities and requirements as fundamental criteria. There are three categories, as follows:

- DBSs, which serve as aerial nodes with some or all functionalities of BS. DBSs have various moving patterns (hovering, rotating, floating, lending, etc.) depending on environmental conditions, communication requirements and machinery.

- Drone relays (DRs) which can be deployed by users or operators. In the first case, unlicensed spectrum can be used for the links between the user and AR. The AR acts as a piece of UE for mobile networks. Operator-deployed DRs can be more sophisticated and should be integrated.

- Drone backhaul/fronthaul providers, where they are used to form an aerial transport network. The interest in this solution is increasing permanently (Mehta and Prasad, 2019) because both (licensed and unlicensed) solutions are possible and hybrid ones seem to be the most efficient.

The integration of drones into contemporary and prospective mobile networks is implied in the ongoing 3rd Generation Partnership Project (3GPP) standardization activities. The most comprehensive study regarding deployment scenarios, channel models, performance requirements and metrics of MEDs implementation can be found in 3GPP TR 36.777 (2018), while specifications regarding field trial results, system-level and mobility evaluations together with fast-fading models are going to be provided in future versions. In comparison with performances of the typical 5G services, MEDs require 100 times less reliability than ultrareliable low-latency communications and 200 times fewer data rates than enhanced mobile broadband cases.

Network slicing enables service-oriented configuration of mobile systems in a flexible and agile manner (Foukas et al., 2017). Hence, 5G networks are able to support different drone services, such as a slice configured for application links of MEDs or a slice to isolate the traffic of WIDs. There are multiple trade-offs when considering splitting options for WIDs. Lower-layer split increases bandwidth requirements and decreases latency tolerance in the fronthaul link, compared with upper-layer splits. It also increases the transmission complexity, especially for the physical layer. In this case, the lower-layer split significantly reduces the computational complexity. That can increase airtime if fronthaul links with large bandwidth and high signal-to-interference-plus-noise ratio (SINR). Finally, the lower-layer split makes centralization more effective and increases the number of served UEs.

The main reason for 5G core to establish a service-based architecture is to exploit advantages offered by SDN and network function virtualization (NFV) (Bojkovic et al., 2018). Increased support for virtualization and slicing makes it easier to integrate new functions, specifically supporting MEDs and WIDs. Since MED traffic varies from latency-tolerant telemetry data to bandwidth-demanding live video streaming, it can be served over multiple slices. A key issue for carrying WID traffic is the creation of an independent slice with some shared control plane functions and nonshared user plane functions, in order to avoid the influence on other services. However, the implementation of only one slice may not be enough since there may be UEs with different services. Since the management of a DBS is complex and expensive, the objective is to utilize it for as many services as possible. Therefore, multiple 5G-core slices per WID are need. Because complexity increases exponentially with the integration of WIDs fleet (Bor-Yaliniz et al., 2019), an efficient management system is required.

In order to support mobile network densification without scaling transport domain, integrated access and backhaul solution is invoked by 3GPP (TR 38.874, 2018). There are two main characteristics of this approach. Firstly, a DR can utilize backhauling natively supported by 5G networks. Moreover, a WID as an intermediate node can reduce the number of hops, providing topology flexibility due to LoS paths and high mobility.

Unlicensed spectrum is one of the most valuable resources. Licensed-assisted access to unlicensed spectrum can be beneficial for drone communications having in mind the maximum transmit power regulations and range of 150 m. Utilization of wideband operations, larger subcarrier spacing and minislot-based transmissions lead to increased throughput but, on the other hand, render less energy per symbol for the same transmit power and hence reduced coverage. Despite limitations for MEDs application, especially above 7 GHz, unlicensed new radio (NR), as well as the wireless local area network aggregation options, can provide numerous diverse integration options for WIDs while exploiting broad free spectrum.

10.4 Drone Communication at Physical Layer

As previously stated, drone-assisted communication networks are applied in the case of unexpected and temporary events. The scenario of drones as DBSs serving a targeted area is shown in Figure 10.2. Each drone contains wireless transceivers to communicate with ground UE as well as other drones.

Drones are equipped for receiving, processing and transmitting the corresponding signals, in order to complement preexisting cellular systems. In that way, the additional capacity to hot spot area in the case of temporary events is provided. At the same time, the communication infrastructure becomes reinforced. This is of importance for emergency and safety reasons. The key technologies are multiple-input multiple-output (MIMO), mmWave communication, NOMA transmission and cognitive radio (CR).

10.4.1 3D Multiple-Input Multiple-Output and Millimeter Wave Communications for Drone-Assisted Networking

There has been significant interest in the application of 3D MIMO, also known as full-dimension MIMO, by exploiting both the vertical and horizontal dimensions in next-generation cellular networks (Nam et al., 2013; Cheng et al., 2014). 3D beamforming enables the creation of separate space beams at the same time, thus reducing intercell interference. Compared with the conventional MIMO, 3D MIMO technology can provide increased

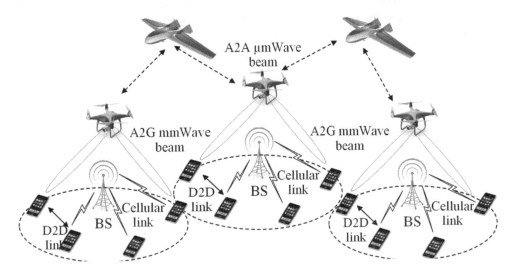

FIGURE 10.2
Drones as aerial BSs. A2A, aerial-to aerial; A2G, aerial-to-ground; BS, base station; D2D, device-to-device.

overall throughput while supporting higher number of users. This solution is more suitable for scenarios in which the number of users is high and they are distributed in three dimensions with different elevation angles with respect to their serving BS.

Due to the high altitude of DBSs, ground UEs can be easily detectable at different altitudes and elevation angles with respect to the drone. LoS channel conditions in A2G communications enable effective beamforming in both azimuth and elevation domains. Furthermore, the high mobility and flexibility of drones allow effective mechanical beamsteering in any direction. Therefore, DBSs are suitable candidates for employing 3D MIMO (Mozaffari et al., 2019b).

mmWave communications are one of the enabling technologies for new generation mobile networks (Bojkovic et al., 2015). mmWave is advanced in multigigabit transmission and beamforming. Conceptually, a large amount of licensed and unlicensed frequency bands is potentially available for use in drone-assisted cellular networks, e.g., 28 GHz licensed band and 60 GHz unlicensed band. Significantly shorter wavelength makes it feasible to design the physically smaller modules, circuits and antennas, which enables to construct the beamforming antenna arrays with dozens to hundreds of antenna elements packed on a tiny chip. It is an ideal solution for short-range drones with limited payloads. Moreover, low interference and increased security benefit from the inherently directional mmWave beams and the narrow beamwidth with high immunity to various attacks, e.g., jamming, eavesdropping and so on.

On the other hand, the short wavelength makes mmWave easily blocked by obstacles. In order to bypass these obstacles, relay nodes are widely required. It is obvious that drones can enable mobile relays in real applications (Kong et al., 2017). Rational path planning and trajectory optimization are mandatory for drone-assisted networking considering the typical constraint of the limited onboard power and thus the flight time. Having in mind that the hovering elevation of the mmWave DBS is comparable with the height of the regular BSs, the air-to-ground channel model with the terrestrial channel model is suitable for the first-order approximation (Gapeyenko et al., 2018).

The beamforming technique can be exploited to construct a narrow directional beam and overcome the high path loss or additional losses caused by atmospheric absorption and scattering. Compared with terrestrial static BSs, DBSs are characterized by dynamic time-varying positions and altitudes. Hence, more efficient beamforming training and tracking are needed, and the channel Doppler effect needs extra consideration, while the drone position and user discovery are intertwined (Le et al., 2019). Achieving dynamic beam alignment between drones and ground UEs with lower overhead and computational complexity is a challenging issue. To reduce the beamforming training time and mmWave beam search overhead, Xiao et al. (2016) proposed a hierarchical beam search scheme based on the tree-structured beamforming codebook which covers the whole search space in angle domain.

With astonishing opportunities, the design, control and optimization of drone-assisted networks with mmWave communications become very challenging and complex. The significant challenges require configuration, management and control of networks in a smarter and more agile manner. Inspired by the trends, machine learning is needed to enable adaptive learning and intelligent decision-making, due to its capability to achieve the convergence of computing power, algorithm improvement and data (Zhang et al., 2019a).

10.4.2 Nonorthogonal Multiple Access for Drone-Assisted Transmission

Multiple access techniques are of great interest while integrating drones into 5G networks and beyond. On this road, NOMA has been regarded as an effective solution to improve

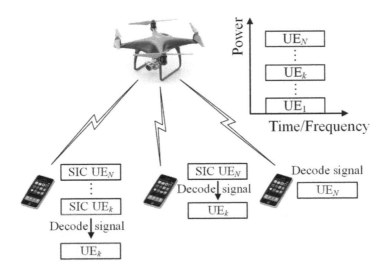

FIGURE 10.3
NOMA transmission in drone-assisted network. NOMA, nonorthogonal multiple access; SIC, successive inter-ference cancellation; UE, user equipment.

spectral efficiency while serving a multitude of UEs simultaneously (Ding et al., 2017). The goal is to serve multiple users on the same resource block but with different power levels. The basis of NOMA implementation relies on the difference in channel conditions among users. An illustrative example of the NOMA transmission in drone-assisted network is presented in Figure 10.3. A drone serves multiple ground users where the signals of UE_1 to UE_{k1} are canceled at the UE_k, whereas the signals of UE_{k+1} to UE_N are received as interference. NOMA uses superposition coding at the transmitter such that the successive interference cancellation (SIC) receiver can separate the users in uplink and downlink channels.

However, the successful operation of NOMA in drone-assisted communications invokes some associated challenges and constraints. First of all, NOMA with improved spectral efficiencies requiring a sophisticated SIC technique at the receiver side is used. SIC solely relies on the channel state information at both the receivers and the transmitters to determine the allocated power for each receiver and the decoding order, which needs to be estimated relatively accurately. Next, more efforts are needed to further eliminate the interlayer interference caused by multiplexing in the power domain. Finally, when considering the high mobility of drones in practice, the communication distance between the drone and ground UEs would vary constantly based on the real-time requirements; thereby the SIC decoding order determined by the received signal strength varies with the drone's location.

10.4.3 Cognitive Drone-Assisted Communications

One of the crucial predicaments faced by drone-enabled networking is the scarcity of the radio spectrum, because of the very intense competition of heterogeneous wireless technologies and dramatic growth of their users. Therefore, it is necessary to obtain further spectrum access by dynamic utilization of the existing frequency bands. Many academia and standardization groups have proposed the incorporation of CR and drone communication systems to increase the spectrum opportunities (Saleem et al., 2015). This concept constitutes a promising network architecture that allows the coexistence of drones with terrestrial mobile devices operating in the same frequency band. In this case, drone-assisted communications can cause severe interference to terrestrial UEs, since drones usually have strong LoS links with ground stations.

Traditional deployments of drones assume a fixed spectrum assignment. Due to the high mobility and high costs of licensed spectrum, it is not effective for drones to operate in this band.

Therefore, generally, drones operate in unlicensed spectrum bands. As mentioned earlier, unlicensed spectrum is becoming overcrowded; therefore, CR technology provides dynamic spectrum access to drones in order to operate in licensed and unlicensed spectrum and maximize network performance (e.g., higher throughput, lower delay, interference with other operating devices and so on).

Zhang and Zhang (2017) proposed an underlay spectrum sharing method between the drone small cells network and traditional cellular network. Using stochastic geometry theory, they derived the explicit expressions for the drone small cells coverage probability and achieved the optimal density of DBSs for maximizing the throughput. Obtained analytical and numerical results show that throughput increases almost linearly with the increase of the drone cell outage constraint. In order to protect the cellular users, the throughput of the drone cell user stops increasing when it meets the cellular network efficiency loss constraint. Similarly, Sboui et al. (2017) proposed the integration of underlay CR into a drone system where the drone as a secondary transmitter opportunistically exploited and shared the primary spectrum for the A2G transmission. The objective was to maximize the energy efficiency of the drone unit, thereby ensuring effective long-time operations. Huang et al. (2018) proposed jointly optimized drone's trajectory and transmit power allocation with the aim of achieving the maximum throughput while restricting the interference imposed at primary receivers below a tolerable level.

10.5 Drone-Assisted Communications from Networking Perspective

The next-generation mobile networks should intelligently and seamlessly integrate multiple nodes to form a multitier hierarchical architecture, including the drone cell tiers for large radio coverage areas, the ground small cell tiers for small radio coverage areas, the user device tiers with device-to-device (D2D) communications and so on. In contrary to the conventional terrestrial networks, aerial networks based on drones accommodate air-mounted transceivers. Drones usually operate according to specific objectives and regulation restrictions, and their trajectory can be more easily predicted. In this respect, Oubbati et al. (2017) focus on drone-based systems and several position-based routing protocols with regard to their performance targets, benefits and drawbacks. Since terrestrial networks often struggle in areas with obstacles and highly mobile and dispersed nodes, drone-assisted networks have a significant chance of achieving LoS links with respect to the ground users, thus complementing all types of ground networks and enhancing coverage and connectivity, especially in highly mobile scenarios. High mobility is a major characteristic of drone systems, which results in new issues for networking. At the same time, drone-assisted networks can support high throughput via very large-scale arrays and sufficient bandwidth through mmWave frequencies. Nevertheless, the successful deployment and long-term operation of drone-assisted networks requires effective interference and radio resource management, as well as coordination and interoperability between heterogeneous wireless systems (Nomikos et al., 2020).

Future aerial networks will be heterogeneous and comprise different types of nodes, namely high-altitude long-range drones, medium-altitude drones and low-altitude short-range drones. The multitier aerial networks are much affected by the density of users and services and can be constructed by utilizing several drone types, which is similar to terrestrial heterogeneous networks (HetNets) with macro/small/picocells and relays. As an initial study, Mehta and Prasad (2017) proposed the concept of aerial HetNets to offload the data traffic from the congested terrestrial BSs, where a fleet of small drones with variable operational altitudes is deployed as an ad hoc network. Sekander et al. (2018) investigated the feasibility of multitier architecture over a single-tier drone network in terms of spectral efficiency in downlink transmission and identified the relevant challenges such as energy consumption of drones, interference management and so on. The influences of different urban environments are analyzed, too.

10.5.1 Drones and Device-to-Device Communications

D2D communications as a new network architecture are becoming increasingly popular, because of significant improvements in network capacity by offloading traffic from BSs, when two neighboring nodes communicate with each other. Together with small cells, D2D communications will form a new low-cost tier with the goal to increase coverage, offload backhaul, as well as to provide fallback connectivity (Bojkovic et al., 2015). Direct communications are typically deployed using underlay transmission links which reuse existing licensed spectrum resources, while the drone can be a good candidate to promptly construct the D2D-enabled network by introducing a new dimension. The operations of drones, side by side width D2D communications over a shared spectrum band, will also introduce important interference management challenges; thus the impact of drone's mobility on D2D and network performance should be analyzed.

Deployment of a drone as a DBS used to provide the fly wireless communications to a given geographical area is analyzed by Mozaffari et al. (2016). In particular, the coexistence between the drone, which is transmitting data in the downlink, and an underlaid D2D communication network is considered in terms of different performance metrics. Both key scenarios, namely, static and mobile drones were considered, respectively. Obtained results show that, depending on the density of D2D users, optimal values for the drone altitude which lead to the maximum system sum rate and coverage probability exist. Moreover, if drones intelligently move over the target area, the total required transmission power while covering the area of interest can be minimized. Finally, in order to provide full coverage, the trade-off between the coverage and delay, in terms of the number of stop points, is discussed.

The concept of spectrum sharing for full-duplex drone relaying systems with underlaid D2D communications was introduced by Wang et al. (2018). Here, the transmit power and drone's trajectory were jointly designed to achieve efficient spectrum sharing.

10.5.2 Software-Defined Drone Networking

SDN paradigm has demonstrated a powerful performance in dealing with complex and dynamic automation networks, such as data centers and backbone networks. Specifically, these networks have a centralized architecture, where a control center is responsible for network management. The SDN controller is deployed in the control plane, and it is aware of the global network conditions to optimize the network operations, by dynamically configuring the data plane routes. Compared with traditional networking, SDN has better controllability and visibility for network components, which enables better management by using the common controller.

In applications of drone cells, the mobile network must be configured efficiently for seamless integration/disintegration of drones, such as changing protocols and creating new paths. Based on the SDN architecture, drones can perform as SDN switches in the data plane for collecting context information in a distributed way, while the ground BSs are controllers gathering data and making control decisions on network functions and resource allocation. Assisted by SDN, network reconfiguration and resource allocation among a swarm of drones can be performed in a more flexible way. Here, aerial nodes must collaborate with each other, and their behaviors (both the data transmission and movements) should be controlled by a centralized entity, i.e., an SDN controller.

Network architecture, which implements the concept of SDN into the drone network to separate the control and data plane, as well as to provide network programmability by controlling drones' operation parameters, is proposed by Zhao et al. (2019). The focus is on the problem of optimal DR placement for real-time video services, which reduces the impact of node mobility on the quality of experience (QoE). The SDN controller considers the global drone context information to prevent collisions, optimize movements and establish a routing path that determines the relay node deployment to provide video transmission with QoE support. Specifically, the controller considers multiple drone context information for routing and drone positions, movement trajectories and residual energy. Major contributions to SDN with drones include, but are not limited to (Li et al., 2019):

- Drone cell management framework and NFV technologies to assist a terrestrial HetNets.
- Provided handover facilities in the drones that support wireless network with lower latencies.
- The placement of the controller in SDN-based drone network for providing better quality of service (QoS) and QoE.
- End-to-end delay for sharing the control information between nodes and SDN controller.
- Resource allocation of a multidrone network to minimize the operating delay and energy consumption by considering the edge and cloud servers.
- Proactive drone cell deployment framework to alleviate overload conditions caused by heavy traffic.
- Resilient multipath routing framework for a drone network, while SDN controller serves to determine the preferred routes in the case of a security attack.

10.5.3 Drone-Assisted Mobile Edge Computing

The MEC proposed by European Telecommunication Standard Institution (ETSI, 2016) is an architecture evolving from 5G that intensely integrates mobile access networks and services. The MEC adds computing, storage, processing and other functions on network side. Beside optimal resource utilization and user experience improvement, the MAC can also reduce computing power of the mobile edge node. The fixed BSs and their limited computing resources cannot meet the communication requirements of mobile devices. A type of mobile relay communication device needs to be defined as the edge computing node (Wua et al., 2020). A drone has become a potential choice for edge computing nodes in the new architecture of MEC.

In drone-assisted networks, the resource-constrained mobile devices are able to offload their intensive computation tasks to a drone with high computing ability and flexible connectivity at the network edge, thereby reducing traffic load and energy consumption at the fixed cloud servers. However, the embedded processor usually may not have enough computing resources due to the small size of drone. This puts a restraint on efficient execution of complex applications. The computation and battery performance of drone can be virtually enhanced by leveraging computation offloading to a remote cloud server or nearby edge servers via terrestrial BSs. As for drone node, it has a choice to either process application using its own resources or distribute its computation tasks to edge server or a remote cloud for processing according to QoS requirements. In this case, it is very important to design an energy-efficient drone-assisted MEC system (Liu et al., 2020). The computation efficiency for the drone-enabled MEC network is studied by Zhang et al. (2019b) by jointly optimizing the offloading times, the central processing unit frequencies, the transmit powers of the user and the trajectory of the drone. Although many energy-efficient methods are proposed, the battery-powered drones still restrict the MEC network performance because of the limited battery capacity.

Wireless power transmission (WPT) technology is expected to extend the energy supply and prolong the drones' operation duration. Particularly, microwave power transmission is one of the important WPT technologies (Strassner and Chang, 2013), which can provide sufficient power for the drone without landing. In this technology, a large disc-shaped rectifying antenna attached to the fuselage of the drone is responsible for energy harvesting from the microwave antenna array on the ground. Microwave power transmission can still keep the advantage of the drone-enabled MEC (Figure 10.4) to provide quick and convenient computation services for ground users.

Here, the drone is powered by the carry-on battery and can be recharged by the microwave power station without landing. The drone can approach the users' locations in the air to simultaneously provide communication and computation services. To achieve that, the drone has an onboard computation processor, wireless communication module and

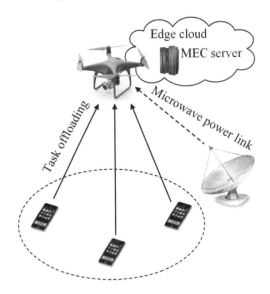

FIGURE 10.4
Drone-assisted MEC network with microwave power transmission. MEC, mobile edge computing.

interface as well as the charging units which consist of rectenna array, power management circuit and rechargeable batteries. This scenario can be formulated as an optimization problem in order to maximize the service utility of the drone by finding the optimal trajectory, computation offloading decisions and offloading duration, as proposed by Liu et al. (2020).

10.6 Further Research Challenges for Drone Communications

Apart from the previously discussed prospects, there are still many open issues related to the practicality of performing drone communications in 5G systems. For instance, in certain use cases, there can exist some dissipations and barriers between the drone and ground station; thus a more realistic A2G channel model that incorporates various atmospheric and topographic parameters is an interesting problem worth future research efforts. The drone-based antenna array system is another footprint for providing high data rates and low service time, since the number of antenna elements is not limited by space constraints. In drone-enabled multiuser NOMA systems, the optimal user clustering and user-pairing algorithms are underexplored fields. Furthermore, an efficient management system is necessary to safely handle the high density of low-altitude drone traffic, which is responsible for the cooperative path planning and collision avoidance.

In order to provide seamless communication service to ground users over a significantly wide area, a swarm of drones constructs a multihop network. However, due to the high mobility and the need to maintain close communication links, the connections with the neighboring drones are interrupted frequently. In this case, the application of traditional routing protocols is not prominent in drone-assisted networking. Additionally, when multiple drones collaborate, collision avoidance also becomes a significant development for safe operation. On the other hand, satellite-to-drone channel models lack detailed propagation effects and thus remains an open topic for future research.

Energy constraint is the bottleneck in any drone-assisted communications scenario. As recent developments in battery technologies (i.e., hydrogen fuel cells), energy harvesting is used to extend the flight times by utilizing green energy sources, such as solar energy. However, the efficiency of energy harvesting is relatively lower due to longer distance and contingent energy distribution. Emerging energy transmission techniques, such as energy beamforming through multiantenna systems and distributed multipoint WPT, are of great interest when considering charging efficiency.

The cybersecurity is still a significant challenge to be overcome in the real utilization of drones. Therefore, designing timely strategies and countermechanisms is required to counteract malicious cyberattacks. Blockchain technology (i.e., aerial blockchain) is expected to be a new paradigm to securely and adaptively maintain the privacy preferences of drone communications.

10.7 Concluding Remarks

The drone technology is becoming more and more mature, which owns advantages on high reliability, flexible deployment, cooperative operation and low cost. By alleviating the pressure on the terrestrial networks and reducing the cost of densely deployed small cells, drones become a significant component of 5G and beyond networks. In this chapter, some of the issues, challenges and future research directions, which arise when drone systems are integrated into 5G mobile networks, are identified and discussed. The integration issues, presented from the physical layer and networking perspective, are important to consider because of the intrinsic characteristics of emerging 5G technologies and drone systems.

The A2A and A2G channels are much persistent than the ground-to-ground channels. In many cases, they can establish LoS links. For these channels, mmWave transmission becomes practicable for drone-assisted communications, which provides a much higher data rate. Furthermore, drones can be combined with 3D MIMO and CR technologies to meet the ever-increasing data demand. Considering the device miniaturization and cost reduction, it is more feasible than ever before to deploy drones as a relay to realize range extension in cellular communication systems.

However, there still exist many challenges in this emerging technology of drone communications and networking, such as interference management, security and safety, beam tracking and alignment, trajectory and placement optimization and so on. Furthermore, drone-enabled networking also requires great effort to reshape the current ground systems for satisfactory performances.

In the context of multidrone deployments, the SDN paradigm has a great potential of meeting network reliability and security requirements. With the emergence of diverse mobile applications, the QoE is greatly limited by computation capacity and finite battery lifetime. MEC and WPT solutions are promising to address this issue.

References

3GPP TR 36.777 V.1.1.0 (2018). Enhanced LTE support for aerial vehicles.

3GPP TR 38.874 V.16.0.0 (2018). Study on integrated access and backhaul.

Al-Turjman, F., Abujubbeh, M., Malekloo, A., and Mostarda, L. (2020). UAVs assessment in software-defined IoT networks: An overview. *Computer Communications*, 150, 519–536.

Bojkovic, Z., Bakmaz, M., and Bakmaz, B. (2015). Research challenges for 5G cellular architecture. *Proceedings of 12th International Conference on Advanced Technologies, Systems and Services in Telecommunications TELSIKS* (pp. 215–222). Nis, Serbia: IEEE.

Bojkovic, Z., Bakmaz, B., and Bakmaz, M. (2018). Principles and enabling technologies of 5G network slicing, in Trestian, R., Muntean, G-M. (Eds.), *Paving the Way for 5G Through the Convergence of Wireless Systems*, Hershey, PA: IGI Global.

Bor-Yaliniz, I., Salem, M., Senerath, G., and Yanikomeroglu, H. (2019). Is 5G ready for drones: A look into contemporary and prospective wireless networks from a standardization perspective. *IEEE Wireless Communications*, 26(1), 18–27.

Ding, Z., Lei, X., Karagiannidis, G. K., Schober, R., Yuan, J., and Bhargava, V. K. (2017). A survey on non-orthogonal multiple access for 5G networks: Research challenges and future trends. *IEEE Journal on Selected Areas in Communications*, 35(10), 2181–2195.

ETSI GS MEC 003 (2016). Mobile Edge Computing (MEC); Framework and Reference Architecture, v. 1.1.1.

Ferranti, L., Cuomo, F., Colonnese, S., and Melodia, T. (2018). Drone cellular networks: Enhancing the quality of experience of video streaming applications. *Ad Hoc Networks*, 78, 1–12.

Foukas, X., Patounas, G., Elmokashfi, A., and Marina, M. K. (2017). Network slicing in 5G: Survey and challenges. *IEEE Communications Magazine*, 55(5), 94–100.

Gapeyenko, M., Bor-Yaliniz, I., Andreev, S., Yanikomeroglu, H., and Koucheryavy, Y. (2018). Effects of blockage in deploying mmWave drone base stations for 5G networks and beyond. *Proceedings of International Conference on Communications Workshops* (pp. 1–6). Kansas City, MO: IEEE.

Hayat, S., Yanmaz, E., and Muzaffar, R. (2016). Survey on unmanned aerial vehicle networks for civil applications: A communications viewpoint. *IEEE Communications Surveys & Tutorials*, 18(4), 2624–2661.

He, D., Liu, H., Chan, S., and Guizani, M. (2019). How to govern the non-cooperative amateur drones? *IEEE Network*, 33(3), 184–189.

Huang, Y., Xu, J., Qiu, L., and Zhang, R. (2018). Cognitive UAV communication via joint trajectory and power control. *Proceedings of the 19th International Workshop on Signal Processing Advances in Wireless Communications* (SPAWC) (pp. 1–5), Kalamata, Greece: IEEE.

Kong, L., Ye, L., Wu, F., Tao, M., Chen, G., and Vasilakos, A. V. (2017). Autonomous relay for millimeter-wave wireless communications. *IEEE Journal on Selected Areas in Communications*, 35(9), 2127–2136.

Li, B., Fei, Z., and Zhang, Y. (2019). UAV communications for 5G and beyond: Recent advances and future trends. *IEEE Internet of Things Journal*, 6(2), 2241–2263.

Liu, Y., Qiu, M., Hu, J., and Yu, H. (2020). Incentive UAV-enabled mobile edge computing based on microwave power transmission. *IEEE Access*, 8, 28584–28593.

Mehta, P. L. and Prasad, R. (2017). Aerial-heterogeneous network: A case study analysis on the network performance under heavy user accumulations. *Wireless Personal Communications*, 96(3), 3765–3784.

Mehta, P. L. and Prasad, R. (2019). Distributed dynamic backhauling in aerial heterogeneous networks. *Wireless Personal Communications*, 109(1), 621–643.

Mozaffari, M., Saad, W., Bennis, M., and Debbah, M. (2016). Unmanned aerial vehicle with underlaid device-to-device communications: Performance and tradeoffs. *IEEE Transactions on Wireless Communications*, 15(6), 3949–3963.

Mozaffari, M., Kasgari, A. T. Z., Saad, W., Bennis, M., and Debbah, M. (2019a). Beyond 5G with UAVs: Foundations of a 3D wireless cellular network. *IEEE Transactions on Wireless Communications*, 18(1), 357–372.

Mozaffari, M., Saad, W., Bennis, M., Nam, Y., and Debbah, M. (2019b). A tutorial on UAVs for wireless networks: Applications, challenges, and open problems. *IEEE Communications Surveys & Tutorials*, 21(3), 2334–2360.

Nam, Y.-H., Ng, B. L., Sayana, K., Li, Y., Zhang, J., Kim, Y., and Lee, J. (2013). Full-dimension MIMO (FD-MIMO) for next generation cellular technology. *IEEE Communications Magazine*, 51(6), 172–179.

Nomikos, N., Michailidis, E. T., Trakadas, P., Vouyioukas, D., Karl, H., Martrat, J., Zahariadis, T., Papadopoulos, K., and Voliotis, S. (2020). A UAV- based moving 5G RAN for massive connectiv-ity of mobile users and IoT devices. Vehicular Communications, In press.

Oubbati, O. S., Lakas, A., Zhou, F., Güneş, M., and Yagoubi, M.B. (2017). A survey on position- based routing protocols for flying ad hoc networks (FANETs). *Vehicular Communications*, 10, 29–56. Saleem, Y., Rehmani, M. H., and Zeadally, S. (2015). Integration of cognitive radio technology with unmanned aerial vehicles: Issues, opportunities, and future research challenges. *Journal of Network and Computer Applications*, 50, 15–31.

Sboui, L., Ghazzai, H., Rezki, Z., and Alouini, M.-S. (2017). Energy- efficient power allocation for UAV cognitive radio systems. *Proceedings of the 86th Vehicular Technology Conference (VTC-Fall)* (pp. 1–5), Toronto, Canada: IEEE.

Sekander, S., Tabassum, H., and Hossain, E. (2018). Multi-tier drone architecture for 5G/B5G cellular networks: Challenges, trends, and prospects. *IEEE Communications Magazine*, 56(3), 96–103.

Strassner, B. and Chang, K. (2013). Microwave power transmission: Historical milestones and system components. *Proceedings of the IEEE*, 101(6), 1379–1396.

Wang, H., Wang, J., Ding, G., Chen, J., Li, Y., and Han, Z. (2018). Spectrum sharing planning for full-duplex UAV relaying systems with underlaid D2D communications. *IEEE Journal on Selected Areas in Communications*, 36(9), 1986–1999.

Wua, G., Miao, Y., Zhang, Y., and Barnawi, A. (2020). Energy efficient for UAV-enabled mobile edge computing networks: Intelligent task prediction and offloading. *Computer Communications*, 150, 556–562.

Xiao, Z., Xia, P., and Xia, X. (2016). Enabling UAV cellular with millimeter-wave communication: Potentials and approaches. *IEEE Communications Magazine*, 54(5), 66–73.

Zhang, C. and Zhang, W. (2017). Spectrum sharing for drone networks. *IEEE Journal on Selected Areas in Communications*, 35(1), 136–144.

Zhang, L., Zhao, H., Hou, S., Zhao, Z., Xu, H., Wu, X., Wu, Q., and Zhang, R. (2019a). A survey on 5G millimeter wave communications for UAV-assisted wireless networks. *IEEE Access*, 7, 117460–117504.

Zhang, X., Zhong, Y., Liu, P., Zhou, F., and Wang, Y. (2019b). Resource allocation for a UAV-enabled mobile-edge computing system: Computation efficiency maximization. *IEEE Access*, 7, 113345–113354.

Zhao, Z., Cumino, P., Souza, A., Rosário, D., Braun, T., Cerqueira, E., and Gerla, M. (2019). Software-defined unmanned aerial vehicles networking for video dissemination services. *Ad Hoc Networks*, 83, 68–77.

Section III

Deployment Scenarios

11

Integration of Multienergy Systems into Smart Cities: Opportunities and Challenges for Implementation on 5G-Based Infrastructure

Vladimir Terzija and Dragan Cetenovic
University of Manchester

Dragorad A. Milovanovic and Zoran S. Bojkovic
University of Belgrade

CONTENTS

11.1 Introduction

The concept of a smart city (SC) is about connecting the physical infrastructure, the information and communication technologies (ICTs) and the social infrastructure. Smart solutions can provide various kinds of services, including smart grid (SG) [1–4]. The continuous need for cost-effectiveness, and high level of reliability and security, has pushed toward new strategies inspired from research fields of fifth-generation (5G) wireless communication and Internet of things (IoT). SG services are transforming cities by improving an integrated multienergy infrastructure [5,6]. Multienergy systems (MES) whereby electricity, heat, gas, hydrogen, cooling, fuels, transport and so on should optimally interact with each other at city level in order to increase technical, economic and environmental performance relative to energy systems (ESs) whose sectors are treated separately, leading to suboptimal planning, operation and maintenance. Accurate MES modeling is a complex task, due to different characteristics and time scales of the phenomena involved, the increase in the solution space awarded by the multinetwork layout and the energy conversion and storage possibilities [7,8].

The third-generation (3G) power grid was launched at the beginning of the 21st century, featured by centralized intelligence and efficient use of renewable energy sources (RESs). The electricity delivery infrastructure consists of power plants, transmission grids and distribution systems. The SG is a new concept intending to include bidirectional communication infrastructure to conventional power grids in order to enable ICTs at any stage of generation, transmission, distribution and even consumption sections of utility grids [9], and consequently to improve the efficacy of the operation of the entire power system. The SG infrastructure is implemented to provide a data communication medium in order to carry several signals for measurement, monitoring, management and control purposes. The essence of a SG is real-time situational awareness, traditionally considered as monitoring real-time control, which requires two-way communications. Fast, secure and reliable two-way communication between different parts of the grid is an integral part of SG. Innovations in sensing and measurement are fundamental for the realization of a more aware and reliable SG. This objective can only be achieved by radically upgrading the sensing, measurement and metering infrastructure throughout the grid. The most recent improvements in sensor technologies have advanced smart sensors used in SG applications. The widespread regulations and advances in energy policies required several developments in power networks and wide area monitoring, protection and control (WAMPAC) issues [10,11]. The smart metering and measurement devices that are advanced with recent developments in SG applications are phasor measurement units (PMUs), intelligent electronic devices (IEDs) and smart meters. PMUs play the key role in monitoring long transmission lines and other system components, functional integration between transmission and distribution networks, and also design of advanced control and protection concepts, supporting integration of RESs into the existing power grid. Integration of widely deployed PMUs with communication and advanced computations will also help in monitoring the state of power grid in a better way and closer to the real time, also called *just in time*. Improved monitoring of the system state makes it more aware of its actual operating state. IEDs are used to detect, selectively eliminate and locate faults, to undertake event recording, measurement, substation control and automation, as well as to provide SCADA/microSCADA type of services in power networks. The smart meters, namely watt-hour meters, are based on voltage and current measurements due to related sensor networks. Smart metering application allows a distribution system operator (DSO) to remotely read the energy consumption of its customers to detect energy theft and to perform some operations on the meter. Information obtained from smart meters can be further used in applications related to forecasting system behavior.

IoT concept considers a massive physical infrastructure in which different types of devices and smart environments such as SGs and SCs are connected to a single cyber-physical system. In order to accomplish such a concept, there is a high bandwidth requirement to be handled, which can be provided by developing and implementing 5G wireless networks. These are some issues needed to be overcome in the current situation of IoT systems. These problems can be classified as automatic sensor configuration, system management modeling, context sharing and security issues. Since the IoT system covers massive, distributed and heterogeneous elements, realization of this platform is very complicated. In addition, cloud systems that provide storage, networking and computing facilities can be combined with the IoT devices. It is also foreseen that the IoT concept will progressively convert the present Internet platform into the machine-to-machine (M2M) communication concept enabled with 5G communication networks.

The development of wireless network generations is mainly affected by progress on wireless devices, higher data rate demand and better system performance expectations. Recently, the 5G mobile wireless systems is achieved the first milestones toward finalization and deployment by 2020. 5G networks will have a significant influence on the development of the SC infrastructure. For example, 5G networks provide higher and faster transmission capabilities, larger bandwidth and higher data capacity. With the introduction of 5G, we are going to combine network slicing, cloud radio access network, various cell sizes and different kinds of supported traffic (M2M). In summary, there is a symbiotic relationship between two of the modes in 5G, massive machine-type communication (mMTC), ultrareliable and low-latency communication (uRLLC) and the SG communication, since the former provides the technology and the latter represents one of the most important use cases that encompass both mMTC and uRLLC. The mentioned advancements will directly contribute to development of novel applications relying on novel sensing and communication technologies, supporting optimal operation of future SCs and power grids in general terms.

11.2 Smart City Infrastructure

The SC refers to a new deployment of information and communication technologies in the combination and integration with urban functions. The term can also be described as the convergence of ICT and support facilities within urban and residential environments. SC evolution includes 20-year development of traditional applications (power grid, traffic, healthcare). SC started to grow as a new feasible and sustainable concept in the 1990s, based on available computing power, introduction of web services and mobile radio networks. It became clear that it was possible to connect and integrate remote devices with processing platform in almost any place and at any time. People and devices become fully connected in the late 2000s based on broadband mobility, initial steps in cloud computing (CC) and sensors in traffic control, smart meters and navigation. Smart applications have become everyone's need (smart home, SG, smart healthcare, smart transport, smart government) [12–14].

Today, almost all cities intend to be more or less *smart* with regard to a different level of new technologies. Consideration will be completed when monitoring and integration of the critical infrastructure, optimization of resources, planning maintenance activities, monitoring of security as well as maximizing services to the citizens are included. The infrastructure consists of the following parts:

- physical (power, water, roads, bridges, rails, airports, seaports) and ICT infrastructure (telecom networks, data centers, access points, terminal devices, available services) and
- social (hospitals, media, social clubs, social networks) and business infrastructure (companies, campuses).

ICT infrastructure has a dominant status in SC. The following four technological layers of SC architecture are subject to changes and transformation:

- sensing layer (sensor or terminal devices),
- network layer (telecommunication, media and associated services),

- operation, administration, maintenance and provisioning, security and
- application layer.

The challenge of sustainable city development and its improvement to become smart is a large amount of effective and efficient investment in infrastructure. There are numerous initiatives focused on the analysis of process conception, implementation methods and SC outcomes (see Table 11.1). Understanding the advantages and disadvantages is necessary not only for planners, managers and project directors but also for people living in SCs.

The energy infrastructure is arguably the single most important feature in any city. If unavailable for a significant enough period of time, all other functions will eventually stop. SG modernizes power systems through self-healing designs, automation, remote monitoring and control and establishment of microgrids. It informs and educates consumers about their energy usage, costs and alternative options to enable them to make decisions autonomously about how and when to use electricity and fuels. SG provides safe, secure and reliable integration of distributed and renewable energy resources. All these add up to an energy infrastructure that is more reliable, more sustainable and more resilient.

The SG is an electrical network that integrates the actions to efficiently deliver sustainable, economic and secure electricity supplies. An SG employs innovative products as services together with monitoring, control, communication and self-healing technologies. The main goals when introducing SGs are as follows:

- accommodate a wide variety of distributed generation and storage systems in order to significantly reduce any impact of the environment, increasing at the same time the efficiency of the electricity supply chain,
- provide consumers with relevant information so that they can make their choices and provide services such as integration of electrical vehicles,
- improve existing standards of reliability and quality of supply,

TABLE 11.1

Sectors and Core Issues for ICTs in a Smart City

Sectors	Key Services	Technologies
Energy and electricity	Automation of transmission and distribution Optimization, management and reduction of accommodated energy	• Various communication standards • Essential part of smart grid systems • One of the biggest potential markets in IoT technologies
Architecture and building	Building management Building automation Home automation	• Different building preferences • IT solution or telecom service providers with newly built building
Automation and transportation	Remote parking management Business fleet management Vehicle telematics	• Autonomous or remote control services • Management systems using IoT technologies • Utilizing individual mobile devices and distributing connected automobiles
Healthcare and monitoring	Smart healthcare Smart hospital	• Rapidly expanded markets • Tracking medical teams and facilities located in hospitals • Electronic medical records and communication systems

ICT, information and communication technology; IoT, Internet of things.

- allow new business models and
- help optimize the utilization and maintenance of assets in an SC.

Modern sustainable SC energy infrastructures generally are integrated multifueled ESs that incorporate energy efficiency, renewable energy and demand-side management. Specific technologies and practices utilized in today's sustainable urban ESs include energy-efficient buildings and transportation systems, district energy (heating, cooling) systems, distributed ESs and SGs (Figure 11.1).

With the current development of mobile ICTs, the 5G wireless communication and IoT application scenarios defined by the 3GPP (3rd Generation Partnership Project) provide three communication modes: enhanced mobile broadband (eMBB), mMTC and uRLLC. However, the heterogeneity of IoT applications and devices leads to higher requirements for low cost, energy efficiency, wide coverage, low delay and reliable communication. To satisfy complex and diverse user computing tasks and content requests, operators deploy CC technology to overcome the limitations of storage and computing in IoT devices. Furthermore, the deployment of artificial intelligence (AI) technology in the remote cloud and edge cloud is an effective way of implementing intelligent services. Characteristics of 5G technology for city benefits are as follows:

- 5G technology provides better coverage and performance in open spaces and buildings (densely populated city areas, stadiums and public transport).
- Reliable high data flows enable consumers to download high-resolution multimedia and city offices can download high-resolution videos.
- 5G represents an alternative to fiber optics, enabling the wireless network to reliably cover areas where today's price is too high for fiber optic applications.
- Future communications networks are adaptive with programming capability for best support of applications.
- Energy efficiency of the network is high so that the battery life of the IoT devices of low power reaches 10 years, reducing the cost of maintenance and replacement of batteries.

FIGURE 11.1
Specific technologies utilized in SC infrastructure. ICT, information and communication technology; SC, smart city.

- Reducing the response time or delay in the 5G network at 1 ms, combination of real-time communication with streaming and capacity for multimedia can become a reality.

- Combining wireless networks deals with integration gateways, licensed or unlicensed networks to be managed as a single network.

- When speaking about quality of experience, 5G culminates into great reliability, while the overall experience for the citizens is improved.

11.2.1 New Concept of Multienergy System

Energy networks exist primarily to exploit and facilitate temporal and spatial diversity in energy production and use and to exploit economies of scale where they exist. The energy trilemma (energy security, environmental impact and social cost) presents many complex interconnected challenges. These challenges vary considerably from region to region due to historical, geographic, political, economic and cultural reasons. As technology and society changes so do these challenges, and therefore, the planning, design and operation of energy networks need to be revisited and optimized. Current energy networks research does not fully embrace a whole systems approach and is therefore not developing a deep enough understanding of the interconnected and interdependent nature of energy network infrastructure. The energy networks community would strongly benefit from a more diverse, open, supportive community with representation from many disciplines beyond traditional engineering (such as computing science, statistics, anthropology, geography, economics and applied mathematics) to help implement a whole systems approach.

Traditionally, energy sectors have been decoupled from both operational and planning viewpoints, whereas tight interactions have always taken place and are increasing. For instance, electricity, heat/cooling and gas networks interact in many cases through various distributed technologies such as CHP (combined heat and power), electric heat pumps, circulation pumps, gas-fired boilers, gas turbine generators, air conditioning devices, trigeneration or combined cooling, heat and power and so on. Similarly, interactions between electricity, the fuel chain and the transport sector are more and more envisaged or already taking place by means of EV (electric vehicles) and biofuels and hydrogen-based transport. In this outlook, a key aspect to evolve toward a cleaner and affordable ES is to better understand and develop integrated energy systems (IESs) or MESs, whereby electricity, heat, cooling, fuels, hydrogen and transport optimally interact with each other at various levels (for instance, within a district, or a city, or at a country level). Electrification of heating and transport and the need to support it through SG options and in case development of suitable distributed energy markets are a tangible example of the need for developing an MES framework [7,8].

IES/MES offer better perspectives for achieving a sustainable energy supply and supporting the energy transition than traditional mono-ES approaches. Such systems consider all relevant energy carriers (electricity, gas, heating, cooling and hydrogen) and services (lighting, cooking and transportation) and consequently create increased degrees of freedom. Because the different energy carriers have different characteristics and applications, the entire solution space becomes larger, increasing the chance of finding better solutions. For example, electricity is quite flexible in its end-use applications and easily produced sustainably, but it is difficult to store. Heat is more easily stored but can only be used for low-grade energy applications. Gas is easiest to store, especially long-term, but (as of yet) difficult to produce in large quantities in a renewable fashion. In other words, short-term carriers can be converted into long(er) term carriers, and carriers can be adjusted based on the required energy service. Especially for a system in transition; an IES can smoothen the path from mostly fossil to mostly renewable.

Accurate MES modeling is a complex task, due to the different characteristics and time scales of the phenomena involved, and the increase in the solution space awarded by the multinetwork layout and the energy conversion and storage possibilities.

- The electricity network at the described geographical scope spans from the high-voltage (HV) network all the way down to the low voltage (LV) levels. Incorporating all of this in an integrated fashion is computationally infeasible. As such, the model is simplified and proposes to merely consider the medium-voltage (MV) (distribution) network: from the HV/MV stations to the MV/LV substations.

- The district heating network is regarded at the same level as the electricity network. Generally, this means a combination of high temperature heat and lower temperature heat; however, it differs greatly per city. It will be modeled as a single network and can include connections to any of the demand sectors specified.

- Similarly, the gas network scope corresponds to the other two network scopes. The distribution network level of natural gas is often at "medium" pressure (corresponding to MV levels).

To set up an integrated multienergy network model and determine the potential of future ES integration at a city level, generally three questions arise:

- Scenario development: given projected demand, relating technology penetration rates (EVs, heat pumps, etc.) and climate goals, what supply mix would be required and feasible?

- System design and operational optimization: how much and which type of conversion and storage would be required and how would it be operated?

- Investment optimization/planning strategy: what is the optimal path toward that optimal system, starting with today's ES and taking into account the prescribed or desired time horizon?

Within the city, many different demand sectors can be defined in three main categories: residential, commercial and industrial demand.

For residential and commercial demand, many different profiles are publicly available. Industrial demand profiles are more challenging, given the heterogeneous nature of industrial processes. Statistical methods could be used to generate profiles using limited data available, for example, as was done here for residential profiles.

The energy supply mix follows from future scenarios specific for a particular city, combined with its climate goals. Within the energy supply mix, there is a difference between intermittent (renewable) supply and controllable (renewable and nonrenewable) supply. The first type is considered given and can be modeled using wind and solar profiles and total installed capacity. If desired, this could be further specified by locations and orientations. The second type can be adjusted based on the needs of the ES. These would generally be defined by capacity limits and ramp rates. In addition to energy supply from within the system, it could be possible that certain energy imports (or exports) are still required.

To enable an IES, energy conversion technologies are required. Currently, several technologies exist which convert one energy carrier to another possible within the energy networks:

- gas to electricity and heat (via a CHP plant),
- electricity to heat (via a heat pump),
- electricity to gas (via power-to-gas technologies).

For all three energy carriers, the main difference between the storage options is the time frame in which they can operate. Electricity storage and battery ES are most effective at a short term (hours to days). Heat (and cold) storage and aquifer thermal energy storage can also be used effectively between seasons. Storing natural gas is also effective at a long term (years to multiple years). Besides the time frame, the effective size of the storage options also differs. For example, batteries can be applied at a small scale (within an EV) or at quite a large scale (provide flexibility for network operators).

However, disturbances of external factors such as nature, social and economic environment may affect the operation of ESs. Conversely, the security and the stability of ES will affect the external environments as well. Therefore, the interactions between ES and external environments should be taken into consideration and studied at the context of macroenergy system (Macro-ES). The macroenergy thinking, regarding electricity as a hub between energy production and consumption, breaks down the physical barriers among power system, primary ES and end-use ES. The big data thinking regards various data resources as fundamental elements of production rather than simple process objects. The integration of the macroenergy thinking and the big data thinking will make the big data on power become the foundation of an extensively interconnected, openly interactive and highly intelligent Macro-ES. Key elements of this integration include the acquirement, transmission and storage of wide area power data with different timescales, the data from related domains, as well as the fast and in-depth knowledge extraction from the multi-source heterogeneous data and its applications.

11.2.2 Research and Standardization in Energy Informatics

Physical layer of SC energy infrastructure can be abstracted to a logical model including network, source, load, storage and conversion elements (Figure 11.2). Network includes electrical network, gas or hydrogen network, heating network, cooling network and communication network. Source node includes power plant such as wind power plant, solar power plant, heat exchange station and so on. Load node includes industrial, building, commercial, residential load, and so on. Storage includes electricity storage, heat storage, and so on. Conversion includes heat pumps and electric boilers. Planning and operations across interdependent domains are necessary to achieve the most efficient, flexible and reliable ES.

ESs present new requirements for interconnection at the physical energy layer. Learning from the successful concepts and technologies pioneered by the Internet and establishing an open interconnection protocol are basic methods. Internet of energy is an integration of energy infrastructure and advanced Internet technologies. Both the Internet and the electrical grid are designed to meet fundamental needs, for information and for energy, respectively, by connecting geographically dispersed suppliers with geographically dispersed consumers. Similarities and differences between the Internet and the electrical

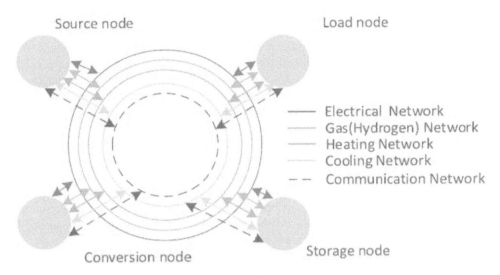

Source node

Load node

—— Electrical Network
——— Gas(Hydrogen) Network
—— Heating Network
········ Cooling Network
— — Communication Network

Conversion node

Storage node

FIGURE 11.2
Interoperability of smart energy physical layer.

grid are identified, and several specific aspects where Internet concepts and technologies can contribute to the development of an SG are proposed.

- Communication network is essential to support the bidirectional flow of information. Based on the data rate, transmission range and operation domain, communication networks in SG can be classified as customer premises area network (home area network, building area network, industrial area network), neighborhood area networks, field area network and wide area network. Through the decoupling of data and control planes enabled by software-defined networking (SDN), a virtual network layer can be built on the top of physical infrastructures, which allows multiple tenants and applications to reuse the same communication infrastructure without being tied to hardware details.

- The large number of programming languages and modeling formalisms developed in the past years are currently spread among several disciplines and communities such as hybrid systems and control, AI-based planning and scheduling, embedded systems design and verification and constraint solving and optimization. However, the rapidly developing computer science technologies are not tailored so far for applications within the energy domain. It is essential to build appropriate simulation frameworks suitable for energy grids and their components, including prosumer behavior. Furthermore, computing algorithms for autonomous and distributed power generation management while incorporating self-healing capabilities, security aspects, and so on are urgently required.

- Ubiquitous information exchange and better awareness of energy availability and load profiles are essential to realize the next-generation ES. First of all, the incorporation of new technologies from different vendors raises the concern of interoperability. Secondly, how to efficiently utilize the big data related to every aspect of ES for performance improvement remains unclear. Thirdly, data security and privacy pose a new challenge since the leakage of the detailed energy usage information recorded in near real-time increases the tension between data access and data privacy.

The development of novel theoretical models and standards is challenging yet exciting research direction of energy informatics (EI) [15,16]. Current IEC (International Electrotechnical Commission) standard defines SG as an electric power system that utilizes information exchange and control technologies, distributed computing and associated sensors and actuators for purposes such as to integrate the behavior and actions of the network users and other stakeholders or to efficiently deliver sustainable, economic and secure electricity supplies. The SG architecture model proposed by the European Union is 3D framework described in Figure 11.3.

The x-axis is composed of five domains including generation, transmission, distribution, distributed energy resource and customer premises. The y-axis described in IEC standard IEC 62 357 has six zones including process, field, station, operation, enterprise and market. The z-axis described in IEC 62 264 is composed of five interoperability layers representing business objectives and processes, functions, information exchange and models, communication protocols and components. Each interoperability layer spans the electrical domains and information management zones.

Current EI research is comprised of a number of interdependent streams of research (Figure 11.4). Most of it focuses on the potential of ICT to realize either smart energy-saving systems or SGs. In the future, we expect to see more EI research in two major areas [17,18]:

- explicit quantification of the trade-off between ICT deployment and the achieved benefits in economic and environmental terms and

- development of more comprehensive ICT solutions and their evaluation based on realistic simulations of cyberphysical systems.

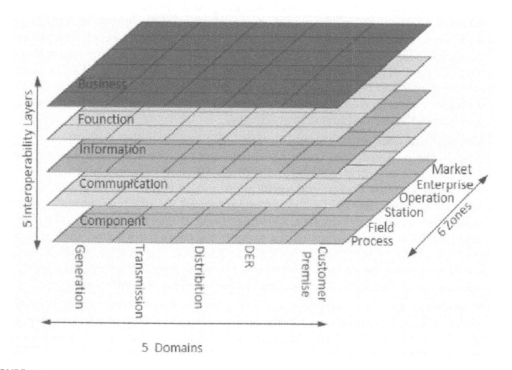

FIGURE 11.3
IEC smart grid architecture model. DER, distributed energy resource; IEC, International Electrotechnical Commission.

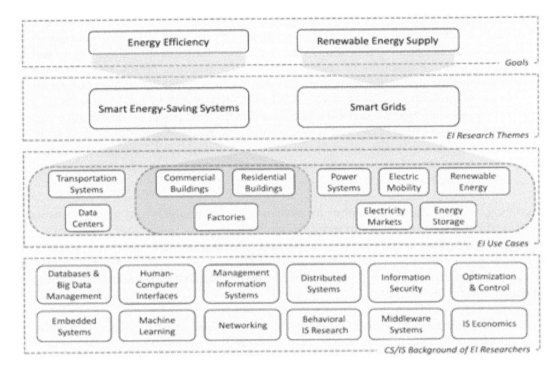

FIGURE 11.4
Scope of energy informatics (EI) research.

- Most of the EI research that has been conducted in the area of SGs so far deals with the role of ICT in demand response (DR), usually with a focus on certain types of electric loads. Since physical SG test beds are not accessible for most EI researchers, they focus on creating new knowledge by applying a model-based research method. Compared with other IS/CS (information systems/computer science) research fields, a lot of data are publicly available, for instance, data about electricity demand, market outcomes and prices, renewable energy production, and so on. Even well-studied optimization problems in the power system context, such as unit commitment and optimal power flow, require significant computational resources. Therefore, EI researchers have begun to investigate the application of distributed optimization techniques to SG control problems.

- Apart from solving optimization problems, EI researchers are also interested in the design of information system infrastructure capable of processing data received from potentially millions of physical resources in real time. If SGs are to be become a reality, the reliability and security of its ICT infrastructure will be vital. A highly active EI use case in the recent past has been the potential of plug-in EVs for grid support. Another use case that EI researchers have begun to explore is the distributed control of loads with intrinsic thermal energy storage. These loads are able to store energy in the form of heat or cold and can be powered electrically. Therefore, the power consumption of loads such as HVAC (heating, ventilation and air conditioning) systems, water heaters and refrigerators can be controlled within certain bounds.

- EI research focusing on smart energy-saving systems has already addressed a large spectrum of related research questions. On the first ICT integration level, a number of recent publications deal with the challenge to collect, store and exchange large amounts of energy-related data. Another highly relevant topic related to the first ICT integration level that has received a lot of attention so far is the identification of small-scale loads based on aggregated measurements. The availability of effective software solutions serving this purpose would allow for obtaining the status of devices as well as highly granular consumption data without having to separately measure the power use of each load. Significant research effort went into designing and evaluating technologies for visualizing energy usage at the end-user level.

Modeling and simulation algorithm for system behavior analysis and energy consumption in SG environment are important research challenges. Since system behavior and energy consumption constitute very important information, it is necessary for modeling various SG scenarios in an efficient and realistic way. Research projects pay effort to develop and mature the next generation of competitive technologies and services. Topics in this domain are different:

- technologies for the storage of energy in the distribution network and their integration and exploitation in the SG context, including decentralized storage at user premises or at substation level,
- synergies between energy networks, tools and technology validation for DR forecast, profiling, segmentation, load forecasting, innovative and user-friendly services for customers based on smart metering,
- tools for the optimization of the distribution grid,
- technologies for autonomous and self-healing grids, energy management and control systems,
- technologies for advanced power electronics and for enhanced observability, e.g., real-time situational awareness,
- secured communications in the SG in particular cybersecurity and big data analytics.

11.3 5G Internet of Things–Based Infrastructure

SC is a complex ecosystem characterized by the intensive use of ICT, with the aim of making cities more sustainable as a unique place for innovation. The main participants are application developers, service providers, citizens, government and public services, the research community and development platform. The innovation cycle consists of numerous technologies, development platforms, maintenance and sustainability, citizen applications as well as technical, social and economic performance indicators. IoT systems have a fundamental role in the implementation of mass heterogeneous infrastructures. The development of 5G wireless networks enables E-IoT (energy IoT), G-IoT (green IoT) and I-IoT (intelligent IoT) technology for SC.

Intelligent applications based on IoT infrastructure are categorized by the type of network, scalability, coverage, flexibility, heterogeneity, repeatability and end-user engagement. Applications can be grouped into personal and household, communal, mobile and business. Utilities include smart energy grid, smart metering/monitoring, water network monitoring and video surveillance.

IoT applications in an SC enforce many demands. For example, IoT is expected to offer low prices, low energy consumption, high quality of service (QoS), broad coverage, greater flexibility, high security and privacy, ultradense implementation and interoperability of multiple manufacturers. IoT deployment is based on a number of network topologies to achieve a completely autonomous environment. Capillary IoT networks offer short distance services. Examples include wireless local area network (WLAN), body area network (BAN) and wireless personal area network (WPAN) wireless networks. Fields of application include e-health services, home automation and street lighting. On the other hand, applications such as intelligent transport system, mobile e-health and waste management use WAN (wide area network), MAN (metropolitan area network) and mobile communication networks. These networks have different characteristics in terms of data, size, coverage, latency and capacity requirements. In such an environment, IoT systems have got a huge role in the deployment of large-scale infrastructures. In addition, the integration of IoT and big data has invoked new challenges such as processing and storage for future smart cities. The objective is very clear, to promote better management of various sectors.

The energy infrastructure is the most important feature in any city. Smart cities depend on an SG to ensure resilient delivery of energy to supply their many functions, improve efficiencies and enable coordination between infrastructure operators. Advanced metering infrastructures (AMIs) and advanced data analytics gather, assess and formulate essential information to improve operational decision-making. For improving energy efficiency and saving energy cost, intelligent management of energy allocation and consumption (smart metering, smart lighting and SG) has lots of benefits. In that way, the application of RESs (solar and wind energy) is promoted.

Recently, it has been proved that applying IoT can optimize the management of the renewable energy resources by enabling sensors and data analytics. Indeed, the renewable ESs based on IoT can facilitate the communication between the renewable energy appliances to optimize the energy efficiency. For example, the energy regeneration will be switched from solar to wind turbine in a particular cloudy day with a lot of wind. On the other side, IoT analytics have a significant effect on renewable energy outperforming. For instance, extracting data from solar panels may help to speed up the maintenance process. Many new ideas and research results in these fields continue to emerge.

With introduction of the 5G technology, mobile cellular networks are being developed into SG platform for acquisition, communication, storage and processing. At the same time, 5G will provide on its side, services for real-time applications in SGs. Mobile edge computing (MEC) and localized processing and management are dramatically decreasing service response latency. The concept is not in collision with mobile CC; they complement each other in building flexible and configurable 5G network using a network slicing approach, where different services may be easily instantiated using different virtualized architectures on top of the high-performance MEC host nodes. Instrumental to the development of flexible core network are reconfigurable architectures based on SDN (SDN) and network function virtualization.

In a networking architecture scenario for IoT-based SC, it is all about integration of technology with sensing networking, information processing and control in real time in order to efficiently utilize public infrastructure.

11.3.1 Challenges and Opportunities in Internet of Things for Energy Applications

Internet of Things for energy applications (E-IoT) will become ubiquitous and enable new automated energy management platforms [19]. There are several benefits of deploying IoT in intelligent electric power systems, such as [20]:

- enhanced reliability, resiliency, adaptability and energy efficiency,
- reduced number of communication protocols,
- networked operation and enhanced information operation capabilities,
- improved control over home appliances,
- enable on-demand information access and end-to-end service provisioning,
- improved sensing capabilities,
- enhanced scalability and interoperability,
- reduced damages from natural disasters and physical attacks.

E-IoT is not without its challenges:

- most E-IoT devices will have and/or require an Internet connection,
- E-IoT devices need to work together to perform different functions across the electric supply and demand value chain,
- new market participants such as aggregators, prosumers, DERs (distributed energy resources) and microgrids will emerge,
- A large quantity of data will be generated and stored or processed in real time.

A mechanism to store, manage and secure the data collected in real time is necessary to protect the interests of all in SC. Although the convergence of the cyber, physical and economic aspects of grid operations poses a challenge, it provides an opportunity for collaboration across various layers of the electricity grid.

Sensing technology plays an indispensable role in providing situational awareness within an E-IoT. As such, sensors exist at the periphery of a communication network to relay data and information from the physical grid to a control or decision-making center. Given the tremendous heterogeneity in the number, type and input of physical E-IoT devices, the measurement role of network-enabled sensing technologies increases immensely. Fortunately, there has been significant innovation in sensing technologies to accommodate these needs. Such advancements include miniaturization, wireless data transfer and decreasing implementation costs.

The most recent improvements in sensor technologies have advanced smart sensors used in SG applications. The widespread regulations and advances in energy policies require several developments in power networks and WAMPAC issues. The smart metering and measurement devices that are advanced with recent developments in SG applications are PMUs, IEDs and smart meters. The IEDs are used to detect faults, protection relaying, event recording, measurement, control and automation aims in power network. The PMUs play a quite important role to monitor long transmission lines and integration to distribution networks. A sinusoidal voltage or current waveform is generally represented by a complex number called phasor having magnitude and angle parts. When the measurement of phasors in a system is synchronized with the help of a precise time stamping technique, these phasors are referred to as synchrophasors. Integration of widely deployed PMUs with communication and advanced computations will help in monitoring the state of power grid in a better way. Better monitoring of state of the grid will make it more aware of its operating state. PMU is an electronic device that measures AC waveforms to provide the measurement of phasors. The phasor estimation is implemented by digital signal processing techniques taking into account the system frequency. These measurements are synchronized through global positioning system (GPS). The measured signals are sampled and processed by a recursive phasor algorithm to generate voltage and current phasors. Frequency is one of the most significant parameters in the SG systems. Thus, accurate frequency estimation becomes an essential task for monitoring, controlling and protecting a real-time SG system. The smart meters, namely watt-hour meters, are based on voltage and current measurements due to related sensor networks. The measured and conditioned signals are converted to digital signals to generate measurement data that are used to be stored and transmitted to remote monitoring and operation centers. Smart metering application allows a DSO to remotely read the energy consumption of its customers, which is usually referred to as automatic meter reading, to obtain power quality information, to detect energy theft (usually referred to as meter tampering) and to perform some operations on the meter (e.g., connection/disconnection, meter programming). Smart metering requires bidirectional communication links to the meters [21–23].

The need for situational awareness also motivated the development of sensor networks. These sensing challenges in the transmission system have motivated the deployment of PMUs (synchrophasors) [11]. Phasor measurements provide a dynamic perspective of the grid's operations because their faster sampling rates help capture dynamic system behavior. PMUs measure voltage and current phasors and can calculate watts, vars, frequency and phase angles 120 times per power line cycle. PMU data immediately enhance topology error correction, state estimation (SE) for robustness and accuracy, faster solution convergence and enhanced observability. Simulations and field experiences also suggest that PMUs can drastically improve the way the power system is monitored and controlled. However, a completely observable system requires a large number of PMUs, which is still very expensive ($40,000–$180,000 per PMU installation [24]). This is why utilities usually install PMUs incrementally.

Recent studies have explored algorithms for optimal placement of PMUs to minimize the number of PMUs required to collect sufficient information. PMU-based wide area monitoring systems (WAMSs) use the global position system (GPS) to synchronize PMU measurements. Such synchronized measurements allow two quantities to be compared in the real-time analysis of grid conditions. Through wide-area monitoring and synchronization, PMUs have made great strides in power system stability, which was often hindered by SCADA's slow state updates. The implementation of synchrophasors has also allowed voltage and current data from diverse locations to be accurately time-stamped in order to assess system conditions in real time. Synchrophasors are also available in protection devices, but since requirements for protection devices are fairly restrictive, the full integration of synchrophasors into line protection is still debated. The increasing application of synchrophasors in wide area monitoring, protection and control systems, postdisturbance analyses and system model validation has made these measurement tools invaluable.

While the integration of PMUs into the transmission system will do much to enhance situational awareness in the transmission system, it is by no means sufficient for the grid as a whole. First, PMUs are primarily meant for applications in the transmission system and to a large extent are not feasible in the distribution system. They are even less appropriate for understanding customers' power consumption profiles. In that regard, the emergence of smart meters has fulfilled a much-needed functionality. Second, PMUs only measure voltage and current phasors. As such, they are able to provide much-needed insights into grid conditions but are not able to inform why these conditions exist. As the electric grid comes to depend more on interdependent infrastructure, weather conditions and consumers' dynamic behavior, secondary measurements of these quantities become increasingly important. In that regard, sensors used in other sectors will have an indispensable role in taking secondary measurements.

SE is a key functionality of the electric power grid's energy management systems. SE aims to provide an estimate of the system state variables (voltage magnitude and angles) at all the buses of the electrical network from a set of remotely acquired measurements. The centralized (classical) SE schemes may prove inapplicable to emerging decentralized and dynamic power grids, due to large communication delays and high computational complexity that compromise their ability for real-time operation. Hence, one interest of the community is shifting from centralized to distributed SE algorithms based on more sophisticated optimization techniques beyond the classical weighted least square approaches. In addition to this goal, novel PMU-based approaches and those involving stochastic description of the monitoring processes, as well as robust and outliers insensitive approaches, are proposed in Refs. [25–32].

Exploiting PMU inputs for a robust, decentralized and real-time SE solution calls for novel communication infrastructure that would support the future WAMS. WAMS aims to detect and counteract power grid disturbances in real time, thus requiring a communication infrastructure able to

- integrate PMU devices with extreme reliability and ultralow (millisecond) latency,
- provide support for distributed and real-time computation architecture for future SE algorithms,
- provide backward compatibility to legacy measurements traditionally collected by supervisory control and data acquisition (SCADA) systems.

In the example in the following, illustrated through Figure 11.5, a typical example of a wide area monitoring is given. The figure presents a segment of the Great Britain (GB) network, in which the capacity transfer (of active and reactive powers) over the existing transmission network had to be monitored. For this purpose, PMUs were installed in key power system substations at the 400-kV voltage level. These PMUs provided information about voltage and current phasors to the central data concentrator, at which information obtained was processed. Firstly, the system state was estimated, and after that, the capacity margin, taking into account the weather conditions and information received from the Meteorological Office, has been calculated. The entire process is visualized, as presented in Figure 11.5 and provided available on different types of media for visualization.

Introduction of nonsynchronous renewable energy resources to modern power systems is causing a number of operational problems related to, e.g., frequency control. Here, both monitoring and control algorithms must rely on reliable and secure communication infrastructure [33–38].

From the perspective of prevention of power system blackouts, the role of ICT in designing novel System Integrity Protection Schemes (SIPS) is particularly becoming critical. Significant changes in power system dynamic properties caused by massive penetration of nonsynchronous generation, i.e., inverter connected RESs, are having a negative impact to power system inertia, resulting in faster system dynamics and difficulties when ensuring, e.g., frequency, voltage, or angular stabilities in the system. One of the effective measures

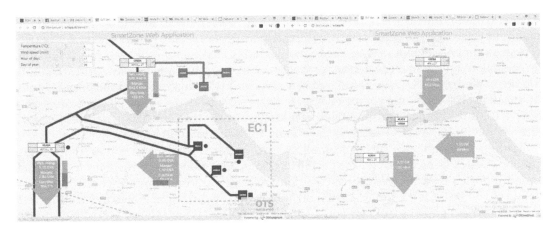

FIGURE 11.5
SmartZone web application (https://svtapp.000webhostapp.com/demo2/).

against potential blackouts caused by such a changed system dynamics are approaches based on intentional and controlled system splitting, also called islanding. Here the questions are how to split the system and when. The existing new concepts are based on utilization of PMUs and ICT, and as such can be described as typical examples of wide area protection, belonging to the broader topic discussed earlier "WAMPAC". Over the last decade, a number of promising concepts for intentional and controlled system islanding are proposed [39–49].

IoT concept considers a massive network structure where different types of devices and smart environments such as SGs in SCs are connected to the network. Similar to other IoT domain, SGs consist of a massive number of sensors and data sources that continuously collect high-resolution data. Managing the large volume of data has been identified as one of the major challenges in IoT. To address this issue, MEC envisions to process the data at the edge of the IoT network close to the embedded devices where the data are collected [50].

The advent of 5G communication networks largely facilitates the provision of the distributed information acquisition and processing services required in WAMSs. As far as information acquisition is concerned, the introduction of mMTC services will allow for a large-scale deployment of AMI. For those measurement devices (PMUs) requiring both very low latency and very high reliability, resorting to uRLLC services will be needed. As for information processing, novel architectural concepts such as MEC will be key for the deployment of the aforementioned distributed SE approaches.

11.3.2 Update on 5G Technology

The advantage of using 3GPP technologies for IoT applications is their almost worldwide coverage and usage of licensed spectrum that, unlike the unlicensed spectrum, provides for higher reliability and interference control. Today's 3GPP technologies are largely optimized for high data rate (human) communications. This makes 3GPP technologies less suitable for wide-area IoT deployments where battery usage and cost might be important requirements [51,52].

5G PPP (Infrastructure Public Private Partnership) is the joint initiative of the EU and the ICT industry, which defines the following groups of uses: densely populated urban areas, broadband access at any location, connected autonomous vehicles, smart office of the future, narrow-band IoT and tactile Internet/automation. For each individual case, KPIs (key perfor-mance indicators) are defined corresponding to performance requirements (in terms of user service experience) supported by the 5G network. 5G PPP classifies the basic functionality on the following parameters: density of the device in a given space, mobility, infrastructure topology, type of traffic, user data flow, latency, reliability, availability and category 5G com-munication. The following competencies are suggested: peak data flow 20 Gbps, user data flow 100 Mbps, latency 1 ms, mobility 500 km/h, density of connection 106/km², energy efficiency 100×4G, spectral efficiency 3×4G and traffic capacity 10 Mbps/m².

Significant advantages of 5G technology are higher speeds and more connections, enabling wireless connectivity of specific locations, faster and more responsive response time supported by time-sensitive applications, as well as ultralow-power connections.

The evolution of telecommunication technology started from narrowband medium for public use, next more broadband and available access, fixed broadband and mobile, to the current mobile broadband access. The implementation of 5G will enable IoT to make the cities much smarter. Billions of devices will have been connected within a very high-speed mobile network by the first phase of 5G to be started in 2020. According to *Statista*, the size of the worldwide IoT market would hit $2.225 trillion in 2020, and by 2025, there will be over 75 billion IoT-connected devices installed worldwide.

IoT is about automation, enabling an intelligent and controllable environment. It is really on the connectivity, data formatting and handling ends. There are different use cases that will demand different IoT protocols, data formats, data handling and interfaces. Currently, there is no one universal standard for IoT. Without a single standard, there is a need for interoperability and possibly open standards. The issue is that IoT application and use cases are too diverse, and industry efforts have been focused and segregated.

The 3GPP consortium develops conditions and requirements for 5G mobile communica-tions in two phases: the first phase was completed by the end of 2018 (Release 15), and the second phase is going to be completed at the end of 2019 (Release 16). 3GPP standardization includes the RAN (radio access network), network backbone (5G core), terminal (CT) and service architecture (SA). The 4G LTE-Advanced technical specifications Releases 13 and 14 were enhancements toward 5G. The purpose of the request is to provide cost-effective connections for a large number of IoT/M2M units with very low consumption and excel-lent coverage of the service zone (Table 11.2).

The 3GPP technologies, especially 3G and 4G technologies, are optimized for high-rate data transmissions. The support of these functionalities at the device increases the com-plexity and performance requirements of its hardware components, resulting ultimately in higher costs and high power consumption. Next, 3GPP systems have relatively large signaling overhead when compared with rather small amounts of user data that are typi-cally transferred in IoT deployments.

- 3GPP machine-type communication (MTC) development has started with the technical specifications Releases 10 and 11.
- The Release 12 specification focuses on the economy and expansion of the service zone. Specification Release 13 introduces narrowband-IoT (NB-IoT), eMTC and EC-GSM-IoT (extended-coverage GSM-IoT) radio access technology that not only relies on LTE technical components but also supports stand-alone operations by introducing individual channel structure.
- The NB-IoT standardization in Release 14 in 3GPP (frozen end of 2017) introduced enhancements of the UE positioning accuracy and multicast transmissions.

TABLE 11.2

3GPP IoT Standardization Process

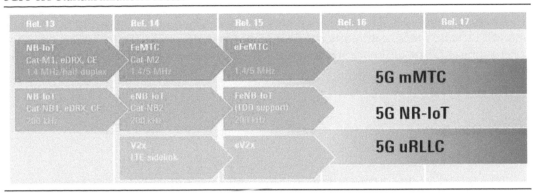

- The further IoT enhancements regarding improved coverage (via multi relaying), scalability and uRLLC are also considered as important topics for 5G standardization in Release 15, which is due end of 2018. mMTC refers to the support of massive number of simple machine-type devices, and it is an evolution of NB-IoT. Specification focuses on completing 5G NR NSA (non–stand-alone mode) standards for eMBB service, as well as setting the foundation for development 5G NR SA (stand-alone mode). The standard completed in December 2017 supports the NSA system that is based on the existing LTE infrastructure when added the new 5G radio network.
- Specification Release 16 by the end of 2019 is going to support new types of service/devices, new implementation/business models and new spectrum sharing types. The uRLLC communication, the use of the unlicensed spectrum and the new spectrum allocation paradigm, communication with autonomous vehicles (5G NR C-V2X) and continuation of 3GPP LPWA (low-power wide area) technologies (NB-IoT/eMTC) are being developed.

Differently from the previous generations of mobile communication systems, 5G will be very much focused on supporting various vertical industries, such as energy, transport, manufacturing and healthcare. The SG communication plays a central role in defining the relationship between 5G and the energy sector. Even more, SG communication represents a large business potential for 5G, both due to the use of uRLLC in applications for energy protection, control and distributed automation because of the importance of mMTC for supporting the future smart metering.

Perhaps the highest novelty in 5G is brought by uRLLC, which aims to provide extremely robust links with guaranteed latency and reliability. The existence of uRLLC links can significantly change the design approach to distributed cyberphysical systems, and one of the main use cases is related to SG distribution automation. Specifically, 3GPP is considering the use of uRLLC in wide area monitoring and control systems for SGs (3GPP TR22862 *Feasibility study on new services and markets technology enablers for critical communications*) [53]. Performance requirements are as follows:

- throughput from 200 to 1521 bytes reliably (99,999 %) delivered in 8 ms,
- one trip time latency between any two communicating points should be less than 8 ms for event-triggered message that may occur anytime,

- device density in dense urban (hundreds of UEs/km^2), urban around (15 UEs/ km^2), populated rural (max 1 UE/km^2).

In summary, there is a symbiotic relationship between two of the modes in 5G (mMTC and uRLLC) and the SG communication, since the former provides the technology and the latter represents one of the most important use cases that encompasses both mMTC and uRLLC.

11.4 Conclusions

SCs, like the SG, evolve slowly, but surely. They will more fully integrate and utilize information to be shared between departments, infrastructure operators, and citizens. Cities will partner to create integrated solutions, and the SG will become only a part of a greater, more responsive urban ecosystem. SCs depend on an SG to ensure resilient delivery of energy to supply their many functions, present opportunities for conservation, improve efficiencies and, most importantly, enable coordination between infrastructure operators. The core of such evolution is the integration of a fast and reliable ICT network, in order to provide accurate SG monitoring and interconnection between the power grid and the intelligent information processing systems.

This chapter reviews recent advances in SG development including the role of 5G-IoT networking, their requirements and challenges. The important features of 5G network connectivity are high data rate transmissions and very low latencies. There are still much more to be done in order to develop integrated solutions and create novel applications. The final objective is to improve human and citizen social life including economic growth. IoT is one of the important ICTs, along with connectivity, data analytics and AI. Many countries have initiated SC projects and invested billions into the development of SCs. For example, in China alone, more than 300 pilot SC projects have been initiated.

There is a need not only to establish a new system planning and system design principles and norms in international, national and other levels but also to reconstruct the flat organization system oriented to multi-energy integration (IES), which is the key of top-down approach. MES can feature better technical, economic and environmental performance relative to independent or separate ESs and at both the operational and the planning stage. It is essential to build an intelligent information system (or platform) to promote the interoperability and coordination of different ESs and ICT planes involved in EI standardization.

Much of the technology already exists to take the first steps. Significant levels of automation, communications and information technology are already being brought to bear on the electrical distribution systems of many utilities to improve reliability. There is a clear movement toward driving more intelligence into substations and field equipment to make faster decisions on fault isolation, location and restoration, feeder reconfiguration and voltage and reactive power management. At the same time, a growing penetration of renewable generation means new needs for managing adaptive protection equipment and extending related substation designs. More than ever, implementation of equipment and protection schemes that support defined standards, principles of integrating information and operations technologies is becoming an essential element in an effective overall design.

5G technology will provide an ideal arena for the development of future distributed SG services. These services will rely on massive and reliable acquisition of timely information from the system, in combination with large-scale computing and storage capabilities, providing a highly responsive, robust and scalable monitoring and control solution for future SGs.

References

1. K.Geisler, *The relationship between smart grids and smart cities*, IEEE Smart Grid, 2013.
2. A.Phan, S.T.Qureshi, *5G impacts on smart city 2030*, Project: 5G Communications, 2017.
3. V.Levi, G.Williamson, J.King, V.Terzija, "Development of GB distribution networks with low carbon technologies and smart solutions: Methodology", *International of Journal of Electrical Power & Energy Systems*, vol.119, 2020. doi: 10.1016/j.ijepes.2020.105833
4. V.Levi, G.Williamson, J.King, V.Terzija, "Development of GB distribution networks with low carbon technologies and smart solutions: Scenarios and results", *International of Journal of Electrical Power & Energy Systems*, vol.119, 2020. doi: 10.1016/j.ijepes.2020.105832
5. S.McClellan, J.A.Jimenez, G.Koutitas (Eds.), *Smart cities: Applications, technologies, standards, and driving factors*, Switzerland: Springer International Publishing AG, 2018.
6. E.Kabalci, Y.Kabalci (Eds.), *Smart grids and their communication systems*, Switzerland: Springer International Publishing AG, 2019.
7. I.Van Beuzekom, L.A.J.Mazairac, M.Gibescu, J.G.Slootweg, "Optimal design and operation of an integrated multi-energy system for smart cities", *IEEE ENERGYCON* 2016, pp.1–7
8. P.Mancarella, "MES (multi-energy systems): An overview of concepts and evaluation models", *Energy*, vol.65, pp.1–17, 2014.
9. J.J.Nielsen et al., *5G and cellular networks in the smart grid*, Chapter 3 in (Eds. H.T.Mouftah, M.Erol-Kantarci, M.H.Rehmani) Transportation and power grid in smart cities: Communication networks and services, Hoboken, NJ, USA: John Wiley & Sons, 2019.
10. S.Chakrabarti, E.Kyriakides, Bi Tianshu, C.Deyu, V.Terzija, "Measurements get together", *IEEE Power and Energy Magazine*, vol.7, no.1, pp.41–49, 2009. doi: 10.1109/MPE.2008.930657
11. V.Terzija et al., "Wide area monitoring, protection and control of future electric power networks", *Proceedings of IEEE*, vol.99, no.1, pp.80–93, 2011. doi: 10.1109/JPROC.2010.2060450
12. L.Yaqoob et al., "Enabling communication technologies for smart cities", *IEEE Communications Magazine*, vol.55, no.1, pp.112–120, Jan. 2017.
13. Y.Mehmood et al., "Internet-of-Things-based smart cities: Recent advances and challenges", *IEEE Communications Magazine*, vol.55, no.9, pp.16–24, Sept. 2017.
14. Y.Qian, D.Wu, W.Bao, P.Lorenz, "The Internet of Things for smart cities: Technologies and applications", *IEEE Network*, vol.33, no.2, pp.4–5, 2019.
15. B.Huang et al., "Energy informatics: Fundamentals and standardization", *ICT Express*, no.3, pp.76–80, Elsevier 2017.
16. C.Goebel et al., "Energy informatics-current and future research directions", *Journal Business & Information Systems Engineering*, vol.6, no.6, pp.25–31, Springer 2014.
17. I.A.Tøndel, J.Foros, S.S.Kilskar, P.Hokstad, M.G.Jaatun, "Interdependencies and reliability in the combined ICT and power system: An overview of current research", *Applied Computing and Informatics*, vol.14, pp.17–27, 2018.
18. N.Karagiorgos, K.Siozios, *A survey of research activities in the domain of smart grid systems*, Chapter 13 in (Eds. K.Siozios, D.Anagnostos, D.Soudris, E.Kosmatopoulos) IoT for smart grids: Design challenges and paradigms, Springer 2019, pp.253–282.
19. S.O.Muhanji, A.E.Flint, A.M.Farid, *eIoT - The development of the energy Internet of Things in energy infrastructure*, Switzerland: Springer International Publishing AG, 2019.

20. G.Bedi, G.K.Venayagamoorthy, R.Singh, R.Brooks, K.C.Wang, "Review of Internet of Things (IoT) in electric power and energy systems", *IEEE Internet of Things Journal*, vol.5, no.2, pp.847–870, 2018.
21. S.Reka, T.Dragicevic, P.Siano, S.R.S.Prabaharan, "Future generation 5G wireless networks for smart grid: A comprehensive review", *Energies*, vol.12, pp.1–17, 2019.
22. M.Cosovic, A.Tsitsimelis, D.Vukobratovic, J.Matamoros, C.Antón-Haro, "5G Mobile cellular networks: Enabling distributed state estimation for smart grid", *IEEE Communications Magazine*, vol.55, no.10, pp.62–69, 2017.
23. D.Ziouzios, A.Sideris, D.Tsiktsiris, M.Dasygenis, *Smart grid modelling and simulation*, Chapter 3 in (Eds. K.Siozios, D.Anagnostos, D.Soudris, E.Kosmatopoulos) IoT for smart grids: Design challenges and paradigms, Springer 2019, pp.43–54.
24. DOE, "Factors affecting PMU installation costs" [Online]. Available: https://www.energy.gov/oe/downloads/factors-affecting-pmu-installation-costs-october-2014
25. S.D.Ahmad, S.Azizi, A.Mohammad, V.Terzija, "Linear LAV-based state estimation integrating hybrid SCADA/PMU measurements", *IET Generation, Transmission & Distribution*, vol.14, no.8, pp.1583–1590, 2020.
26. A.Dubey, S.Chakrabarti, A.Sharma, V.Terzija, "Optimal utilization of PMU measurements in power system hybrid state estimators", *IET Generation, Transmission & Distribution*, vol.13, no.21, pp.4978–4986, 2019. doi: 10.1049/iet-gtd.2019.0010
27. J.Zhao et al., "Power system dynamic state estimation: Motivations, definitions, methodologies, and future work", *IEEE Transactions on Power Systems*, vol.34, no.4, pp.3188–3198, 2019. doi: 10.1109/TPWRS.2019.2894769
28. A.S.Dobakhshari, S.Azizi, M.Paolone, V.Terzija, "Ultra fast linear state estimation utilizing SCADA measurements", *IEEE Transactions on Power Systems*, vol.34, no.4, pp.2622–2631, 2019. doi: 10.1109/TPWRS.2019.2894518
29. D.N.Ćetenović, A.M.Ranković, "Optimal parameterization of Kalman filter based three-phase dynamic state estimator for active distribution networks", *International of Journal of Electrical Power & Energy Systems*, vol.101, pp.472–481, 2018. doi: 10.1016/j.ijepes.2018.04.008
30. Z.Jin, P.Dattaray, P.Wall, J.Yu, V.Terzija, "A screening rule-based iterative numerical method for observability analysis", *IEEE Transactions on Power Systems*, vol.32, no.6, pp.4188–4198, 2017. doi: 10.1109/TPWRS.2017.2660068
31. G.Valverde, V.Terzija, "Unscented Kalman filter for power system dynamic state estimation", *IET Generation, Transmission & Distribution*, vol.5, no.1, pp.29–37, 2011. doi: 10.1049/iet-gtd.2010.0210
32. G.Valverde, S.Chakrabarti, E.Kyriakides, V.Terzija, "A constrained formulation for hybrid state estimation", *IEEE Transactions on Power Systems*, vol.26, no.3, pp.1102–1109, 2011. doi: 10.1109/TPWRS.2010.2079960
33. M.Kheshti et al., "Toward intelligent inertial frequency participation of wind farms for the grid frequency control", *IEEE Transactions on Industrial Informatics*, 2019. doi: 10.1109/TII.2019.2924662
34. D.Wilson, J.Yu, N.Al-Ashwal, B.Heimisson, V.Terzija, "Measuring effective area inertia to determine fast-acting frequency response requirements", *International of Journal of Electrical Power & Energy Systems*, vol.113, pp.1–8, 2019. doi: 10.1016/j.ijepes.2019.05.034
35. M.Sun et al., "On-line power system inertia calculation using wide area measurements", *International of Journal of Electrical Power & Energy Systems*, vol.109, pp.325–331, 2019. doi: 10.1016/j.ijepes.2019.02.013
36. M.Kheshti, L.Ding, M.Nayeripour, X.Wang, V.Terzija, "Active power support of wind turbines for grid frequency events using a reliable power reference scheme", *Renewable Energy*, vol.139, pp.1241–1254, 2019. doi: 10.1016/j.renene.2019.03.016
37. R.Azizipanah-Abarghooee, M.Malekpour, M.Paolone, V.Terzija, "A new approach to the online estimation of the loss of generation size in power systems", *IEEE Transactions on Power Systems*, vol.34, no.3, pp.2103–2113, 2019. doi: 10.1109/TPWRS.2018.2879542
38. N.Shams, P.Wall, V.Terzija, "Active power imbalance detection, size and location estimation using limited PMU measurements", *IEEE Transactions on Power Systems*, vol.34, no.2, pp.1362–1372, 2019. doi: 10.1109/TPWRS.2018.2872868

39. M.Naglic, M.Popov, M. van der Meijden, V.Terzija, "Synchronized measurement technology supported online generator slow coherency identification and adaptive tracking", *IEEE Transactions on Smart Grid*, accepted on 13/12/2019. doi: 10.1109/TSG.2019.2962246

40. M.Naglic et al., "Synchronized measurement technology supported AC and HVDC online disturbance detection", *Electric Power Systems Research*, vol.160, pp.308–317, 2018. doi: 10.1016/j.epsr.2018.03.007

41. I.Tyuryukanov, M.Popov, M. van der Meijden, V.Terzija, "Discovering clusters in power networks from orthogonal structure of spectral embedding", *IEEE Transactions on Power Systems*, vol.33, no.6, pp.6441–6451, 2018. doi: 10.1109/TPWRS.2018.2854962

42. M.Naglic, M.Popov, M. van der Meijden, V.Terzija, "Synchro-measurement application development framework: An IEEE standard C37.118.2–2011 supported MATLAB library", *IEEE Transactions on Instrumentation and Measurement*, vol.67, no.8, pp.1804–1814, 2018. doi: 10.1109/TIM.2018.2807000

43. L.Ding, Y.Guo, P.Wall, K.Sun, V.Terzija, "Identifying the timing of controlled islanding using a controlling UEP based method", *IEEE Transactions on Power Systems*, vol.33, no.6, pp.5913–5922, 2018. doi: 10.1109/TPWRS.2018.2842709

44. J.Quirós-Tortós, P.Demetriou, M.Panteli, E.Kyriakides, V.Terzija, "Intentional controlled islanding and risk assessment: A unified framework", *IEEE Systems Journal*, vol.12, no.4, pp.3637–3648, 2018. doi: 10.1109/JSYST.2017.2773837

45. L.Ding, Z.Ma, P.Wall, V.Terzija, "Graph spectra based controlled islanding for low inertia power systems", *IEEE Transactions on Power Delivery*, vol.32, no.1, pp.302–309, 2017. doi: 10.1109/TPWRD.2016.2582519

46. J.Quirós-Tortós, R.Sánchez-García, J.Brodzki, J.Bialek, V.Terzija, "Constrained spectral clustering-based methodology for intentional controlled islanding of large-scale power systems", *IET Generation, Transmission & Distribution*, vol.9, no.1, pp.31–42, 2015. doi: 10.1049/iet-gtd.2014.0228

47. J.Quirós-Tortós, P.Wall, L.Ding, V.Terzija, "Determination of sectionalising strategies for parallel power system restoration: A spectral clustering-based methodology", *Electric Power Systems Research*, vol.116, pp.381–390, 2014. doi: 10.1016/j.epsr.2014.07.005

48. L.Ding, P.Wall, V.Terzija, "Constrained spectral clustering based controlled islanding", *International of Journal of Electrical Power & Energy Systems*, vol.63, no.1, pp.687–694, 2014. doi: 10.1049/iet-gtd.2014.0228

49. L.Ding, F.Gonzalez-Longatt, P.Wall, V.Terzija, "Two-step spectral clustering controlled islanding algorithm", *IEEE Transactions on Power Systems*, vol.28, no.1, pp.75–84, 2013. doi: 10.1109/TPWRS.2012.2197640

50. F.Samie, L.Bauer, J.Henkel, *Edge computing for smart grid: An overview on architectures and solutions*, Chapter 2 in (Eds. K.Siozios, D.Anagnostos, D.Soudris, E.Kosmatopoulos) IoT for Smart grids: Design challenges and paradigms, Switzerland: Springer International Publishing AG, 2019, pp.21–42.

51. Z.Bojkovic, D.Milovanovic, "5G Connectivity technologies for the IoT: Research and development challenges", in *Proc. ELECOM2018*, Springer LNEE, vol.561, pp. 362–371, 2018.

52. D.Milovanovic, Z.Bojkovic, "5G Ultra reliable and low-latency communication: Fundamental aspects and key enabling technologies", in *Proc. ELECOM2018*, Springer LNEE, vol.561, pp. 372–379, 2018.

53. 3GPP TR 22862 v14.1.0, *Feasibility study on new services and markets technology enablers for critical communications*, Release 14, Sept. 2016.

12

Enabling Massive IoT in 5G and Beyond Systems: Evolution and Challenges

Chaochao Yao, Jia You, Gongpu Wang, and Yigang Cen

Beijing Jiaotong University

CONTENTS

12.1 Introduction

In March 2019, 3GPP (3rd Generation Partnership Project) announced the completion of its first 5G standard version, with one of the goals of realizing large-scale Internet of things (IoT). It claims that the 5G characteristics, including high bandwidth, low latency and reliable transmission, can enable everything to be connected intelligently, i.e., enable physical objects to interact and exchange data with each other. One of the new and key scenarios of 5G is massive machine-type communications (mMTC), which can provide communication services for a large number of microsensors and actuators in the future intelligent systems. Furthermore, the future sixth-generation mobile networks propose a more ambitious goal: smartly connecting everything.

The IoT depends on the convergence of multiple technologies, including access, communications, networking, security and architecture. In the first part of the chapter, we focus on two key technologies: multiple access and backscatter communications. Multiple access technology aims to address the problem how to decide who can use the channel. Extensive studies about radio backscatter have been performed in the past 10 years. Recently, new backscatter technologies are proposed and attract both academic attentions and industrial interests. Here, we briefly introduce three types of backscatter communications: traditional, bistatic and ambient. In the third part, we discuss the development and deployment of IoT, which currently face many technical and economical challenges.

12.2 Evolution of IoT

The definition of the IoT has evolved due to the convergence of multiple technologies.

- The concept of IoT first appeared in the book *The Road Ahead,* written by Bill Gate in 1995, where it claims the *camera connected to a personal computer or other video equipment will enable us to hold video conferences more conveniently.* It is the first form of IoT.
- Later in 1999, Dr. Kevin Ashton, cofounder of the Auto-ID Center at Massachusetts Institute of Technology (MIT), used the term IoT for the first time. He considers IoT as the connection between the real world and the Internet through sensors and RFID technology, enabling intelligent interconnection and management.
- In 2005, the International Telecommunication Union (ITU) released the *ITU Internet Report 2005: Internet of Things* at the World Summit on the Information Society (WSIS) in Tunisia. Since then, the concept of the IoT has been widely utilized in academic and industrial circles (Figure 12.1).

Subsequently, the IoT is regarded as the third wave of the information world after the invention of computers and Internet. Many governments have started to develop IoT technologies.

- In November 2008, the IBM published a report entitled *Smart Earth: The Next Agenda of Leaders* in New York, United States, which proposed the concept of Smart Earth. The report indicated that the next generation of information technologies would be applied to all aspects of life. This idea has received a positive response from the former US President, Barack Obama, who has incorporated *Smart Earth* into his national strategy.

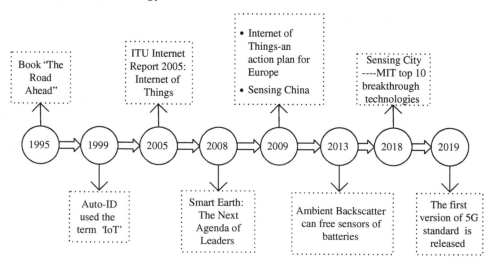

FIGURE 12.1
Milestone events in the IoT development. IoT, Internet of things.

- On June 8, 2009, the European Commission published the *Internet of Things – an action plan for Europe*, which proposed 12 actions to accelerate development of IoT. It implies that the European Union has put the realization of IoT on the agenda.
- On August 7, 2009, Chinese Premier Jiabao Wen proposed the concept of sensing China during his visit in Wuxi, a south city in China. In the following year, he explicitly included the development and promotion of IoT into his government annual report.
- In 2013, researchers from the University of Washington suggest ambient backscatter, a green technology that utilizes ambient wireless signals to enable sensors to communicate with other nodes. Ambient backscatter can liberate sensors from maintenance-heavy batteries and thus has significant commercial value.
- In 2018, sensing city is selected as one of the top ten breakthrough technologies of this year by MIT. The aim of sensing city is to make urban areas more affordable, livable and environmentally friendly based on a huge IoT with massive sensors.
- In March 2019, 3GPP announced the completion of its first 5G standard version.

12.3 Technologies for Internet of Things

IoT depends on a host of technologies, including access, communications, networking, security and architecture. In this chapter, we focus on two key technologies: multiple access and backscatter communications.

12.3.1 Multiple Access

Multiple access technology aims to address the problem: when there are several users, e.g., sensors, to communicate with the central node over one channel, how to decide who can use the channel. This problem can be solved in physical layer and medium access control (MAC) layer. Difference of access technologies between physical layer and MAC is that physical layer deals with bits within single packet, while MAC copes with each packet. Multiple access can be divided into two types: grant-based and grant-free. Grant-based multiple access approaches require that the user obtain authorization from other users or base stations (BSs) before transmission. Widely used grant-based multiple access algorithms include carrier-sensing multiple access (CSMA) and random access channel (RACH) in long-term evolution (LTE) technology (LTE RACH). Grant-free algorithms allow users to start transmission freely without any grant. Clearly, grant-free algorithms can have low control overhead but may suffer from collisions. Therefore, the key challenge to solve is collision between data transmission from different users. Typical grant-free algorithms are nonorthogonal multiple access (NOMA) and multiuser shared access (MUSA) in physical layer and ALOHA[1] and coded slotted ALOHA (CSA) in MAC layer (Figure 12.2).

---

[1] ALOHA is a classical algorithm of random access suggested in 1970s.

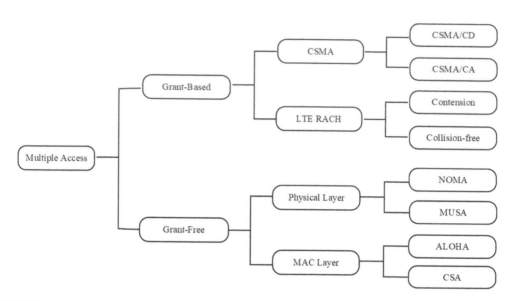

FIGURE 12.2
Multiple access algorithms.

Nonorthogonal multiple access. The basic idea of NOMA is to actively introduce inter-ference information to enable a nonorthogonal transmission with different transmit pow-ers. The receiver uses serial interference cancellation (SIC) to eliminate interference and realize orthogonal demodulation [1]. When using NOMA in orthogonal frequency-division multiplexing (OFDM) systems, the subchannels are still orthogonal and do not interfere with each other. However, one subchannel is no longer assigned to only one user; it can be shared by multiple users. Then, nonorthogonal transmission between different users on the same subchannel will cause the problem of interference, which can be addressed by SIC at the receiver so that multiuser detection is realized. At the transmitting end, dif-ferent users on the same subchannel transmit with various powers. These transmitting powers are specially designed so that they are different when arriving at the receiver. The receiver then performs SIC in a certain order to achieve correct demodulation and thus distinguish different users. SIC is of vital importance in the receiver for NOMA transmis-sion. The SIC receiver can eliminate the interference between users at the same frequency and sequentially recover user data. In this way, the same subchannel is shared by multiple users so that the spectrum efficiency of NOMA is enhanced [2]. The diagram of NOMA implementation is shown in Figure 12.3.

Consider the simplest case of a downlink NOMA system with one BS and two users. The Shannon capacity is

$$C = \sum_{i=1,2} B_i \log_2 \left(1 + \frac{P_i}{I_i + N_i}\right) \tag{12.1}$$

where C represents channel capacity, B indicates bandwidth, P is signal power, I denotes interference power and N is noise power.

The BS uses overlay coding to transmit two users signals x_1 and x_2 at the same frequency but with different powers P_1 and P_2, respectively. Thus the transmitted signal is

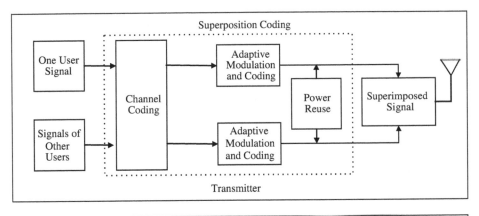

Superposition Coding / Transmitter

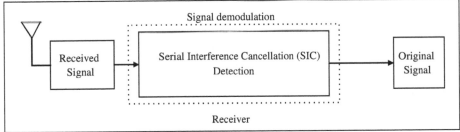

Signal demodulation / Receiver

FIGURE 12.3
The diagram of downlink NOMA implementation. NOMA, nonorthogonal multiple access.

$$x = \sqrt{P_1}\, x_1 + \sqrt{P_2}\, x_2 \tag{12.2}$$

Suppose the channel between BS and the ith user is h_i. Then the signal received by the ith user can be given as

$$y_i = h_i x + w_i \tag{12.3}$$

Without loss of generality, assume $P_1 > P_2$, and the receiver first recovers signal x_1 from y_i and then substracts the recovered x_1 from y_i so as to detect x_2. In the case of perfect SIC, the channel capacity for user 1 and 2 can be found as

$$C_1 = \log_2\left[1 + \frac{P_1|h_1|^2}{N_{0,1}}\right], \quad C_2 = \log_2\left[1 + \frac{P_2|h_2|^2}{\left(P_1|h_2|^2 + N_{0,1}\right)}\right] \tag{12.4}$$

In orthogonal frequency-division multiple access (OFDMA), the available bandwidth is orthogonally allocated to two users. Suppose the bandwidth for user 1 is α and user 2 is $(1-\alpha)$. It can be readily checked that the channel capacities for user 1 and user 2 are

$$C_1 = \alpha\log_2\left[1 + \frac{P_1|h_1|^2}{\alpha N_{0,1}}\right], \quad C_2 = (1-\alpha)\log_2\left[1 + \frac{P_2|h_2|^2}{(1-\alpha)N_{0,1}}\right] \tag{12.5}$$

Calculating and comparing the equations (12.4) and (12.5) will show that NOMA improves spectral efficiency by 30%–40% over OFDMA [3].

Multiuser shared access. MUSA is a new multiaccess technology for 5G IoT. It is non-orthogonal multiple-access technology based on the code domain, which is regarded as an improvement scheme of code-division multiple access (CDMA) technology. Through special design of multivariate code sequence in the complex domain and the multiuser detection of SIC, MUSA allows multiple times of user access on the same time–frequency resource. The basic elements of a MUSA system with N simultaneous users are shown in Figure 12.4. First, the signal S_i of the ith user is encoded into the bits C_i. Then the bits C_i are modulated into the symbol m_i using a complex domain multivariate code sequence with low cross-correlation l_i that can be easily differentiated by SIC receivers. After that, the extended symbols R_i of user i can be transmitted over the same time–frequency resource. Finally, the receiver motivates linear signal processing and SIC to recover the original bits of each user. MUSA allows a large number of users to share access without requiring grants, which are usually obtained through complex control processes such as resource application and scheduling confirmation. That is, MUSA can be grant-free. Besides, MUSA can relax or even eliminate the strict uplink synchronization process and only needs to implement simple downlink synchronization. MUSA can support scheduling-free access, eliminating the resource scheduling process and simplifying synchronization, power control and other related process. These features can save system signaling overhead, reducing access delays, simplifying terminal implementation and reducing terminal energy consumption, which makes MUSA suitable for the 5G IoT.

Access methods based on ALOHA. ALOHA along with time-division multiple access (TDMA) was one of the first protocols used in radio networks [4]. The basic idea of ALOHA system is simple: whenever the user has data to send, it will be transmitted immediately. It will certainly cause collision in many cases. Therefore, the transmitter needs to check collision. In pure ALOHA algorithm shown in Figure 12.5a, when the collision problem occurs,

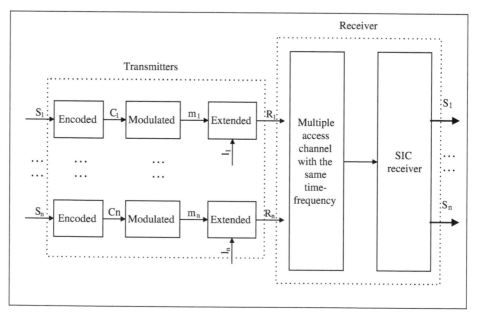

FIGURE 12.4
Basic elements of a MUSA system. MUSA, multiuser shared access.

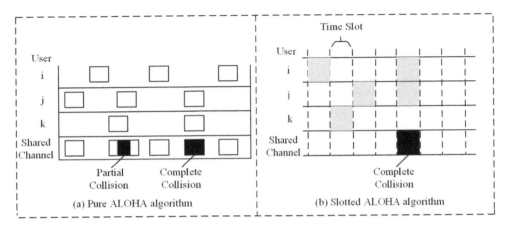

FIGURE 12.5
Two algorithms based on ALOHA: (a) pure ALOHA algorithm and (b) slotted ALOHA algorithm.

the received frame will be corrupted, and the transmitter needs to send this frame after a random time period. Although pure ALOHA allows multiple users to share the same channel, the corresponding throughput is not large due to often collision. Consequently, two improved versions based on ALOHA have been proposed: slotted ALOHA [5] and coded slotted ALOHA (CSA) [6].

- Slotted ALOHA, as depicted in Figure 12.5b, unlike pure ALOHA, divides time into multiple equal slots, which is slightly larger than a frame. Data can only be transmitted at the beginning of each time slot so that there is no partial collision. This algorithm can reduce the collision period and improve channel utilization.

- CSA is an improved version of slotted ALOHA. It separates original data packets into k small data packets before transmission. After that, linear block codes (n,k) will be used to encode k small data packets. All the processed data packets will be randomly placed into n time slots within a frame and then sent out [6]. At the receiver, successive interference cancellation (SIC) technology and maximum a posterior (MAP) of linear block codes will be used to solve the collision problem and recover the original data. Figure 12.6 illustrates main steps of CSA algorithm. Suppose there are three users A, B and C and each frame contains eight slots. We assume that user A chooses (4,2) linear block codes, whereas user B and user C select (3,2) linear block coding. The data packet of each user will be first divided into two parts P_1 and P_2 that will go through the encoder of the linear block coding and then generate some new data packets. Assume three users will generate 4, 3, 3 new packets, respectively. These new generated data packets will be transmitted in randomly selected time slots to the receiver. The corresponding receiver will check the data packets in every time slot for each frame. In this example, the receiver will find that there is only one packet and no confliction at the second, fourth, fifth and eighth time slots. Accordingly, these data packets can be recovered successfully. Next the receiver can find there exist collisions at the first, third and sixth slots. It can also be readily checked that if the data packets in the third slot can be recovered, then the data of user B can be fully recovered. Fortunately, MAP and SIC can aid in detection of the data of user B

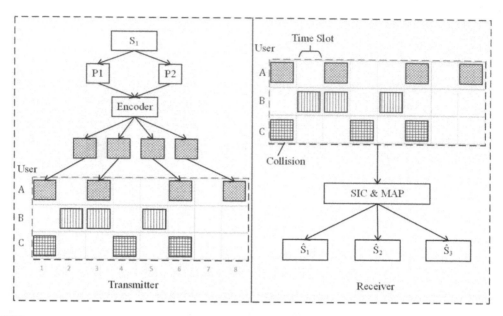

FIGURE 12.6
CSA algorithm. CSA, coded slotted ALOHA.

at the third slot. With full information of the data of user B, the data of user A at the third slot can also be obtained with SIC. Next, due to the redundancy of (4,2) linear block codes of user A, its data can also be detected. Finally, the receiver can cancel the interference of the user A to user C at the first and sixth slots so as to recover the data of user C. This example shows that CSA provides the possibility for massive access to IoT.

12.3.2 Backscatter Communications

It may be less strange if we mention radio frequency identification (RFID) than backscatter communications. Actually, RFID systems employ backscatter technology for communications.

A typical RFID system mainly consists of a reader (also known as an interrogator) and multiple tags (also known as transponder). Depending on the power supply for tags, RFID systems can be active, passive or semipassive. For passive RFID, the reader first generates an electromagnetic wave, and the tag receives and backscatters the wave with modulated information bits to the reader.

Backscatter communications originated from World War II. At that time, radio signals were designed and transmitted by some radar and then backscattered from an on-coming airplane so as to identify it as *friend or foe*. The first paper about backscatter communication was given by Harry Stockman in 1948 [7]. Since then, it has been continuously studied, and RFID products have been developed for identification/supply chain applications.

From 1990 to 2000, a famous and successful application of RFID systems was electronic toll collection (ETC). ETC can enable vehicles on the high speedways to pay the toll fee without stopping. After the 1990s, the rapid progress in integrated circuits decreased the

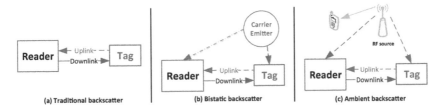

FIGURE 12.7
Backscatter communications: (a) traditional, (b) bistatic and (c) ambient.

cost of tags, which enabled the widespread usage of RFID products and also aroused great interests on further investigation of the backscatter technology.

Extensive studies about radio backscatter have been performed in the past 10 years. Recently, new backscatter technologies are proposed and attract both academic attentions and industrial interests. Here, we briefly introduce three types of backscatter communications shown in Figure 12.7: traditional, bistatic [8] and ambient [9,10].

Traditional backscatter. As shown in Figure 12.7a, the traditional backscatter is utilized in the RFID system where the reader first generates a continuous carrier wave, and then the tag receives and remodulates the wave. There exists a round-trip path loss during the signal transmission from the reader to the tag and then back to the reader, which will impose a limit on the communication distance between the reader and the tag [10].

Bistatic backscatter. To further increase the field coverage and communication range or avoid the round-trip path loss, bistatic scatter is proposed [8]. As shown in Figure 12.7b, bistatic scatter locates a carrier emitter close to the tag. The tag remodulates the signals transmitted by the carrier emitter and backscatters them to the reader. By dislocating the carrier emitter from the receiver of reader in traditional RFID systems, bistatic scatter can achieve long ranges of up to 130 m with 20 mW of carrier power [8], significantly enhancing the communication range.

Ambient backscatter. Ambient backscatter utilizes ambient radio frequency (RF) signals, such as television (TV) radio, cellular signals and wireless fidelity (Wi-Fi), to enable battery-free tag to communicate with the reader.

As shown in Figure 12.7c, the BS communicates with the mobile user. The signal transmitted by the BS can be utilized by the tag. The tag receives these signals, remodulates and backscatters them to the reader.

Suppose the channels between the tag and the BS, between the tag and the reader and between the BS and the reader are h_{tB}, h_{tr} and h_{Br}, respectively. When the BS transmits signals $s(t)$, the tag will receive $h_{tB}\,s(t)$, modulate its own signal $B(t)$ onto the received signals and then backscatter the combined signals $h_{tB}\,s(t)\,B(t)$ to the reader. The reader will obtain

$$y(t) = h_{Br}s(t)B(t) + w(t) \tag{12.6}$$

Clearly, the reader aims to recover $B(t)$ from $y(t)$. It is worth noting that in most cases, the reader has little knowledge about the source signals $s(t)$. Therefore, special design of the tag signals $B(t)$ and the detector for the reader are of vital importance.

Ambient backscatter is a new communication technology that utilizes ambient wireless signals to enable the tag to communicate with the reader. It can also enable battery-free devices to communicate mutually or connect to the Internet. Accordingly, ambient backscatter can liberate tags and sensors from maintenance-heavy batteries and thus has significant commercial value.

12.4 Challenges and Open Problems

Despite the ambitious aims of 5G and 6G, the development and deployment of IoT currently face many technical, economical and even political challenges. As far as technologies are concerned, the following challenges exist: interference control and optimization, massive access methods, synchronization, security, energy source for sensors, interconnection architecture and internetworking protocols. In this section, we introduce these challenges and discuss the corresponding researches.

Interference control and optimization. The central nodes collect information from widely deployed sensors. When many sensors transmit wireless signals to the central nodes, interference will definitely incur. Besides, signals between central nodes and from other BSs using the same frequency bands can result in strong interference. It is worth noting that symbiotic radio (SR) [11], a new paradigm, has been proposed as one potential solution to interference control and resource optimization. It integrates secondary devices with a primary communication system, and the transmitters and receivers are jointly designed to optimize transmission resources.

Massive access. A large number of connections, up to 300,000 in a single cell, require a set of completely different access methods from those for existing LTE systems. Nonorthogonal access approaches, such as NOMA and MUSA, can provide a larger number of users over orthogonal methods. It is worth noting that the gain is obtained at the cost of computational complexity, which makes NOMA and MUSA more applicable for uplink. Besides, spread spectrum and sparse signal processing may be candidate technologies to address the challenge of massive access. Apart from huge number of connections, low energy consumption is another challenge for realizing mMTC in 5G. Low energy consumption requires new low signaling overhead access protocols because less frequent and shorter transmissions preserve energy. Therefore, short packet transmission and corresponding protocol design are also worth investigation.

Synchronization. Many sensors for future IoT will be small batteryless devices that do not have powerful computing resources and also can only have limited or few training symbols for synchronization. Therefore, simple but reliable synchronization protocols or methods are needed for these small devices when synchronization is necessary, especially for battery-free backscatter schemes in future IoT.

Security. Security problem is born with wireless communication systems [12]. Wireless channels are broadcasting, and thus the wireless signals can be easily acquired. Eavesdroppers close to the transmitter can have stronger signals than licensed users. In addition, active attack can also lead to strong interference in wireless transmission. Encryption algorithms and protocols are traditional methods to guarantee security. Physical layer methods, such as beamforming and artificial noise, are new approaches for secure transmission. For IoT with a large number of wireless sensors, security studies are on their early stage, especially for physical layer. Security is even more important to IoT systems based on backscatter communications. It is not only because that wireless channels are open, i.e., accessible by any devices closed to the transceivers, but also that the backscattered signals do not belong to the batteryless backscatter communication systems. Therefore, it is of vital importance for backscatter communication systems to take special measures to guarantee their secure transmission.

Energy source. Currently, most sensors and tags in IoT use batteries for energy source. However, batteries have limited service lifetime and additional maintenance costs. When the batteries consume all the energy stored, these small devices need new batteries and also replacement, which is time-consuming and inconvenient. Besides, the sensors may be limited by severe external conditions. For example, it is almost impossible to replace batteries of sensors embedded in walls. For another instance, sensors as well as their batteries along the beach suffer from seawater corrosion. An efficient strategy to address the energy problem is to harvest energy from ambient environment, instead of utilizing time-limited and replacement-required batteries. Existing possible candidates are solar, wind, vibrational and electromagnetic energy.

Interconnection architecture and protocols. Smartly connecting everything requires interconnection between many small IoTs. In reality, there are many small sensor networks that work well locally and self-sufficiently. However, those local sensor networks cannot connect with each other, which seems like "information isolated islands." That is, currently, there are few interconnection protocols or architectures available that can join all local sensor networks together. In the history of information technology, transmission control protocol/Internet protocol (TCP/IP) successfully joined countless local area networks into the world famous Internet. Now such architecture and protocols are also needed by IoT.

12.5 Conclusions

In this chapter, we first provide a brief overview of the IoT history and then mainly introduce two key technologies, multiple access and backscatter communications, and finally suggest open problems for future IoT. We predict that a large number of batteryless sensors will be deployed in IoT and 5G, which will give rise to many open problems for both fundamental theories and practical applications [13,14].

References

1. Z. Ding, X. Lei, G. K. Karagiannidis, R. Schober, J. Yuan and V. K. Bhargava, A Survey on Non-Orthogonal Multiple Access for 5G Networks: Research Challenges and Future Trends, *IEEE Journal on Selected Areas in Communications*, vol. 35, no. 10, pp. 2181–2195, Oct. 2017.
2. S. M. R. Islam, N. Avazov, O. A. Dobre and K. Kwak, Power-Domain Non-Orthogonal Multiple Access (NOMA) in 5G Systems: Potentials and Challenges, *IEEE Communications Surveys & Tutorials*, vol. 19, no. 2, pp. 721–742, Second quarter 2017.
3. Y. Saito, Y. Kishiyama, A. Benjebbour, T. Nakamura, A. Li and K. Higuchi, Non-Orthogonal Multiple Access (NOMA) for Cellular Future Radio Access, *IEEE 77th Vehicular Technology Conference* (VTC Spring), pp. 1–5, 2013.
4. F. Baccelli, B. Blaszczyszyn and P. Muhlethaler, An Aloha Protocol for Multihop Mobile Wireless Networks, *IEEE Transactions on Information Theory*, vol. 52, no. 2, pp. 421–436, Feb. 2006.
5. J. Arnbak and W. van Blitterswijk, Capacity of Slotted ALOHA in Rayleigh-Fading Channels, *IEEE Journal on Selected Areas in Communications*, vol. 5, no. 2, pp. 261–269, Feb. 1987.
6. Z. Sun, Y. Xie, J. Yuan and T. Yang, Coded Slotted ALOHA for Erasure Channels: Design and Throughput Analysis, *IEEE Transactions on Communications*, vol. 65, no. 11, pp. 4817–4830, Nov. 2017.

7. H. Stockman, Communication by Means of Reflected Power, *in Proceedings IRE*, vol. 36. no. 10, pp. 1196–1204. Oct. 1948.
8. J. Kimionis, A. Bletsas and J. N. Sahalos, Increased Range Bistatic Scatter Radio, *IEEE Transactions on Communications*, vol. 62, no. 3, pp. 1091–1104, Mar. 2014.
9. V. Liu, A. Parks, V. Talla, S. Gollakota, D. Wetherall, and J. R. Smith, Ambient Backscatter: Wireless Communication Out of Thin Air, in *Proc. ACM SIGCOMM*, pp. 39–50, Hong Kong, China, 2013.
10. G. Wang, F. Gao, R. Fan and C. Tellambura, Ambient Backscatter Communication Systems: Detection and Performance Analysis, *IEEE Transactions on Communications*, vol. 64, no. 11, pp. 4836–4846, Nov. 2016.
11. R. Long, H. Guo, G. Yang and Y. Liang, Symbiotic Radio: A New Communication Paradigm for Passive Internet-of-Things, *IEEE Internet of Things Journal*, vol. 7, no. 2, pp. 1350–1363, Feb. 2020.
12. Y. Zou, J. Zhu, X. Li and L. Hanzo, Relay Selection for Wireless Communications against Eavesdropping: A Security-Reliability Tradeoff Perspective, *IEEE Network*, vol. 30, no. 5, pp. 74–79, Sept. 2016.
13. G. Wang, Q. Liu, R. He, F. Gao and C. Tellambura, Acquisition of Channel State Information in Heterogeneous Cloud Radio Access Networks: Challenges and Research Directions, *IEEE Wireless Communications*, vol. 22, no. 3, pp. 100–107, June 2015.
14. C. Xing, Y. Jing, S. Wang, S. Ma and H. V. Poor, New Viewpoint and Algorithms for Water-Filling Solutions in Wireless Communications, *IEEE Transactions on Signal Processing*, vol. 68, pp. 1618–1634, 2020, doi:10.1109/TSP.2020.2973488.

13

Intelligent Transportation System as an Emerging Application in 5G

Natasa Bojkovic, Marijana Petrovic, and Tanja Zivojinovic
University of Belgrade

CONTENTS

13.1 Introduction

Integration of transport engineering with telecommunications, electronics and information technology marked a new era in transportation systems. In the core of the present and future development are innovative transport solutions known under the term intelligent transport systems (ITSs). The backbone of ITSs is the use of information and communication technologies (ICTs), entailing ICT integration into transport infrastructure and vehicles.

Real-time information, improved decision-making and adaptive responses to changing traffic conditions are altogether characteristics that make transport systems intelligent. It is actually about upgrading transport processes, in a way that recognizes information as a resource for increasing safety, efficiency and quality of service. The users benefit from new technologies by improving their driving experience and personal mobility in general.

It is noticeable that most attention is paid for ITS solutions in road sector. In support of that is the European Action Plan for the Implementation of ITS, adopted in 2010, which envisages ITS development in road traffic, as well as for the connection of road transport with other modes. This position is explained by the fact that in other modes of transport, advanced information and in traffic management systems have been strategically and

247

operatively developed. Unlike rail or air transportation modes, where communications and centralized management are often a condition for traffic establishment, road sector is characterized by easy market access and a high degree of decentralization. In other words, a large number of individual decisions left to drivers make traffic management and safety more complex. For this reason, road traffic and transportation is the most demanding one, and at the same time, it opens up a broad field for ITS application.

When it comes to the development of ITS, the literature mainly elaborates on two phases that preceded the present-day technological rise (Figueiredo et al., 2001; Nowacki, 2012; Miles, 2014): initial development and the creation of technical foundations. The beginnings of ITS date back to the 1970s, when the first systems of navigation, that is, the dynamic routing of vehicles in the changing traffic conditions (ARI – *Auto-fahrer Rundfunk Information* in Germany, and CACS – *Comprehensive Automobile Control System* in Japan), have been introduced. Due to the limited development of computer and communication technologies at the time, these systems could not have a wider practical application. In the next phase, from the 1980s to the mid-1990s, in the most developed countries of the world, a number of programs and projects for basic technological development were launched. In this period which is called a "feasibility study," many ideas have been tested, and market opportunities have been evaluated.

An important event, and the "catalyst" of ITS development, is the first World ITS Congress, held in Paris in 1994, where the term ITSs has been officially introduced. In the mid-1990s, along with the intensive advancement of technological support, ITSs had become a conceptual heading for a wide range of technological and organizational solutions, applications and services.

This chapter provides the review of the current developments in ITS and new services that are yet to be introduced. The aim is to understand potentials and limitations of the implementations and to understand the communication requirements for the deployment of new ITS services. The focus is on "vehicle to everything" (V2X) communication which is a breakthrough technology.

The content of the text is organized in five sections. In the next two sections, the fundamentals of ITSs are exposed. The current domains of applications and technologies in use are outlined in Section 13.2, while the next, technologically more demanding generation of ITS is the subject of Section 13.3. The advances of 5G over current communication technologies as well as enabling ITS solutions are discussed in Section 13.4. Some of 5G advances that will support vehicular communications are also given in this section. The chapter ends with topics related to 5G ITS ecosystem including converging technologies, stakeholders and business models under consideration.

13.2 Basic Functional Areas of Intelligent Transportation System

There are numerous ITS solutions, of which many are already largely in commercial use. The basic functional areas according to which ITS is commonly classified include the following:

Advanced payment system (APS). APS refers to various electronic payment solutions for tolls, public transport tickets and parking fees. Electronic toll collection enables transactions by establishing a short-range communication system between transponders and external units installed at a pay point. In addition to this widespread application today, advanced payment technology is also used in cities to charge for the entry of vehicles into central city areas. This kind of restriction is known as congestion charging. Regardless of the purpose and position of the charge (on highways or in cities), the basic components of technology are automatic vehicle identification, automatic vehicle classification, transaction processing and system for violations identification.

Advanced transportation (traffic) management system (ATMS). ATMS provides a dynamic approach to road information and traffic signaling systems. This includes the introduction of centralized road traffic management from the traffic control center (TCC), which operates similarly to flight control or signaling centers in rail traffic. The data are collected and transmitted for processing in TCC through sensors and cameras. Sensors are built into the surface of the road or placed on equipment alongside the road, whereas cameras are set up to allow for a wide visual display of the road section. The result is real-time multimedia information on the state of roads and traffic, driving recommendations or service information (which may include environmental aspects, e.g., tunneling pollution or meteorological conditions, etc.). An important domain of proactive management is adaptive traffic signal control system. It adjusts the operating mode of the signal devices to traffic conditions (number of vehicles on the road). Sensors detect vehicles, process data and anticipate the size of the incoming vehicle flow. Accordingly, instead of a static timing plan, the signal status is continuously tuned.

Advanced traveler information system (ATIS). ATIS includes a range of applications that provide travelers with different travel information, important for planning and making decisions on a trip. What *advanced* information systems differs from the previous ones are personalized assistance and real-time decision support systems, such as dynamic navigation that allows customization of the proposed route to the current traffic conditions. ATIS is intended for drivers as well as users of public transport. Information can be transmitted to drivers within the vehicle or through the traffic management centers. For public transport users, a more advanced form of information is multimodal travel planner, which provides information on the optimal way of moving from the starting point to the destination.

Advanced vehicle control and safety system (AVCSS). AVCSS supports driving in many different ways. It includes speed and distance control, unwanted traffic lane departure prevention, parking assistance, etc. These systems are evolving toward active support, which implies not only the reception of information and warnings but also systems for an adequate response. For example, the adaptive cruise control system allows automatic speed adjustment to maintain a safe distance from the vehicle in front. Advanced detection systems record these distances and transmit information to the braking system. When the vehicle moves farther away, the predefined speed is automatically restored. Lane keeping system (LKS) is a proactive technology that not only carries on track marking but also recognizes whether the driver deliberately leaves the traffic lane or not. In case of unintentional deviation, LKS undertakes corrective actions by acting on the vehicle's steering system. These and other advanced driver support technologies are already in commercial use.

13.3 Cooperative Intelligent Transport Systems

A new ITS generation transforms vehicles from autonomous into cooperative systems. Cooperative intelligent transport systems (C-ITS) establish communication processes between vehicle and its environment. A constant message exchange between vehicles interacting with each other (V2V), vehicles and infrastructure (V2I) or vehicle and pedestrians (V2P) enables vehicles to be aware of surroundings, beyond the visible (Figure 13.1). In addition, other types of technological areas are considered. For example, connections between vehicle and network (V2N) or vehicle and grid (V2G) are also possible. The former enables access to cellular infrastructure for infotainment or navigation purposes. The latter connects electric vehicles (EVs) with power grid for bidirectional energy exchange. V2G is seen as crucial technology for the prevention of power grid overloading possibly caused by EV market growth.

Collectively, this is called V2X – "vehicle to everything" concept, meaning that the vehicle is connected to any entity that affects its movement, regardless of distance. With all relevant information, driving becomes safer and smarter. The first wave of V2X mostly includes various warnings being forwarded to drivers. More proactive is adding V2X communication to the advanced in-vehicle controlling systems such as steering or braking, enhancing functionality of these systems. With higher levels of automation, and the development of artificial intelligence (AI), the tendency would be to achieve the machine–machine interaction. This will exclude the driver as information recipient, leading to self-driving cars eventually. For the time being, many solutions for connecting vehicles with each other and with the roadside units (RSUs) (V2V and V2I) are demonstrated within pilot projects, or they are in a testing phase.

There are various possible scenarios where connecting vehicles prove to be vital for performing adequate driving tasks. We mention some of the applications that are currently seen as the most beneficial for safety and traffic efficiency. Depending on the purpose and objectives, we distinguish applications intended for traffic safety and management at intersections and in the traffic flow as well as for parking operations (Figure 13.2).

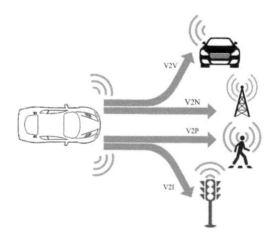

FIGURE 13.1
V2X concept. V2I, vehicle to infrastructure; V2N, vehicle to network; V2P, vehicle to pedestrian; V2V, vehicle to vehicle; V2X, vehicle to everything.

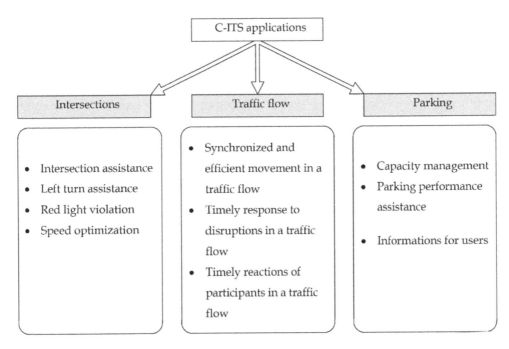

FIGURE 13.2
Cooperative ITS applications. ITS, intelligent transportation system

13.3.1 Intersections-Related Applications

At intersections, safety can be enhanced both by V2V and V2I communications. Intersection movement assistance and left turn assistance are V2V applications which warn the drivers of unsafe situations when entering intersections or maneuvering – turning left across traffic. With V2V, a vehicle becomes aware of the presence and the speed of other vehicles including those out of the sight. The red light violation warning (RLVW) is V2I application that gives the driver information of possible red signal violation. The system combines the data on vehicle movement (speed and acceleration) with intersection geometry and signal phase and timing to estimate if the driver would have sufficient time to stop at red light.

Another issue of interest that takes advantage of V2V and V2I communication is the adjustment of vehicle speed with the signal phase and timing at signalized intersections and vice versa. Based on vehicle to infrastructure message interchange, and knowing the vehicle location and kinematics, the onboard processor calculates and recommends the optimal speed before entering intersection. In this way the vehicle encounters green light, avoids stops and hence improves energy efficiency.

13.3.2 Traffic Flow–Related Applications

Several V2V and V2I applications focus to traffic flow–related safety and efficiency. Emergency electronic brake light uses V2V to alert of sudden braking of the vehicle(s) ahead, enabling timely reaction to avoid collision. Traffic flow–related warnings include not only sudden braking but also other conditions that atypically slow down the speed of the flow (such as traffic accidents, weather conditions, road hazards, road works, etc.). The related information can be delivered by both V2V and V2I communication. When it comes to traffic efficiency, the topic is synchronized movement in a traffic flow. By supplementing adaptive cruise control with V2V, the system becomes cooperative (CACC), meaning that information are instantly exchanged among all vehicles in a flow, independently on sensing range. The benefit of such extension is in maintaining optimal distance between vehicles without unnecessary braking or accelerating. This saves energy, reduces pollution and increases road capacity.

The emergency vehicle warning system is another V2V application which enables a proper behavior in case of appearance of emergency vehicles. Thanks to information interchange between vehicles, it is possible to know ahead of emergency vehicle approaching, even before the sound of the siren. Thus the vehicles can be safely and timely removed from the traffic lane.

13.3.3 Parking-Related Applications

A considerable volume of traffic is generated from searching for available parking spaces. According to some estimates, cruising accounts for 30% of traffic (Shoup, 2006). Time waste, congestion and air pollution increase generalized transport costs, which is why efficient management of parking operations is needed. New, smart-based parking solutions use sensors, technologies and applications to disseminate information about parking lots occupancy and guide vehicles to free parking space. To provide distributed, decentralized parking information system, both V2V and V2I two-way communication can be used. In V2V applications, available parking spaces are recorded by vehicles and reported to other vehicles, while in two-way V2I, the information comes from the equipment installed on parking lots. In V2I cellular-based communication, the requests for parking and reservation are sent, and thereafter the information on parking availability, together with allocation of parking lot, driving direction and payment instructions are forwarded to driver (V2I).

In choosing the appropriate technology, protection of open parking lots and high capital costs have to be taken into account. In that sense, machine vision and neural networks are among promising, lower cost technologies for object detection, suitable for on-street parking (Paidi et al., 2018). Another field of application is parking performance assistance where V2V and communication with parking controllers is used to occupy as little parking space as possible.

13.4 V2X from Technological Perspective

13.4.1 Dedicated Short-Range Communication versus Cellular Technology

There are currently two rival technologies for V2X support. One is Wi-Fi technology, based on IEEE 802.11p wireless standard, which enables direct ad hoc intervehicle and vehicle-to-infrastructure communication. The IEEE 802.11p standard laid the foundation for DSRC (dedicated short-range communication) technology in North America and its European counterpart ITS – G5, developed by ETSI. The other is cellular-based technology, C-V2X. It is supported by the 3rd Generation Partnership Project (3GPP) technical specifications, introduced to meet the requirements of both direct short-range communications (V2V, V2I and V2P) and indirect, longer-range, network-based applications (V2N).

As for the current state of play, there are strong arguments to support Wi-Fi-based V2X communications. First, DSRC is tested for more than a decade. It is proved, validated and, as a mature technology, ready for widespread deployment. It has dedicated spectrums in 5.9 GHz frequency band, in most of world for more than a decade, which is considered as one of the greatest assets. There have been massive investments in DSRC infrastructure worldwide – in RSUs installation and in on-board units equipment. Some leading automakers such as General Motors, Toyota and Volkswagen started deploying DSRC equipment in their vehicles. Another advantage of DSRC is that it is distributed, peer-to-peer system, operating without relying on network-based transmission, and hence independent from single-point deterioration and/or out-of-coverage problem. For the time being, it is the only commercial available technology that enables stable communication and ubiquitous, uncoordinated access.

Although IEEE 802.11p well supports many ITS applications, a question is whether it can technically advance to maintain the quality of service for massive V2X communications. On the other side, we are witnessing dramatic improvements of cellular networks, and that can be a decisive factor for C-V2X superiority in the future. The proponents of C-V2X point out that cellular communication will be necessary to support self-driving vehicles. Cellular technology is also the key enabler of new business models in transportation, like different models of shared mobility. Beside autonomous vehicles, shared mobility and every form of mobility personalization are seen as the future of transportation industry.

Another topic of interest is the technology for V2P and P2V communication. The majority of existing solutions are based on traditional sensor-based detection methods (RADARs, computer vision, LASER scanners, etc.), which are dependent on line of sight detecting. Besides the capability to support cooperative/collective sensing (360° non–line-of-sight awareness), especially in automotive driving mode, in favor of cellular-based solutions is also the possibility to utilize existing infrastructure and devices. Namely, smartphones are not by default DSCR-enabled devices, and efforts to make them operative in the framework of IEEE 802.11p are still in their infancy with scarce commercial tests (Subramanian, 2013). On the other hand, in the case of cellular communication solutions for V2P, there is no need to install new hardware (e.g., reliant on the RSUs), while the lack of cellular connectivity of the cars can be overcome by relying on driver's smartphone. Also, cellular-based V2P offers high mobility support, as well as high bit rate, communication range and capacity, compared with utilizing Wi-Fi (which suffers from limitations posed by interference, range only up to 100 m, sensitivity to Doppler effect, etc.), and is more easily aligned with cloud-based computation (Bagheri et al., 2014). Some specific problems related to cellular-based V2P deployment have been resolved for some time. Examples of specific problems related to cellular-based V2P deployment are prediction algorithms for warning pedestrians via smartphone (Hussein et al., 2016)[1] and battery power consumption (Bagheri et al., 2014). Currently, the most advanced cellular technology is long-term evolution (LTE)-V2X, which is a bridge standard on the road to 5G-based V2X. With the introduction of proximity services (ProSe) feature in 3GPP Release 12, the device-to-device communication is enabled. Starting from LTE Release 14, 3GPP adapts to V2X services by enhancing resource allocation with centralized (mode 3) and distributed (mode 4) mechanisms. In mode 3, cellular network is used to manage radio resources in V2V communications, whereas in mode 4, vehicles manage their radio resources and communicate without cellular network support. Direct communication, which is crucial for many safety applications with low-latency requirements, is implemented over LTE PC5 interface. A long-range communication is supported by wide area network LTE Uu interface and facilitates vehicles to make use of Internet and network services (V2N).

LTE Release 14 supports basic safety use cases. With the further evolution of LTE V2X in Release 15, and NR-based V2X, which is a part of Release 16 and represents a global standard for 5G, the way to 5G V2X is traced. The 5G network communication capabilities open up for a range of advanced ITS applications (Section 13.4.2). 5G V2X will not replace LTE V2X but will still rely on it for safety-critical use cases. It will consolidate LTE and new radio technology to address basic safety as well as enhanced/advanced V2X services.

13.4.2 5G V2X Applications

The 5G vehicular network is envisioned to meet the communication requirements for new services exposed by automotive industry: platooning, advanced driving, remote driving and extended sensors. These are more stringent use cases, whose functions go beyond safety. They are considered as autonomous driving applications and are waypoint to full automation. 5G-based solutions are needed for these use cases, commonly referred as "enhanced" V2X (eV2X), because of ultrareliable low-latency and high-throughput communication requirements (Table 13.1).

Platooning. Platoon is a convoy of vehicles formed to move in a coordinated way, close following each other. Supplied with a cooperative adaptive cruise control (CACC), a group of vehicles is acting as a road train while in the platoon. In an autonomous vehicular platoon, one leading vehicle is followed by autonomously driven vehicles at small distances. Without unnecessary braking and accelerating and with reduced air drags, a significant potential to cut fuel consumption is gained. Apart from fuel savings, low intervehicle distance improves the use of road capacity. High level of automation with driver controlling only the leading vehicle makes platooning less dependent on human decisions and hence safer and more productive.

The stability and safety of the platoon is provided by control system, which integrates information from V2V links and data collected from sensing technologies. Due to operational interdependencies between communication and control system, the stability of the system should not be deteriorated by wireless transmissions delay (Zeng et al., 2019). The complexity of platooning stems from string requirements that should be simultaneously achieved. Fundamental issues of concern in platooning are related to topology, i.e., longitudinal and lateral control for the following vehicles and the delivery requirements for cooperative messaging (Nardini et al., 2018). V2V communication allows for exchange information about kinetics and intentions among platoon members and also supports joining and leaving a platoon. Besides, platooning operations include indication on the

TABLE 13.1

Performance Requirements for Vehicles Platooning, Extended Sensors, Remote Driving and Advanced Driving (ETSI, 2018)

	Degree of Automation	Max End-To-End Latency (ms)	Reliability (%)	Data Rate (Mbps)	Minimum Required Communication Range (m)
Vehicles platooning	Lowest-low	25–20	90–/	/	/–350
	Highest-high	10–20	99.99–/	65	80–180
Extended sensors	Lower	100	99	/	1000
	Higher	3–50	99–99.999	10–1000	50–1000
Remote driving	/	5	99.999	Uplink – 25; downlink – 1	/
Advanced driving	Lower	25–100	90	10–30	700
	Higher	10–100	99.99	50–53	360

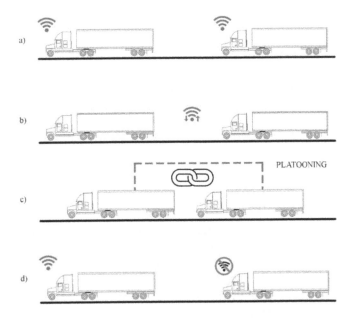

FIGURE 13.3
Different scenarios of V2V communication: (a) vehicles with radar technology; (b) V2V wireless communication; (c) V2V wireless linking – platooning; (d) disconnection from platooning network. V2V, vehicle to vehicle.

presence of a platoon to other vehicles approaching, so that they can join, bypass or avoid it. Different scenarios of V2V communication are shown in Figure 13.3.

Both simulations and real-world testing have shown that current V2V communication fails to effectively support platooning under dense traffic conditions and/or with more vehicles in the platoon formation. Rashdan et al. (2016) analyzed three performance metrics in a simulation scenario of 10 platoon vehicles at 8-lane highway: end-to-end delay (E2E) for cooperative awareness messages (CAMs) transmission, packet delivery ratio (PDU) from the leading vehicle to platoon members, and update delay (UD) that indicates the distribution of packet loss. With ETSI G5, the average E2E delay increases when communication channel is used by platoon members and nearby vehicles at the same time. Similar is found for PDU – the channel congestion and packet loss occurs even without background traffic, due to message collisions. This reduces PDU values at the end-of-line vehicles. The simulation results for UD show the occurrence of more consecutively lost packages. The probability of such events rises with increasing distance between the leading vehicle and a platoon member.

One of the problems that need to be solved in cellular-based platooning is the handover. With the increase of the platoon size, the data transfer requirements could be compromised when traversing different cell sites. As pointed in Nardini et al. (2018), this should be addressed with some coordination mechanism between neighboring nodes and also with handover solutions that enable all vehicles to execute handovers at the same time (bulk handover).

Remote driving. Remote driving enables control over the vehicle's operations by driver/operator in the control center. It not only supports driving in normal traffic conditions but also makes use of human knowledge and experience in dangerous environment, where fully autonomous vehicles still have to "learn" how to cope with (for example, bad weather conditions). Remote control will be needed even after the deployment of self-driving vehicles, to serve as a backup for unsafe situations (Kang et al., 2018). Except that remote driving can be utilized by people with disabilities, it is important for the development of car rentals and car sharing markets. Remote driving makes fleet management easier since cars can efficiently be repositioned to be accessible to users and also detour can be performed.

In 2017, the first field trials for remote driving powered by 5G ultrahigh bandwidth and ultralow latency took place in Shanghai. The multi-HD video data coming from a vehicle was transmitted to the operator with 50 Mbps uplink throughput. It gives the operator/remote driver detailed sensory perception. The control signals from the operator back to the vehicle were transmitted with E2E latency under 10 ms. Some simulations have, however, shown that in loaded conditions with high traffic density scenario, larger bandwidth, dedicated V2N carrier or other techniques such as MIMO (multiple-input multiple-output) would be required (Edwertz, 2017).

Advanced driving. This category refers to higher levels of automation where vehicles share information about their movement and intentions to improve safety. Information is collected from local sensors and RSUs. Messages are generated aperiodically with the size depending on the use case. Typical examples are cooperative collision avoidance (CoCA), and more specific, cooperative intersection collision avoidance system (CICAS), emergency trajectory alignment (ETrA) and cooperative lane change (CLC). With ETrA messages, a vehicle can share information of unsafe situation with neighboring vehicles so that they can cooperatively align their trajectories (Bian et al., 2017). Cooperative lane change application facilitates the exchange of information of lane change operations, which is often unpredictable and difficult to recognize (Li et al., 2018).

The aforementioned enhanced V2X services are in a testing phase, while a number of architectural issues have to be solved. For example, as pointed out in Garcia-Roger et al. (2019), the quality-of-service requirements for advanced V2X applications are hardly to be achieved with a single communication interface. The architecture for simultaneous use of cellular-based Uu and PC5 V2V-based communications and their dynamic selection have been developing. Among issues to be solved is how and when to select radio access technology for PC5, given that it can be LTE- or NR based. The topics are also network slicing for eV2X, support of edge computing, interworking of 5G V2X with evolved packet core (EPC) V2X (Husain et al., 2018), etc.

13.4.3 5G Cutting-Edge Technologies in Vehicular Communications

Many ideas have been explored to discover what technologies will put 5G into practice. Some of them are particularly about making vehicles fully connected, i.e., moving from in line-in-sight sensing to 360° non–line-of-sight awareness and paving the way to super-advanced applications such as adaptive platooning or fully automated driving. The core issues to be resolved are related to ultrahigh data throughput, delay-insensitive data transmission and mitigating interference in high mobility circumstance. A mix of 5G enabling technologies: *millimeter* waves (mmWaves), beamforming, small cells, massive MIMO and full duplex (FD) is envisioned as the key support for cooperative sensing of vehicles. These 5G building blocks may be argued as the most discussed/researched, but the list of upcoming solutions is not nearly exhausted, since new ideas are emerging every day. Among fresh breakthroughs are network slicing and mobile edge computing. Elasticity, efficiency and functionality of 5G network, increased through virtualization of functions, will help to fulfill diverse and complex requirements of multitenant V2X services.

5G enabling technologies for vehicular communications. According to predictions, about 90% of new US vehicles will have cellular modems by 2025 (Lu et al., 2014), while the average number of sensors is projected to be around 200 by 2020 (Choi et al., 2016; Massaro et al., 2017). At the same time, it is expected that connected cars will send out 25GB of data every hour (55Mbit/s) and autonomous vehicles up 1 TB of data per hour of driving, while a four-lane highway in normal conditions will require a throughput of tens of Gbit/s per kilometer (Buzzi, 2018).

Existing DSRC and 4G cellular communications do not support the gigabit-per-second data rates (terabyte-per-hour) needed for raw sensor data exchange between vehicles. One of the key technologies envisioned for the next-generation V2X networks and autonomous vehicles is mmWaves. This is for the following potentials: gigabit-per-second data rates, broad bandwidth, wide field of view sensing, precise localization capabilities and apprehending raw sensor data exchange among vehicles and infrastructure. mmWaves exploit spectrum between 30 and 300GHz, use large arrays of antennas and highly-directive beams to recompense the propagation effects in high frequency bands and offer additional gains in terms of spatial sharing. They are already used by automotive radars – e.g., RaDAR for applications such as adaptive cruise control and collision avoidance (Hasch et al., 2012; Kumari et al., 2015; Coll-Perales et al., 2018). As one of the key pillars of future 5G networks, mmWaves are foreseen as keystone of 5G vehicular communication with already developing solutions introduced in 5G-MiEdge and 5GCAR projects. 5G currently operates as LTE at frequencies up to 6GHz, but international organizations such as 3GPP are working to manage the allocation of mmWave spectrum for 5G cellular applications – move to 28GHz 5G, 39GHz 5G and E band 5G with peak rates 1.5 Gbps, 3Gbps and 24 Gbps, respectively.

mmWaves are envisioned as an enabler predominantly for V2I (e.g., infotainment, traffic monitoring, etc.,) and V2V (e.g., platooning) having that V2N (e.g., traffic management services) has less stricter latency or reliability requirements (except in case of teleoperated vehicles) (Boban et al., 2017). In terms of mmWaves vehicular technology, V2I is more in focus, but solutions for V2V are emerging (e.g., answers to reducing the beam alignment introduced in Va et al., 2016; Choi et al., 2016; Loch et al., 2017). For example, MacHardy et al. (2018) extracted directionality (line-of-sight [LoS] connection) and unsuitability of a car to serve as carrier for antenna with 360 coverage, as the key challenges for 5G V2V mmWave deployment. However, some of the mmWave challenges can be perceived as specific in V2X applications. For example, as argued by Karmoose et al. (2018), directionality characterizing mmWaves can help platooning cars to create encrypted shared keys with rates up to 166 Mbps. This is important having that platooning is seen as one of the most vulnerable applications of vehicular networks in terms of safety and security. Also, blockage by objects, a general problem associated with mmWave technology, is also argued to be particular in terms of V2V communications. As discussed by Choi et al. (2016), blockage of signals can be even seen as beneficial for V2V communications if only vehicles in proximity share raw sensor data. Referring to Guha et al. (2012), they explain that blockage effect when combined with narrow mmWave beams can reduce intervehicle interference.

When it comes to standards, mmWave vehicular communication is already available through adaptation of IEEE 802.11ad – WLAN standard for mmWave communications at 60GHz, which supports directional multi-gigabit channel access and beamforming protocols (like MAC – medium access control). Mitigating interference, i.e., beamforming with narrow beams is a central issue of mmWaves deployment with specific challenges in vehicular communications posed by dynamic topology of networks and high mobility. Developed initially for indoor exploitation, IEEE 802.11ad standard is limited in supporting high mobility along with intermittent connectivity in vehicular environments (Choi et al., 2016). Apart from IEEE 802.11ad, two recognized streams of solution for mmWave for vehicular communications are 5G mmWave cellular and a new standard (e.g., on the basis of ISO/DIS 21216 standard devised in 2012) (Choi et al., 2016; Coll-Perales et al., 2018). ISO standard for vehicular application at the 60 GHz band is not a complete system and only provides several parameter values of the physical layer (Va et al., 2016).

Besides beamforming, small cells, massive MIMO and FD are also seen as supportive of 5G mmWaves vehicular communications. 5G vehicular small-cell networking (V-SCN) is seen as a prominent solution for growing demand of high volumes of mobile data in vehicular communications. Small cells, i.e., low-powered radio access points accommodate exchanging delay-insensitive data in vehicular communications related to map updates, infotainment, collecting traffic and sensor information, etc. Interference management and handover in case of high mobility are seen as the most challenging issues for V-SCN while solutions based on mmWaves and MIMO antenna technology are envisioned as promising. For example, Qian et al. (2017) followed the experiences with OFDMA in 4G and LTE and proposed to use nonorthogonal multiple access–enabled 5G V-SCNs to avoid excessive handovers and improve spectrum and energy efficiency in vehicular communications. However, as they point out, 5G V-SCNs are far more dynamic than usual small cell networks, and very dense use of small cells is needed to make moving vehicles join vehicular wireless networks. This leads to more frequent handovers and cochannel interference, i.e., more technological hurdles to overcome. With theoretical potential to double the spectrum efficiency through concurrent downlink/uplink transmission inside the same frequency band, FD in a dense small cell scenario is upcoming 5G radio access technology. Challenges in mitigating self-interference associated with FD are additionally complex in fast time-varying V2X propagation environment and call for new design of MAC protocols. However, vehicle onboard units are seen as good candidates for FD transceivers due to power supply and processing characteristics (Campolo et al., 2017a). Although IEEE 802.11 capacity is limited by the available unlicensed spectrum, FD has been disclosed for 802.11-based V2X communications with similar considerations for 5G V2X communications. When it comes to C-V2X, FD can help for rising the number of vehicles in the range in terms of cooperative awareness service based on LTE-V2V (Bazzi et al., 2017).

Improved 5G network architecture for diverse V2X use cases. The idea of new 5G network architecture is grounded in software-defined networks and virtualization of network functions. It is built up with the aim to bring elasticity to 5G, save money and energy and increase efficiency and functionality. Namely, the emergence of new services raises the interest for better use of physical resources by linking them into the virtual network using software solutions. The result is tailored made (customized) network for particular users, applications, devices, operators, etc.

From the perspective of vehicular communications, this is of great importance having the diverse requirements posed by specific use cases. For example, autonomous driving is more about latency than a throughput while the opposite is for some infotainment services (e.g., streaming a video in a moving car). Business bundles, i.e., different types of network slices packaged as a single product tailored for business customers can meet the specific vehicular communications demand – e.g., vehicle may need simultaneously a high bandwidth slice for infotainment and an ultrareliable slice for assisted driving. This comes from the key feature in 5G network slicing for V2X labeled as multitenancy. Tenants refer, for example, to the car, the driver, the occupant, the road authority, etc. The situation is complicated by the fact that each of these tenants requires different services often simultaneously. All these services are mapped to different types of slices offered by different providers and provided through multiowned infrastructure. Simply explained, it is about one car that is in the process of remote diagnostic, with a driver who is in safety data exchange and with a passenger who is watching HD video.

Although still in its infancy, network slicing is paving a way to V2X applications with first field tests in South Korea as well as several propositions by scholars (e.g., a model for abstracting vehicles and RSUs as software-defined networking switches proposed by He et al. (2016) or configuration of the predefined set of slices introduced in Campolo et al. (2017b).

Apart from network slicing, a recent study by Shah et al. (2018) pinpoints to proximity services and mobile edge computing (MEC) as driving wheels of 5G vehicular communications. MEC enables mutual optimization of radio and computing resources, which leads to empowering sensors with amplified computational capabilities to fulfill sophisticated requirements in vehicular communications such as low latency, reliability, increased bandwidth and processing large amount of sensed data. MEC is also referred as distributed cloud or fog computing that will allow cars to leverage local mobile terminals/cloud and perform computations without gaining the additional latency (MacHardy et al., 2018). Giust et al. (2018) extracted relevance of MEC for V2X safety use cases following 5GAAA experience and highlighted intersection movement assist, real-time situational awareness with HD maps, cooperative lane change of automated vehicles and vulnerable road user discovery as the most relevant. An example from filed trial point of view is mmWave and MEC for automated driving included in 5G-MiEdge project. It is about exchanging high-definition maps with focus on applications such as collision avoidance, navigation, city guide and city parking.

13.5 Environment for 5G V2X Deployment

Vehicular communication under 5G is expected to pool together wide area knowledge and multiple technologies. The major impact is expected from the following converging technologies:

- human–machine interface (HMI) for handling human–computer interactions;
- AI whose role would be to facilitate data-driven decisions, such as traffic flow predictions, congestion control at intersections, dynamic data carrier node selection (Ye et al., 2018) and so on;
- cloud computing with vehicular cloud framework that aggregates and shares underutilized vehicular resources in both static and mobile environment;
- big data analytics to manage vast amount of real-time data.

The 5G V2X ecosystem involves mutual cooperation across stakeholder chain. The main components of this chain, interacting with each other, are mobile network operators, vendors of telecommunication equipment, standardization organizations, original equipment manufacturers in automotive industry, road infrastructure operators, regulatory bodies, local authorities and users (Figure 13.4).

Institutions that contribute to the development and standardization of 5G networks are presented in Table 13.2.

The role of 5G V2X stakeholders and inclusion of other parties depends on the service type. For example, for smart parking solutions, parking providers and payment solutions, providers have to be included; car sharing should be supported by renting companies, fleet

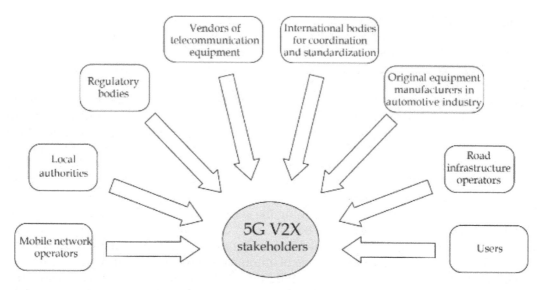

FIGURE 13.4
5G V2X ecosystem. V2X, vehicle to everything.

TABLE 13.2

Overview of 5G Networks Development and Standardization (Storck and Duarte-Figueiredo, 2019)

Institution	Projects and Initiatives	Target	Main Contributions
ITU	International Mobile Telecommunications for 2020 and Beyond (ITM-2020)	Radio regulations; operational aspects; protocols and test specifications; performance, QoS and QoE; security	Recommendations (standards)
3GPP	5G specifications	Radio access network; service and systems aspects; core network and terminals	Releases; technical specifications
ETSI	5G technologies	mmWave transmission; next-generation protocols; MEC; NFV	Technical specifications
NGMN	Next-generation mobile networks (NGMN) 5G initiative	Technology evolution toward 5G	White papers
ATIS	Technical forum	Incubator of new business models	White papers
5G PPP	5G Public Private Partnership (5G-PPP) projects	5G infrastructure; 5G architecture	White papers
IEEE Future Networks	Technical community	Providing practical, timely technical and theoretical content; development and deployment of 5G	Research publications
5G Americas	5G network development on Americas	Support and promote the full development of wireless technology capabilities	White papers
5GMF	5G research and development by industry	5G radio access technologies; network technologies for 5G	5GMF white paper

(Continued)

TABLE 13.2 (*Continued*)

Overview of 5G Networks Development and Standardization (Storck and Duarte-Figueiredo, 2019)

Institution	Projects and Initiatives	Target	Main Contributions
Verizon 5G TF	Forum and technical specifications	Specifications for physical layer, MAC, RLC, PDCP and RRC	5G specifications
5TONIC	Open research laboratory	SDN; NFV; physical and MAC layer	5G technologies
5GAA	Mobility and transportation services	Use cases and technical requirements; system architecture; standards and spectrum; business models	White papers

MAC, medium access control; mmWave, millimeter wave; NFV, network functions virtualization; PDCP, packet data convergence protocol; QoE, quality of experience; QoS, quality of service; RLC, radio link control; RRC, radio resource control; SDN, software-defined networking.

managers, payment solutions providers and mobility operators; e-call strongly depends upon regulators and operation of public safety answering points (PSAPs); for remote diagnostic, workshop service is also needed.

Currently, various business models are considered for C-V2X viability. Generally, new service types such as infotainment generate more possibilities for interactions between stakeholders than safety applications alone. The nature of interactions between parties involved is a matter that has yet to be determined.

For example, as pointed in the 5G-PPP white paper related to the 5G V2X deployment (5G-PPP, 2018), both road infrastructure operators and automotive industry could act as a lessee of 5G connectivity. Accordingly, end users set up their business relationships with road infrastructure operators or automotive industry. Under the term "end users," we assume vehicle owners, which do not necessarily have to be drivers but the mobility service providers as well. For that reason, passengers also fall into this category. Within V2P context, pedestrians act as end users alike.

Regardless of who the end user is, they can be burdened with high costs, which is thereby a concern for the penetration level. The hot topic is how to deliver high mobile data rates in a cost-effective way. Moreover, the onboard units are also at the expense of vehicle owners. For example, according to calculations, for a vehicle traveling at a distance of 100 km at a speed of 100 km per hour, creating HD maps would require a 120 Mbytes data volume (5G-PPP, 2018). A promising offset to these costs is to utilize vehicles' computing capabilities to forward mobile data to nearby vehicles in exchange for various kinds of rewards. This new business opportunity is known as vehicle-as-infrastructure concept. On the other side, it is estimated that the capital and operational expenditures (CAPEX and OPEX) will be too high for a single network provider to bear. To cut off the deployment costs, several infrastructure sharing models have been considered. As reported in BEREC Report on Infrastructure Sharing (BEREC, 2018), the savings are highest when both cellular network base stations and spectrum are shared between operators. In that case, operators transmit on dedicated spectrum while radio access resources interlink with core networks of other operators. The estimated cost savings are 33%–45% CAPEX and 30%–33% OPEX.

Another issue of interest refers to coexistence of V2V technologies in the 5.9 GHz band. The 5GAA advocates open market approach meaning to allow both DSRC and C-V2X to operate on 5.9 GHz band. Because of interference, distinctive channels of 10 MHz should be reserved.

Judging by the number of car OEMs (original equipment manufacturer) in the 5G Automotive Association (5GAA), it is likely that C-V2X will quickly penetrate the auto market, although the industry is fragmented at the moment. The deployment of technology does not have to be limited to luxury vehicles but deployed in all classes, which could achieve the economies of scale. Single chipset solution, which integrates PC5 interface and Uu connectivity and will be commercially available from year 2020, is expected to contribute to greater C-V2X adoption.

Coordinated, connected and automated mobility is a key topic in policy documents, and there is a tendency to mandate it in the coming years. European Strategy on Cooperative Intelligent Transport Systems (C-ITS) (European Commission, 2016) published in 2016, followed by Communication from the European Commission in 2018 (European Commission, 2018), paves the way for building connected vehicle ecosystem in Europe. It addresses specific issues such as safety, cyber security, data protection and access and also gives an insight into possible long-term impacts on society and economy. Despite that the Commission stands for technology-neutral policy, it has budgeted considerable amount for 5G connectivity testing and pilot projects. In the United States, a proposal for the mandatory introduction of V2V and V2I communications for all new vehicles has been prepared since 2017. For the time being, the 5.9 GHz band is allocated to DSRC-based V2X applications. However, NHTSA declares its technological neutrality and is engaged in research and testing C-V2X technologies and devices. As for the current state of play, it is likely that C-V2X will be first commercialized in China, where there is a strong governmental support and tight collaboration of stakeholders for city-wide trials and projects. From 2016, the 20 MHz spectrum in 5.9 GHz band has been reserved for cellular-based ITS trials in major cities. It is planned to establish the preliminary standards for low-level automated driving by 2020 and to develop technology for autonomous driving by 2025.

13.6 Concluding Remarks

The 5G communication should allow for a range of advanced ITS applications, which are a waypoint to autonomous vehicles as a breakthrough that modern society is striving for. Although the alternative DSRC technology supports some safety-critical V2X use cases, technological improvements of cellular networks will make them superior for massive V2X communications, needed for self-driving vehicles. The autonomous driving applications such as platooning, extended sensors, remote driving and advanced driving are currently under development, all of which require ultrareliable, low-latency and high-throughput V2V and V2I communication. To achieve the quality of service, simultaneous use of cellular-based Uu and PC5 V2V-based communications is required. The test fields and simulations show that in high-traffic density conditions, vehicle platooning and remote driving applications are still challenging tasks even with improved cellular technology.

Only synergy of 5G enabling technologies will empower very dynamic vehicular networks to come to practice. mmWave technology will enable to move up along the spectrum to support the ultrahigh data throughput associated with cooperative perception of sensing. This will impose supplementary technological improvements: dense use of small cells to provide delay-insensitive data communication, further leading to the need for new beamforming designs and massive MIMO along with FD to mitigate interference in high mobility circumstances. In addition, new 5G network architecture built up around the idea of network slicing, and mobile edge computing will help to cope with multitenant V2X services.

References

Bagheri, M., Siekkinen, M., Nurminen, J. K. (2014). Cellular-based vehicle to pedestrian (V2P) adaptive communication for collision avoidance. In *Proceedings of the International conference on Connected Vehicles and Expo (ICCVE)*, Vienna, Austria.

Bazzi, A., Masini, B. M., Zanella, A. (2017). How many vehicles in the LTE-V2V awareness range with half or full duplex radios? In *Proceedings of the 15th International Conference on ITS Telecommunications (ITST)*, Warsaw, Poland.

BEREC (2018). BEREC report on infrastructure sharing. Report No. BoR (18) 116. Available at: https://berec.europa.eu/eng/document_register/subject_matter/berec/reports/8164-berec-report-on-infrastructure-sharing

Bian, K., Zhang, G., Song, L. (2017). Security in use cases of vehicle-to-everything communications. In *Proceedings of the 86th Vehicular Technology Conference (VTC-Fall)*, Toronto, ON, Canada.

Boban, M., Kousaridas, A., Manolakis, K., Eichinger, J., Xu, W. (2017). Use cases, requirements, and design considerations for 5G V2X. Available at: arxiv.org/abs/1712.01754.

Buzzi, S. (2018). Principles of millimeter wave communications for V2X [Online presentation]. Available at: https://nms.kcl.ac.uk/toktam.mahmoodi/v2x-summer-school/slides/Day-1/SB-5GCAR%20PhDSchool.pdf

Campolo, C., Molinaro, A., Berthet, A. O., Vinel, A. (2017a). Full-duplex radios for vehicular communications. *IEEE Communications Magazine*, 55(6), 182–189.

Campolo, C., Molinaro, A., Iera, A., Menichella, F. (2017b). 5G network slicing for vehicle-to-everything services. *IEEE Wireless Communications*, 24(6), 38–45.

Choi, J., Va, V., Gonzalez-Prelcic, N., Daniels, R., Bhat, C. R., Heath, R. W. (2016). Millimeter-wave vehicular communication to support massive automotive sensing. *IEEE Communications Magazine*, 54(12), 160–167.

Coll-Perales, B., Gruteser, M., Gozalvez, J. (2018). Evaluation of IEEE 802.11 ad for mmWave V2V communications. In *Proceedings of the Wireless Communications and Networking Conference Workshops (WCNCW)*, Barcelona, Spain.

Edwertz, O. (2017). *Performance evaluation of 5G vehicle-to-network use cases*. Master's Thesis. Gothenburg: Chalmers University of Technology.

ETSI (2018). Service requirements for enhanced V2X Scenarios. Technical Specifications ETSI TS 122 186 V15.3.0. Available at: https://www.etsi.org/standards#Pre-defined Collections

European Commission (2016). A European strategy on cooperative intelligent transport systems, a milestone towards cooperative, connected and automated mobility. Technical Report No. COM (2016) 766; Brussels, Belgium: European Commission.

European Commission (2018). On the road to automated mobility: An EU strategy for mobility of the future. Technical Report No. COM (2018) 283; Brussels, Belgium: European Commission.

5G-PPP Automotive Working Group (2018). A study on 5G V2X deployment. *Online White Paper*. Available at: https://5g-ppp.eu/wp-content/uploads/2018/02/5G-PPP-Automotive-WG-White-Paper_Feb.2018.pdf

Figueiredo, L., Jesus, I., Machado, J. T., Ferreira, J. R., De Carvalho, J. M. (2001). Towards the development of intelligent transportation systems. In *Proceedings of the Intelligent Transportation Systems*, Oakland, CA, USA.

Garcia-Roger, D., Roger, S., Martín-Sacristán, D., Monserrat, J. F., Kousaridas, A., Spapis, P., Zhou, C. (2019). 5G functional architecture and signaling enhancements to support path management for eV2X. *IEEE Access*, 7, 20484–20498.

Giust, F., Sciancalepore, V., Sabella, D., Filippou, M. C., Mangiante, S., Featherstone, W., Munaretto, D. (2018). Multi-access Edge computing: The driver behind the wheel of 5G-connected cars. *IEEE Communications Standards Magazine, 2*(3), 66–73.

Guha, R. K., Chennikara-Varghese, J., Chen, W. (2012). Evaluation of a multi-antenna switched link-based network architecture for quasi-stationary vehicle network. In *Proceedings of the 8th Wireless Communications and Mobile Computing Conference (IWCMC)*, Limassol, Cyprus.

Hasch, J., Topak, E., Schnabel, R., Zwick, T., Weigel, R., Waldschmidt, C. (2012). Millimeter-wave technology for automotive radar sensors in the 77 GHz frequency band. *IEEE Transactions on Microwave Theory and Techniques, 60*(3), 845–860.

He, Z., Cao, J., Liu, X. (2016). SDVN: Enabling rapid network innovation for heterogeneous vehicular communication. *IEEE Network, 30*(4), 10–15.

Hussein, A., García, F., Armingol, J. M., Olaverri-Monreal, C. (2016). P2V and V2P communication for Pedestrian warning on the basis of autonomous vehicles. In *Proceedings of the 19th International Conference on Intelligent Transportation Systems (ITSC)*, Rio De Janeiro, Brasil.

Husain, S., Kunz, A., Prasad, A., Pateromichelakis, E., Samdanis, K., Song, J. (2018). The road to 5G V2X: ultra-high reliable communications. In *Proceedings of the Conference on Standards for Communications and Networking (CSCN)*, Paris, France.

Kang, L., Zhao, W., Qi, B., Banerjee, S. (2018). Augmenting self-driving with remote control: Challenges and directions. In *Proceedings of the 19th International Workshop on Mobile Computing Systems & Applications*, Tempe, AZ, USA.

Karmoose, M., Fragouli, C., Diggavi, S., Misoczki, R., Yang, L. L., Zhang, Z. (2018). Leveraging mm-Wave Communication for Security. Available at: arxiv.org/abs/1803.08188.

Kumari, P., Gonzalez-Prelcic, N., Heath, R. W. (2015). Investigating the IEEE 802.11 ad standard for millimeter wave automotive radar. In *Proceedings of the 82nd Vehicular Technology Conference (VTC Fall)*, Boston, MA, USA.

Li, B., Zhang, Y., Zhang, Y., Jia, N. (2018). Cooperative lane change motion planning of connected and automated vehicles: A stepwise computational framework. In *Proceedings of the Intelligent Vehicles Symposium (IV)*, Changshu, China.

Loch, A., Asadi, A., Sim, G. H., Widmer, J., Hollick, M. (2017). mm-Wave on wheels: Practical 60 GHz vehicular communication without beam training. In *Proceedings of the 9th International Conference on Communication Systems and Networks (COMSNETS)*, Bangalore, India.

Lu, N., Cheng, N., Zhang, N., Shen, X., Mark, J. W. (2014). Connected vehicles: Solutions and challenges. *IEEE Internet of Things Journal, 1*(4), 289–299.

Massaro, E., Ahn, C., Ratti, C., Santi, P., Stahlmann, R., Lamprecht, A., Roehder, M., Huber, M. (2017). The car as an ambient sensing platform [point of view]. *Proceedings of the IEEE, 105*(1), 3–7.

MacHardy, Z., Khan, A., Obana, K., Iwashina, S. (2018). V2X access technologies: Regulation, research, and remaining challenges. *IEEE Communications Surveys & Tutorials, 20*(3), 1858–1877.

Miles, J. C. (2014). *Intelligent transport systems: Overview and structure (History, Applications, and Architectures). Encyclopedia of Automotive Engineering.* John Wiley & Sons, Inc., Hoboken, NJ.

Nardini, G., Virdis, A., Campolo, C., Molinaro, A., Stea, G. (2018). Cellular-V2X communications for platooning: Design and evaluation. *Sensors, 18*(5), 1527.

Nowacki, G. (2012). Development and standardization of intelligent transport systems. *International Journal on Marine Navigation and Safety of Sea Transportation, 6*(3), 403–411.

Paidi, V., Fleyeh, H., Håkansson, J., Nyberg, R. G. (2018). Smart parking sensors, technologies and applications for open parking lots: A review. *IET Intelligent Transport Systems, 12*(8), 735–741.

Qian, L. P., Wu, Y., Zhou, H., Shen, X. (2017). Non-orthogonal multiple access vehicular small cell networks: Architecture and solution. *IEEE Network, 31*(4), 15–21.

Rashdan, I., Muller, F., Sand, S. (2016). ITS-G5 Challenges and 5G Solutions for Vehicular Platooning. In *Proceedings of the WWRF37*, Kassel, Germany.

Shah, S. A. A., Ahmed, E., Imran, M., Zeadally, S. (2018). 5G for vehicular communications. *IEEE Communications Magazine, 56*(1), 111–117.

Shoup, D. C. (2006). Cruising for parking. Transport Policy, 13(6), 479–486.

Storck, C. R., Duarte-Figueiredo, F. (2019). A 5G V2X ecosystem providing internet of vehicles. *Sensors*, 19(3), 550.

Subramanian, S. (2013). Cooperative ITS for all: Enabling DSRC in mobile devices [Online presentation]. Available at: https://docbox.etsi.org/Workshop/ 2013/201302_ITSWORKSHOP/S05_COOPERATIVEITSandFUTUREASPECTS/QUALCOMM_SUBRAMANIAN.pdf

Va, V., Shimizu, T., Bansal, G., Heath, Jr, R. W. (2016). Millimeter wave vehicular communications: A survey. *Foundations and Trends® in Networking*, 10(1), 1–113.

Ye, H., Liang, L., Li, G. Y., Kim, J., Lu, L., Wu, M. (2018). Machine learning for vehicular networks: Recent advances and application examples. *IEEE Vehicular Technology Magazine*, 13(2), 94–101.

Zeng, T., Semiariy, O., Saad, W., Bennis, M. (2019). Joint communication and control for wireless autonomous vehicular platoon systems. *IEEE Transactions on Communications*, 67 (11), 7907–7922.

14

5G Connectivity in the Transport Sector: Vehicles and Drones Use Cases

V. Bassoo, V. Hurbungs, Tulsi Pawan Fowdur, and Y. Beeharry
University of Mauritius

CONTENTS

14.1 Introduction

Transportation is no longer defined as the movement of passengers or goods from one terminal to another. With the evolution of smart cities and broadband connectivity, intelligent transport systems (ITSs) aim to improve traffic, reduce congestion, minimize travel time and control vehicles in order to ease traffic flow. In addition to minimizing traffic problems, ITS increases road safety and makes transport safer with the integration of information and communication technologies. The first generation of ITSs were the electronic toll collection (ETC) systems and vehicle information and communication systems (VICSs); the second generation used dedicated short-range communication (DSRC) systems and collision avoidance as well as the integration of communication systems for vehicular applications. Moreover, drones are being used increasingly in a variety of civilian applications such as delivery of goods, remote sensing and precision agriculture. We are now witnessing a revolution with the development of connected vehicles, fully automated driving systems and drones that would be facilitated by the worldwide deployment of 5G networks in the near future.

14.2 Impact of 5G on Vehicles through Technologies

5G is set to change everything; every industry, business and everyday experience. With faster speeds and more reliable connections, 5G offers a panoply of options to car manufacturers in terms of connected and autonomous vehicles. 5G could eventually be the solution to improve traffic flow, prevent traffic jams and bottlenecks, increase road safety and forecast potential dangers. Smart traffic lights could, for example, talk to vehicles, and the same could happen with street lights, road signs, junctions and roads. The CAR 2 CAR Communication Consortium (C2C-CC), created in 2002, ensures the implementation of cooperative intelligent transport systems and services (C-ITS) in Europe. In 2017, C2C-CC declared its strong support for transmitting C-ITS messages using both direct vehicle communication (V2X-ITS-G5 technology) and mobile radio communication [1]. The formation of the 5G Automotive Association (5GAA) in 2016 is a clear signal for connecting the telecom industry and vehicle manufacturers. The vision of 5GAA is to build the future of connected mobility with the aim of deploying intelligent transport and communication solutions [2]. With the help of 5G, 5GAA aims to pave the path toward new ways of communication that will change the transportation ecosystem.

14.2.1 Vehicle Communication Technologies

Over the years, various technologies have been deployed to manage transportation. The communication standards in vehicular communication are DSRC and IEEE. DSRC is characterized by low communication delays and high data transfers and supports seven channels [3]. DSRC combined with GPS provides a low-cost vehicular communication system that provides an overall view of similarly equipped vehicles. Wireless networking based on IEEE802.11p is now the main standard for vehicle-to-vehicle communication with a supported data rate of up to 27 Mbps [4]. To address challenges such as vehicle speed, traffic patterns and high mobility, the IEEE 1609 standard for wireless access in vehicular environments (WAVE) is being developed and implements the wave short message protocol (WSMP) which is a highly efficient messaging protocol that supports high priority and time-sensitive communication [5,6]. The next-generation vehicles are becoming more intelligent, location aware and more connected to road objects and other vehicles. The connected car technology evolution is illustrated in Figure 14.1[6].

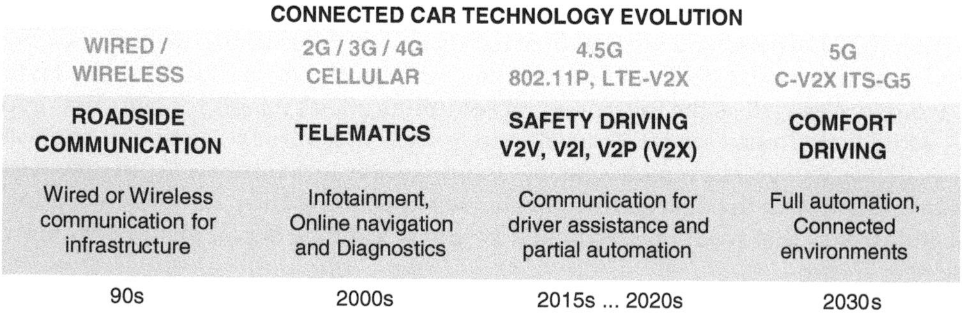

CONNECTED CAR TECHNOLOGY EVOLUTION

WIRED / WIRELESS	2G / 3G / 4G CELLULAR	4.5G 802.11P, LTE-V2X	5G C-V2X ITS-G5
ROADSIDE COMMUNICATION	**TELEMATICS**	**SAFETY DRIVING V2V, V2I, V2P (V2X)**	**COMFORT DRIVING**
Wired or Wireless communication for infrastructure	Infotainment, Online navigation and Diagnostics	Communication for driver assistance and partial automation	Full automation, Connected environments
90s	2000s	2015s … 2020s	2030s

FIGURE 14.1
Connected car technology evolution.

Vehicle-to-everything (V2X) is the technology that allows vehicles to communicate with all possible objects nearby, including cars, cyclists, traffic lights and so on. V2X communications systems ensure vehicle safety by making them aware of surrounding objects to avoid collisions. In addition, weather, nearby accidents and road conditions can also be provided by V2X [7]. V2X allows road safety applications such as follows:

- **Emergency vehicle approach warning**: Attract the attention by sending warnings to other road users as an emergency vehicle comes near.
- **See through**: Visibility beyond other vehicles, buildings, corners and intersections to alert drivers of hidden hazards.
- **Forward collision warning**: Using cameras or radars to scan the road in order to alert the driver if the distance between the vehicle ahead is too short.
- **Lane change warning/blind spot warning**: Assist the driver to avoid a collision by detecting vehicles that are invisible when changing lanes.
- **Emergency electric brake light warning**: Notify a driver of a vehicle braking suddenly ahead.
- **Intersection movement assist**: Warn a driver if another vehicle is making a rapid turn.
- **Roadworks warning**: Alert drivers of upcoming roadworks, giving them time to slow down or change lane.
- **Platooning**: Grouping of self-driving vehicles that can travel very closely together.

V2X encompasses different communication technologies as shown in Figure 14.2.

Vehicle-to-vehicle (V2V). The main objective of the V2V communication model is to avoid possible accidents by transmitting vehicle data and speed. V2V technology provides real-time short-range communication between vehicles and was pioneered by Mercedes-Benz and Cadillac [8]. It uses a wireless protocol, and V2V messages have a range of approximately 300 m and can be used to exchange vehicle heading, speed, braking status, etc. [9]. According to the *Consumer Connected Cars: Applications, Telematics & V2V 2017–2022* report, 50% of new vehicles will be equipped with V2V hardware by 2022 with a compound annual growth rate (CAGR) of 376% [10]. This in-depth examination of OEM of cars by Juniper Research also found that V2V will be among the significant technologies

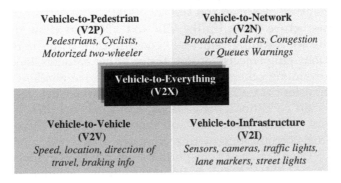

FIGURE 14.2
V2X technologies. V2X, vehicle-to-everything.

in providing autonomous driving systems, and cellular connectivity is essential for the success of V2V communication. The report also recommends the implementation of 5G in OEM such that connected vehicles can benefit from low latency, high bandwidth and wide coverage to enable new services in autonomous systems. With the help of 5G, V2V solutions will enable low-latency and high-range communications between vehicles mainly for the purpose of driver safety [11].

Several V2V and V2I applications focus on traffic flow–related safety and efficiency. Emergency electronic brake light uses V2V to alert of sudden braking of the vehicle(s) ahead, enabling timely reaction to avoid collision. Traffic flow–related warnings include not only sudden braking but also other conditions that atypically slow down the speed of the flow (such as traffic accidents, weather conditions, road hazards, road works and so on). The related information can be delivered both by V2V and V2I communication. When it comes to traffic efficiency, the topic is synchronized movement in a traffic flow. By supplementing adaptive cruise control with V2V, the system becomes cooperative (CACC), meaning that information are instantly exchanged among all vehicles in a flow, independently on sensing range. The benefit of such extension is in maintaining optimal distance between vehicles without unnecessary braking or accelerating. This saves energy, reduces pollution and increases road capacity.

The emergency vehicle warning System is another V2V application which enables a proper behavior in case of appearance of emergency vehicles. Thanks to information interchange between vehicles, it is possible to know ahead of emergency vehicle approaching, even before the sound of the siren. Thus the vehicles can be safely and timely removed from the traffic lane.

Vehicle-to-network (V2N). V2N allows vehicles to exchange real-time information about traffic, routes and road conditions using both broadcast and unicast communications. V2N represents the connectivity between vehicles and a network operator. This type of communication is enabled by the LTE network, and drivers are able receive broadcasted alerts regarding road accidents or congestion warnings. V2N use cases include parking information and management, park and ride information, information on fueling and charging stations, traffic details, zone access control and loading zone management. Further V2N features such as network slicing and edge computing could be enabled by 5G networks. Network slicing solutions involve the creation of subnetworks within a single network to support V2N use cases. For example, autonomous driving would require 10 Mbps downlink and uplink with a reliability of almost 100% [12]. Edge computing could be the answer to the problem of high traffic density during the interaction between vehicles and the road infrastructure by placing computational resources nearer to the end users of the network, thereby reducing network overhead and latency [13,14].

Vehicle-to-infrastructure (V2I). Unlike V2V and V2N that allow intervehicular communication, V2I enables traveling vehicles to communicate with the road system, which may consist of the following objects: sensors, traffic lights, cameras, lane signs, road lights and parking meters among others. V2I communications are wireless and bi-directional and use DSRC to transfer data. V2I sensors can acquire infrastructural data, thus informing drivers about road conditions, congestions, accidents and presence of construction site or availability of parking spaces. Beyond that, V2I can help drivers choose the best route and reduce fuel consumption by providing real-time traffic updates to drivers. In January 2017, the US Department of Transportation released V2I guidelines for the Federal Highway Administration (FHWA) to help transportation system owners/operators as they deploy V2I technology [15]. The aim was to improve safety and mobility and accelerate the implementation of communication systems. The V2I guidelines cover among others

V2I message lexicons, environment considerations and transportation planning for connected vehicles, licensing requirements for roadside units (RSUs) and services that support vehicle – V2I applications. In 2018, 5GAA recommended FHWA to use cellular V2X (C-V2X) to move forward with the new generation of transportation in North America. According to 5G Americas, V2I use cases should have a download link and upload link of 1–5 Mbps, with a latency of 5 ms and mobility in the range of 0–160 km/h [16].

Vehicle-to-pedestrian (V2P). According to the United States of Transportation, 32,719 people lost their lives in motor vehicle crashes in 2013 [17], and nearly 270,000 pedestrians die every year in road traffic accidents globally according to World Health Organization reports [18]. Advances in communication technology now provide pedestrians with traffic information to help them on roads. However, how this information could be easily communicated remains unclear. V2P therefore aims at preventing traffic accidents involving pedestrians by alerting the latter or drivers in time at cross-roads or intersections where accidents occur more frequently. The V2P approach includes not only pedestrians but also other vulnerable road users such as cyclists and motorized two-wheeler (MTW) operators [19]. Pedestrians may walk alone or in groups, and speed varies by age and physical characteristics. A typical pedestrian walking speed is 5 km/h, a cyclist travels at a speed of 15 km/h and a MTW traveling speed is 50 km/h. Telstra and Cohda Wireless have successfully tested the V2P technology over a mobile network in Australia. The trial showed the interaction between vehicles, pedestrians and cyclists on a 4G network [20]. 5G networks would aim at providing five times lower latencies than 4G networks while allowing the more efficient use of the radio spectrum with the help of software-defined networks (SDNs), network slicing and virtualization techniques. According to 5G Americas, V2P use cases should have a download link and upload link of 100 kbps–1 Mbps, with a latency of 1 ms and mobility in the range of 0–160 km/h [16].

14.2.2 Cellular V2X and 5G

C-V2X is a modern technology which is geared toward 5G and designed to connect vehicles to each other. C-V2X is characterized by low-latency, high-speed communication and provides an integrated solution using existing cellular network infrastructure. According to 5GAA, C-V2X will redefine transportation by providing real-time, reliable and information flows to ensure safety and mobility [21]. C-V2X is one cellular technology but operates in two complementary communication modes: sidelink (V2V, V2I, V2P) and uplink/downlink (V2N). V2V, V2I and V2P operates in bands (e.g., 5.9 GHz) independent of cellular network, whereas V2N works in the traditional mobile broadband–licensed spectrum. C-V2X also supports both basic safety and advanced use cases. C-V2X is evolving from LTE-V2X (Rel-14) toward 5G-V2X (Rel-16). LTE-V2X Rel-14 interoperability tests performed in Shanghai in November 2018 illustrates C-V2X industry maturity in terms of triple-level interoperability testing of LTE-V2X applications, multivendor operability of module/device/OEM and access layer implementing 3GPP R14 LTE-V2X PC5 standard [22]. The evolution to 5G-V2X R16 maintains backward compatibility, and the different releases are highlighted in Table 14.1. The security requirements of C-V2X and 5G are illustrated in Table 14.2.

14.2.3 Vehicular Communication and 5G – The Way Forward

According to the 5G Americas report, *Spectrum Landscape for Mobile Services*, suitable spectral range for low-latency and high data rates applications is 3–6 GHz. Ultrahigh-data rate

TABLE 14.1

C-V2X: Evolution to 5G [22]

3GPP Rels.8-13 March 2016	3GPP Rel. 14 March 2017	3GPP Rel.15 June 2018	3GPP Rel.16 December 2019
LTE V2N	Direct Communication LTE V2V/V2	5G NR V2N High bandwidth/low latency	5G NR Direct Communication 5G NR V2V/V2
Hazard warning	V2V safety use case	Enhanced Navigation & Infotainment	Cooperative automated driving

TABLE 14.2

C-V2X: C-V2X and 5G Requirements [23]

Safety	Technology	Description
Basic	C-V2XR14	• Foundation for basic V2X services
Enhanced	C-V2XR14/15	• Longer range and more reliability • Progress towards 5G • Maintaining backward compatibility with C-V2XR14
Advanced	C-V2XR16	• More throughput • More reliability • Wideband ranging and positioning • Lower latency • Backward compatibility with C-V2XR14/15

and ultralow-data rates applications will require spectral range of >24 GHz. To support 5G applications in different vehicular environments, spectra below 6 GHz and above 24 GHz are necessary [24]. C-V2X is intended to operate in the ITS 5.9 GHz spectrum; support for C-V2X in the ITS band was added in 3GPP Release 14, C-V2X uses dedicated spectrum for vehicles to communicate and C-V2X together with 802.11p can cooperate together on different channels in the ITS band. 5G will bring forward new prospects for future autonomous vehicles and aims at providing a unified infrastructure in terms of mobile broadband, critical services and Internet of things. 5G brings new capabilities to V2X communications, for example, orthogonal frequency-division multiplexing numerology, wideband transmissions for positioning, advanced low-density parity check/polar channel coding, self-contained subframe, low-latency slot structure design and massive MIMO. Furthermore, 5G V2X will bring new possibilities for the connected vehicle while ensuring backward technology compatibility. These include sensor and trajectory sharing, wideband ranging and positioning and high-definition maps. 5G technological components will include new carrier frequency including mmWaves and larger bandwidth; massive MIMO enables higher spectral efficiency, hybrid centralized/distributed resource management and ultralean design with reduced fixed overhead [25].

14.3 Impact of 5G on Drones

Drones or unmanned aerial vehicles (UAVs) were initially created for military applications such as reconnaissance and surveillance. Drones can operate in an autonomous fashion or can be remotely piloted. Today, drones are being used in a variety of civilian applications both commercial and personal. In the United States, the Federal Aviation Administration predicted that the market for commercial drones would triple in size between 2019 and 2023 [26]. Goldman Sachs Research has forecasted that the market for commercial and civil government drone applications will grow to around 100 billion dollars by 2020 [27]. The widespread adoption of drones has been made possible due to improvement in the basic technologies. Miniaturization of electronic devices and improvement in battery life, communication links, camera technology, sensors and software have turned drones into viable options for many applications [28]. The regulatory environment for drone operation has also evolved in a positive direction. According to recent rules adopted by the European Commission (EU), most drones will merely need to be registered and have electronic identification. Moreover, the EU is also developing different types of framework to enable intricate drone maneuvers with higher level of automation [29].

14.3.1 Drones and 5G

All drone communications are wireless, and mobile network is the obvious candidate for connectivity given the wide coverage and robustness of the existing mobile networks infrastructure. 5G networks will be even more drone-friendly and substantially diversify potential applications of drones. 5G will enable drones to be more precise, efficient and safer. The main specifications of 5G supporting drone operation are shown in Figure 14.3.

High coverage. Future 5G networks are expected to provide "100% coverage anywhere anytime," and this is being achieved by the densification of the cellular networks. A variety of low-powered small base stations such as picocells and femtocells are being used to improve coverage [30,31]. Given that drones need to be able to travel to any location seamlessly and still be traceable and able to communicate in real time, 5G network is the appropriate candidate to improve drone operation. With the wide coverage provided by the 5G networks, drones can be used in rural or remote locations for a variety of applications such as wildfire monitoring, search and rescue operations, maritime surveillance and wildlife monitoring [32].

Low latency. Another key specification of 5G networks is low latency down to 1 ms, and in order to achieve this specification, research work is being carried out in designing novel network topologies free of hardware components. Concepts such as SDN, network virtualized function, edge computing and caching are being utilized to lower latency [33]. Low latency is crucial for the safe operation of drones as it helps to provide improved control. It is even more important in scenarios where multidrones are deployed and coordination, collision avoidance and quick adaptability are essential [34,35]. Moreover, low latency can enhance "real-time" transmission of high-definition photos and videos.

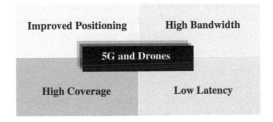

FIGURE 14.3
5G specification supporting drone applications.

High bandwidth. 5G networks are expected to support high data rates of up to 10 Gbps. The currently underutilized millimeter waves (mmWaves) are being considered to provide the high bandwidth as the lower band of the radio spectrum is already congested. A spectrum of almost 100 GHz is present for mobile communication in the mmWave band [36].

Drones are expected to transfer large amount of data in "real time." For applications such as mapping and surveying, high-resolution maps and 3D models are captured by the drones and transferred to the ground station. Moreover, in search and rescue or forest fire detection operations, emergency services use drones to obtain real-time high-quality images and videos of remote locations with difficult access [32].

Improved positioning. Better positioning is another important aspect of 5G networks. Traditionally, global navigation satellite systems (GNSSs) have been used for vehicular positioning for numerous applications ranging from military to personal ones [37,38]. As the name implies, the GNSS uses a series of satellites to provide positioning information globally. Electronic receivers on the device use line of sight signals transmitted from satellites to establish location information in terms of longitude, latitude and elevation. GNSS positioning can lead to inaccuracies of a few meters, and it suffers from latency and low refresh rate, which may cause significant problems for drone navigation. Moreover, GNSS signals are often blocked by dense forest canopies, heavy vegetation and compact urban areas. In 5G, GNSS will be supplemented by new techniques, and precision may be improved by up to 0.1 m. According to recent research work, this may be achieved by using wider mmWave bandwidth and improved calculation of the position based on larger number of sensors, infrastructure and technologies [39]. No fly zone is a key air traffic regulation concept that is applied to prevent drones from entering specific air space regions and is implemented by using geofencing functions. Improved positioning through 5G networks will lead to drones more accurately respecting no fly zones and allow precise tracking by drone operators and authorities [40]. In Dubai, drones are used to autonomously inspect construction sites. High rises are often difficult and dangerous to inspect, but with drones, comprehensive examination can be carried out safely. The drone is fed with GPS coordinates of the construction site to be visited, and once at the location, the drone takes images, videos or 3D scans of the structure. The drone transmits live feed to a server for the inspector to view or flies back to the office and uploads the captured data for viewing at a later time. Accurate positioning and navigation of the drones are essential for this application.

14.3.2 Drones and Vehicular Networks

In many cities around the world, drones are being used to observe traffic flows and provide aerial surveillance. As the technology matures, drones can be employed to support vehicular networks [41]. Such networks require low latency, high coverage area and 100% network availability. Future 5G technology aims at fulfilling all the requirements; however, in some special scenarios, drones can be deployed to assist vehicular networks. In rural areas where there is lack of mobile infrastructure and poor coverage and where the deployment of permanent infrastructure is viewed as uneconomical, drones can be used as relays to provide connectivity. Drones can also form heterogeneous networks where vehicles can access network resources. Moreover, drones have the capability of organizing into swarm networks and dynamically adapting to the required coverage area and topology. Another advantage of using drones is that they offer line of sight connection between ground infrastructure and other drones, therefore providing reliable links [41]. It was reported in Ref. [42] that drone-assisted vehicular networks improve path connectivity and end-to-end delay.

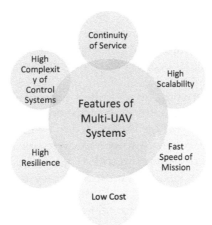

FIGURE 14.4
The main features of a multi-UAV system. UAV, unmanned aerial vehicle.

14.3.3 Multi Unmanned Aerial Vehicle Systems

Using a series of drones, also known as multi-UAV system, swarm of drones or FANETs (flying ad hoc networks) to perform a specific task is becoming more popular. UAV applications are generally mission specific, and the use of multiple UAVs enables a particular task to be completed faster, more efficiently and often at a cheaper cost [32]. The main features of a multi-UAV system are illustrated in Figure 14.4. Multiple UAV systems are particularly interesting as they promote the use of smaller UAVs, which are inexpensive to purchase and maintain. Moreover, multi-UAV systems drastically increase the area of coverage, and therefore, a mission is completed much faster. The system has a higher resilience, as the mission is not terminated by the failure of one UAV. Additionally, in multi-UAV systems, only a few UAVs maintain communication with the ground station, transmitting information gathered by all UAVs in the network, therefore leading to savings in battery life [35,43]. However, the coordination of a number of drones is a complex undertaking, requires extensive bandwidth and is computationally intensive.

The capability of 5G networks to provide high bandwidth and low latency ensures safe operations of drone by facilitating rapid communication to allow the system to reconfigure to changing surroundings.

14.4 Impact of 5G on Air and Maritime Transport

Other popular transportation methods where 5G could prove useful are air and sea travel. For air travel, better in-flight experience and the need for faster Internet connectivity in airport terminals and airplanes fuel the quick adoption of 5G networks for the aviation industry. According to Ref. [44], the aviation 5G market will grow to USD $3.9 billion by 2026 with CAGR of 52.7% from 2021 to 2026. In-flight Internet provider GOGO is developing a 5G network for planes starting in 2021 [45]. The new 5G service will power faster in-flight Wi-Fi on small commercial airlines in the United States and Canada. GOGO is relying on 5G and an increased reliance on satellite Internet to keep its business profitable.

On September 2019, China Eastern Airlines, Beijing Unicom and Huawei Beijing launched the 5G smart travel system at the Beijing Daxing International Airport. This 5G smart travel service system aims at improving passenger airport experience in terms of facial recognition at check-in counters, personalized airport experience and paperless luggage services [46].

Ships equipped with 5G networks are definitely a possibility with the aims of enhancing operations and on-board consumer experience on cruise ships, for example. The 55-km Hong Kong–Zhuhai–Macao bridge is expected to be upgraded from 4G to 5G service in the future with extended coverage to ensure network signal along the bridge [47]. According to Maritime and Port Authority (MPA) of Singapore, maritime use cases using 5G could include autonomous vessels with low latency connectivity for remote operations, ship-shore communications for vessel traffic management and timely operations, and search and rescue with IoT sensors for real-time communications and accurate positioning [48].

14.5 Conclusion

The race to 5G deployment is on the way with the benefits of increased bandwidth, faster speeds, lower latency and increased reliability. Transportation in general could greatly benefit from the next-generation wireless networks. This chapter has highlighted the main considerations of 5G with respect to vehicular communications and drone use cases. For vehicular applications, there is a shift from V2X to C-V2X, which provides the advantages of low-latency and high-speed communication over cellular networks. Moreover, 5G provides improved positioning and coverage, high bandwidth and low latency for a wider range of drone applications. The benefits of 5G in air and maritime transport have also been highlighted in this chapter.

References

1. CAR2Car Communication Consortium, September 2019. [Online]. Available: https://www.car-2-car.org/about-us/.
2. C. Voigt, September 2019. [Online]. Available: https://5gaa.org/.
3. B. Bilgin and V. Gungor, "Performance Comparison of IEEE 802.11p and IEEE 802.11b for Vehicle-to-Vehicle Communications in Highway, Rural, and Urban Areas," *International Journal of Vehicular Technology*, vol. 2013, June 2013.
4. J. Jansons, E. Petersons and N. Bogdanovs, "Vehicle-to-Infrastructure Communication Based on 802.11n Wireless Local Area Network Technology," in *2nd Baltic Congress on Future Internet Communications*, Vilnius, Lithuania, 2012.
5. S. A. M. Ahmed, S. H. S. Ariffin and N. Fisal, "Overview of Wireless Access in Vehicular Environment (WAVE) Protocols and Standards," *Indian Journal of Science and Technology*, vol. 6, no. 7, p. 8, July 2013.
6. L. Hariharan, July 2018. [Online]. Available: https://medium.com/@ResourceLeaders/v2x-and-v2v-a-one-way-road-to-great-business-opportunities.

7. W. Kenton, December 2017. [Online]. Available: https://www.investopedia.com/terms/v/v2x-vehicletovehicle-or-vehicletoinfrastructure.asp.

8. SmartCitiesWorld, September 2019. [Online]. Available: https://www.smartcitiesworld.net/news/news/whats-next-for-v2v-1695.

9. NHTSA, September 2019. [Online]. Available: https://www.nhtsa.gov/sites/nhtsa.dot.gov/files/documents/v2v_fact_sheet_101414_v2a.pdf.

10. Juniper Research, March 2017. [Online]. Available: https://www.prnewswire.com/news-releases/juniper-research-v2v-capabilities-to-feature-in-over-50-of-cars-sold-by-2022-as-autonomous-vehicles-set-to-take-off-622529574.html.

11. FutureIoT, January 2019. [Online]. Available: https://futureiot.tech/5g-will-fuel-demand-for-v2v-communication/.

12. C. Claudia, M. Antonella, I. Antonio and M. Francesco, "5G Network Slicing for Vehicle-to-Everything Services," *IEEE Wireless Communications*, vol. 24, no. 6, pp. 38–45, December 2017.

13. A. Vladyko, A. Khakimov, A. Muthanna, A. A. Ateya and A. Koucheryavy, "Distributed Edge Computing to Assist Ultra-Low-Latency VANET Applications," *Future Internet*, vol. 11, no. 6, p. 128, June 2019.

14. Z. W. Lamb and D. P. Agrawal, "Analysis of Mobile Edge Computing for Vehicular Networks," *Sensors*, vol. 19, no. 6, p. 1303, March 2019.

15. N. Gaffney, January 2017. [Online]. Available: https://www.its.dot.gov/v2i/.

16. 5G Americas, November 2017. [Online]. Available: https://www.5gamericas.org/wp-content/uploads/2019/07/5G_Americas_Whitepaper_Spectrum_Landscape_For_Mobile_Services_1.5.pdf.

17. A. Gold, September 2019. [Online]. Available: https://www.its.dot.gov/research_archives/safety/v2p_comm_safety.htm.

18. Intellias, May 2018. [Online]. Available: https://www.intellias.com/v2x-basics-connected-vehicle-technology/.

19. P. Sewalkar and J. Seitz, "Vehicle-to-Pedestrian Communication for Vulnerable Road Users: Survey, Design Considerations, and Challenges," *Sensors*, vol. 19, no. 2, p. 358, January 2019.

20. Cohdawireless, September 2019. [Online]. Available: https://cohdawireless.com/telstra-and-cohda-wireless-conduct-australian-first-v2p-technology-trial/.

21. 5GAA, September 2019. [Online]. Available: https://5gaa.org/5g-technology/c-v2x/.

22. J. Springer, September 2019. [Online]. Available: https://www.itu.int/en/ITU-T/extcoop/cits/Documents/Meeting-20190308-Geneva/14_5GAA-progress_report.pdf.

23. 5GAA, September 2019. [Online]. Available: https://5gaa.org/5g-technology/paving-the-way/.

24. M. Kratsios, "Emerging Technologies and Their Expected Impact on Non-Federal Spectrumdemand," May 2019. [Online]. Available: https://www.whitehouse.gov/wp-content/uploads/2019/05/Emerging-Technologies-and-Impact-on-Non-Federal-Spectrum-Demand-Report-May-2019.pdf.

25. S. Sorrentino, "Automotive and ITS: On the road to 5G," 2016, https://smartmobilitycommunity.eu/sites/default/files/AI_130916_05-Ericsson%20StefanoSorrentino-5G%20for%20ITS%20130916.pdf [Accessed 30 6 2020].

26. Federal Aviation Administration, "FAA Aerospace Forecast - Fiscal Years 2019–2039," 2019, https://www.faa.gov/data_research/aviation/aerospace_forecasts/media/FY2019-39_FAA_Aerospace_Forecast.pdf [Accessed 30 6 2020].

27. Goldman Sachs Research, 2019. [Online]. Available: https://www.goldmansachs.com/insights/technology-driving-innovation/drones/. [Accessed 24 9 2019].

28. P. Chandhar and E. Larsson, "Massive MIMO for Connectivity with Drones: Case Studies and Future Directions," *IEEE Access*, vol. 7, pp. 94677–94691, 31 7 2019.

29. European Commission, 24 5 2019. [Online]. Available: https://ec.europa.eu/transport/modes/air/news/2019-05-24-rules-operating-drones_en.

30. X. Ge, S. Tu, G. Mao, C.-X. Wang and T. Han, "5G Ultra-Dense Cellular Networks," *IEEE Wireless Communications*, vol. 23, no. 11, pp. 72–79, February 2016.

31. M. Agiwal, A. Roy and N. Saxena, "Next Generation 5G Wireless Networks: A Comprehensive Survey," IEEE Communications Surveys & Tutorials, vol. 18, no. 3, pp. 1617 - 1655, February 2016.
32. S. Hayat, E. Yanmaz and R. Muzaffar, "Survey on Unmanned Aerial Vehicle Networks for Civil Applications: A Communications Viewpoint," *IEEE Communications Surveys & Tutorials*, vol. 18, no. 4, pp. 2624–2661, April 2016.
33. I. Parvez, A. Rahmati, I. Guvenc, A. Sarwat and H. Dai, "A Survey on Low Latency Towards 5G: RAN, Core Network and Caching Solutions," *IEEE Communications Surveys & Tutorials*, vol. 20, no. 4, pp. 3098–3130, May 2018.
34. E. Yanmaz, S. Yahyanejad, B. Rinner, H. Hellwagner and C. Bettstetter, "Drone Networks: Communications, Coordination, and Sensing," *Ad Hoc Networks*, vol. 68, pp. 1–15, January 2018.
35. I. Bekmezci, O. K. Sahingoz and S. Temel, "Flying Ad-Hoc Networks (FANETs): A Survey," *Ad Hoc Networks*, vol. 11, no. 3, pp. 1254–1270, December 2012.
36. S. A. Busari, K. M. Huq, S. Mumtaz, L. Dai and J. Rodriguez, "Millimeter-Wave Massive MIMO Communication for Future Wireless Systems: A Survey," *IEEE Communications Surveys and Tutorials*, vol. 20, no. 2, pp. 836–869, December 2017.
37. H. Wymeersch, G. Seco-Granados, G. Destino, D. Dardari and F. Tufvesson, "5G mmWave Positioning for Vehicular Networks," *IEEE Wireless Communications*, vol. 24, no. 6, pp. 80–86, December 2017.
38. J. Duffy, C. Andrew, L. Debell, C. Sandbrook, S. Wich, J. Shutler, I. Myers-Smith, M. Varela and K. Anderson, "Location, Location, Location: Considerations When Using Lightweight Drones in Challenging Environments," *Remote Sensing in Ecology and Conservation*, vol. 4, no. 1, pp. 7–19, 2018.
39. Z. Chaloupka, "Technology and Standardization Gaps for High Accuracy Positioning in 5g," *IEEE Communications Standards Magazine*, vol. 1, no. 1, pp. 59–65, May 2017.
40. G. Yang, X. Lin, Y. Li, H. Cui, M. Xu, D. Wu, H. Rydén and S. Redhwan, "A Telecom Perspective on the Internet of Drones: From LTE-Advanced to 5G," arXiv e-prints, 3 2018.
41. W. Shi, H. Zhou, L. Junling, W. Xu, N. Zhang and X. Shen, "Drone Assisted Vehicular Networks: Architecture, Challenges and Opportunities," *IEEE Network*, vol. 32, no. 3, pp. 130–137, May 2018.
42. W. Fawaz, R. Atal, C. Assi and M. Khabbaz, "Unmanned Aerial Vehicles as Store-Carry-Forward Nodes for Vehicular Networks," *IEEE Access*, vol. 5, pp. 23710–23718, 2017.
43. L. Gupta, R. Jain and G. Vaszkun, "Survey of Important Issues in UAV Communication Networks," *IEEE Communications Surveys & Tutorials*, vol. 18, no. 2, pp. 1123–1152, November 2015.
44. everythingRF, August 2019. [Online]. Available: https://www.everythingrf.com/news/details/8719-Demand-for-Better-In-Flight-Connectivity-to-Boost-the-5G-Market-for-the-Aviation-Industry.
45. S. O'Kane, May 2019. [Online]. Available: https://www.theverge.com/2019/5/29/18644707/gogo-5g-network-planes-developing-in-flight.
46. Huawei, September 2019. [Online]. Available: https://www.huawei.com/ch-en/press-events/news/2019/9/Eastern-airline-huawei-5G-indoor-unicom.
47. Xinhua, November 2018. [Online]. Available: http://www.chinadaily.com.cn/a/201811/11/WS5be7f986a310eff303287e59.html.
48. Maritime and Port Authority of Singapore, July 2019. [Online]. Available: https://www2.imda.gov.sg/-/media/Imda/Files/Programme/5G-Innovation-and-Grant/Overview-on-Maritime-5G-Trials-20190701.pdf?la=en.

15

5G Smart and Innovative Healthcare Services: Opportunities, Challenges, and Prospective Solutions

Girish Bekaroo and Aditya Santokhee
Middlesex University Mauritius

Juan Carlos Augusto
Middlesex University

CONTENTS

15.1 Introduction

Due to the rapid growth in human population around the world, huge pressures are being exerted on modern healthcare systems. Consequently, healthcare has become one of the most significant issues for governments, particularly due to the significant demand for medical resources (e.g., personnel, equipment and space) (Yang, et al., 2016). This can further deteriorate in the future due to the progressively ageing population as well as the associated increase in chronic diseases (Baker, et al., 2017), since the global population above 60 years is expected to double in 2050 as compared with 11% of same age group in 2000 (Beard, et al., 2015). Subsequently, healthcare demands are expected to outgrow the capabilities of medical practitioners toward providing safe and quality healthcare in a timely manner, where the global needs and the factual amount of health workforce are expected to increase to 81.8 million and 67.3 million by 2030 as compared with 60.4 million and 43 million, respectively, in 2013 (Scheffler, et al., 2016). Taking awareness of such needs, resolutions toward alleviating pressures on healthcare systems are needed while ensuring continuity to deliver high-quality healthcare services to at-risk patients.

Owing to its ability to boost productivity, reduce costs and enhance user satisfaction, technology-enabled healthcare or smart healthcare is widely recognized as a potential solution to reduce pressures on healthcare systems (Latif, et al., 2017; Baker, et al., 2017). It can extend consumption and delivery of healthcare services outside the traditional settings ranging from monitoring patient health at distance to even conducting surgeries remotely. As a result of benefits including increased availability, efficiencies and accuracy of medical treatment, smart healthcare has gained increasing attention during recent years (Banerjee & Gupta, 2015). The proliferation of smart healthcare could also be attributed to advances in different innovative technologies (e.g., biosensors, wearables and mobile applications), in addition to communication networks such as the Internet, Wi-Fi and cellular networks. Among these networks, the fifth-generation (5G) cellular network technology is designed to provide a huge step ahead in terms of bandwidth, connectivity and latency, among others. Following worldwide deployment of 5G, consumption and delivery of healthcare services will be enhanced through novel healthcare solutions toward boosting the digital transformation of healthcare and facilitating an ad hoc orchestration of healthcare services (e.g., medical diagnosis and treatment) through better connecting healthcare stakeholders. As such, in the 5G era, healthcare stakeholders will need to rethink and innovate their concept of healthcare in the process of delivering and consuming related services. For instance, the advent of 5G will enable various innovative solutions in telemedicine, remote surgery and health-based wearable devices, among others, that will offer people with quality care due to enhancements in medical imagery, diagnostics and treatment. On the other hand, different challenges can hinder the proliferation of 5G smart and innovative healthcare solutions, including security and heterogeneous devices (Tehrani, et al., 2014). This chapter presents how 5G will boost digital transformation of healthcare through delivery and consumption of smart and innovative healthcare services, while probing into key hurdles in the process as well as prospective solutions.

15.2 Digital Transformation of Healthcare: A Conceptual Framework

The widespread deployment of 5G is expected to bring a massive boost in the digital transformation of healthcare, with a myriad of innovations within the various healthcare services where possibilities will be boundless. Digital transformation here entails the integration of digital technologies in healthcare in order to provide and support services such as medical diagnosis, treatment and surgery, among others. Through the integration of such technologies, significant improvements in care quality and patient experiences, as well as enhanced operational efficiencies, are possible to better address the needs of the progressively growing human population. With advances and adoption of in-home medical care and telemedicine in the 5G era, health-related issues can be more effectively resolved via remote consultation, treatment and monitoring of patients. Similarly, an assortment of cutting-edge technologies including robotics, Internet of Things (IoT), artificial intelligence (AI) and big data will advance to the next level under 5G toward boosting the digital transformation of healthcare.

In this digital transformation process, various components are involved that add to the complexity of smart healthcare systems. In order to address such complexity, a conceptual framework is recommended for simplistic representation of the constituents and associated links (Demirkan, 2013). Such a conceptual framework is given in Figure 15.1 that depicts the key building blocks entailed in the consumption and delivery of smart and innovative healthcare services in the 5G epoch. Based on the components of this framework, discussions follow about how deployment of 5G is expected to advance key healthcare services.

The conceptual framework is based on the needs of healthcare seekers where smart and innovative healthcare services lie at the very center to cater for such needs, and at both ends, the consumers and providers of healthcare services are present. This also makes it essential to understand key terms to be used throughout this chapter. While healthcare means to provide medical care to individuals or a community in an organized manner, healthcare services entail the provision of medical care by medical professionals, organizations and ancillary healthcare workers to those in need. Examples of healthcare services include medical diagnosis, treatment and surgery, among others, and these are provided by a number of healthcare practitioners and providers including nurses, specialists and pharmacists. Based on these terms, smart and innovative healthcare services involve the delivery and consumption of healthcare services in an innovative, effective, efficient, organized, reliable and timely manner by using emerging and increasingly autonomous technologies while promoting connected entities (e.g., patients and health service providers) that share data and information between themselves.

The delivery and consumption of smart and innovative healthcare services are facilitated by a range of cutting-edge technologies such as IoT, big data, AI, augmented reality (AR) and virtual reality (VR) in addition to communication technologies including Wi-Fi and 5G. Since this chapter focuses on 5G, it is important to briefly comprehend the technical

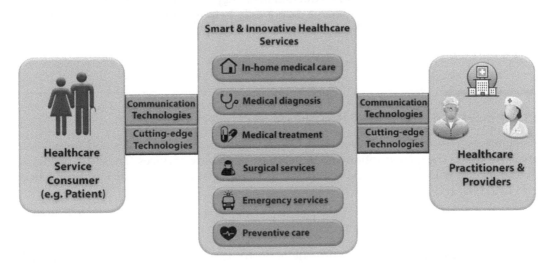

FIGURE 15.1
Conceptual framework for smart and innovative healthcare.

specifications and some insights on the ability of 5G wireless communications systems that can aid in digitally transforming healthcare. As compared with 4G, 5G wireless communications envision a huge step ahead in terms of connectivity, network speed, capacity and scalability in addition to reduction in latency and energy consumption (Agiwal, et al., 2016; GSMA Intelligence, 2014), elaborated as follows:

High data rates. A key technical ability of 5G wireless communications is high data rate, ranging between 1 and 10 Gbps in real networks (Andrews, et al., 2014). This also represents an increase by hundred times as compared with the 4G technology.

Reduced latency. As compared with the 10 ms round trip time within 4G, 5G aims at reducing latency to 1 ms round trip time, thus implying an approximate reduction by 10 times (Agiwal, et al., 2016). Within healthcare services, this reduction in network communication delays is expected to enable remotely controlled robots for medical or first response, in addition to AR and VR applications that necessitate rapid request–response cycles (Agyapong, et al., 2014).

Enhanced connectivity. With improved connectivity, billions of connected devices and sensors are expected to be supported within 5G wireless networks, while also enabling access to zettabytes (ZB) of data (Anwar & Prasad, 2018). This is also a means to realize the vision of IoT toward addressing the needs to provide widespread connectivity of devices, while at the same time ensuring higher bandwidth for longer durations within a particular geographical area (Agiwal, et al., 2016). Typical devices connected within such network include smartphones and smartwatches running health-related applications in addition to wearables that enable health monitoring and fitness tracking through real-time sensing of related information.

Improved availability and increased geographical coverage. The next-generation 5G systems envision a practically 100% available network in addition to complete geographical coverage irrespective of locations of users (Agiwal, et al., 2016). Even though this is not a technical objective of 5G, the new generation of cellular network technology is expected to provide opportunities for businesses and network operators toward improving availability and increasing geographical coverage (GSMA Intelligence, 2014).

Reduced energy use and improved battery life. Essential requirements of the new generation of cellular networks include reduction in energy usage and improved battery life of connected devices. These requirements do not only help to reduce costs but also promote environmental and ecological sustainability.

15.3 Smart and Innovative Healthcare Services

As smart technology is revolutionizing healthcare, numerous advances and innovations are expected in the consumption and delivery of healthcare services. These are discussed in this section with focus on key healthcare services provided by healthcare providers, as also listed in Figure 15.1.

15.3.1 In-Home Medical Care

5G and cutting-edge technologies provide opportunities to decentralize healthcare facilities and extend consumption and delivery of healthcare services outside the traditional settings whereby moving the point of care closer to the patient. This will enable a patient to assume a more active and independent role in taking care of his or her own personal health at home whereby promoting ambient-assisted living. Among the cutting-edge technologies, wearables, implantable medical devices and mobile robotic home assistants have important roles to play in enabling in-home medical care. Wearables and implantable medical devices have sensing or actuation capabilities and can either be worn, carried, implanted in the body or beneath the skin, placed within different parts of the body as intelligent patches or even be running through the blood stream. While sensing devices are meant for measuring certain internal or external parameters related to the human body (e.g., monitoring blood pressure, heart beat and emotions), actuators interact with the human body (e.g., pump correct dose of medicine into body)–based data obtained from sensors or other sources. As such, these devices can sample, monitor, process, control and communicate essential information in real time to the patient or medical staff (Milenković, et al., 2006). In terms of applications, although relevant for all age groups, the use of these devices can provide elderly people with a sense of comfort for monitoring physiological issues and early detection of abnormal health scenarios. Likewise, with the help of baby monitors, mobile applications and other smart health solutions, parents can keep track of baby body's movement, respiratory patterns and even moisture levels in diapers (e.g., through smart diapers). Due to their prospects, various such devices are proliferating at an unprecedented rate (Dunn, et al., 2018) for providing assistance in numerous health and medical issues, namely, cardiovascular and gastrointestinal monitoring, pulmonary health conditions, mental health issues and maternal care. With 5G, there will be opportunities for further diffusion of such medical devices due to the support for a huge number of connected devices that communicate at high bandwidth and ultralow latency in addition to robust mobility support (Lema, et al., 2017). Likewise, there are prospects for the proliferation of mobile robotic home assistants such as the Care-O-bot to extend their services to medical ones in the 5G era. This will enable provision of in-home medical care assistance by robots where key features range from health monitoring to basic in-home treatment and provision of emergency healthcare services. Although meant for all ages, such mobile robotic systems can assist individuals who lack mobility (e.g., elderly people and patients with some disabilities), and if connected to the medical network and healthcare practitioners, further opportunities for in-home medical care will arise.

Moreover, with the expected rapid expansion of telemedicine between 2017 and 2023 at an annual rate of 16.5% (Vaidya, 2017), opportunities arise in the 5G era for the application of telepresence, teleauscultation, teleconsulting and telemonitoring to deliver in-home medical care. With 5G, doctors could be sitting in their clinic or at home while teleconsulting patients located in various parts of the world and benefit from high-definition video streaming within teleconsultation software and video conferencing tools in the 5G era. In this endeavor, reliable diagnostic tools can be promoted by doctors providing remote consultation in order to acquire diagnostic reports at distance. In this endeavor, mobile applications can be helpful. Millions of such applications have been developed for smartphones and tablets since the first appearance of such devices where over 100,000 health-related apps have already been published in platforms including Google Play and Apple App Store. It was also found that one out of five smartphone users have downloaded a medical or health-related app where the most popular ones relate to fitness and diet monitoring (Fox & Duggan, 2012). In the 5G epoch, with further adoption of smart phones in developing countries and rural areas, there are avenues for more mobile applications to address different health issues, languages and diagnostic methodologies. In relevant cases where the patient does not have measurement devices (e.g., blood pressure monitors), alternatives for the use of sensors from mobile devices or mobile applications can be investigated by the research and development community. For instance, mobile application that analyzes selfie videos to measure the blood pressure of individuals can be an alternative for patients who do not have blood pressure monitors. Likewise, mobile AR-based diagnostic applications can also be utilized at distance and provide healthcare providers with diagnostic reports. Such applications relate to examining different parts of the human body through the camera of mobile phones (to scan details) while providing real-time information on the screen of the same device. For instance, diagnostic AR-based applications could be used to detect optical issues, skin problems and dental concerns, among others. In addition, with a plethora of online platforms for discussions on medical issues, people can exchange experiences pertaining to diagnosis and medical treatment or even seek advices from the community. Due to their usefulness, mobile-based diagnostic tools have started to become part of the professional practice of many medical practitioners and healthcare workers with existing 4G networks, and 5G can bring enhanced communication that will further improve diffusion of such diagnostic tools (West, 2016).

In addition, 5G is expected to make significant contributions in the expansion of medical IoT and body area networks (BANs) toward envisioning totally connected healthcare (Jones & Katzis, 2018). With these technologies, rather than relying for patients to report values such as blood pressure as well as how these values varied temporally, doctors can obtain real-time monitoring data presented in visually attractive forms to help early diagnosis of diseases (Al-Turjman, et al., 2019). These technologies have huge prospects in providing smart healthcare services in the 5G era and have the potential to diagnose various life-threatening diseases through real-time patient monitoring such that timely actions could be taken. With more connected entities in the 5G era, smarter and innovative services could proliferate including automatic ordering and delivery of drugs based on treatment records, crowd sensing for medical supports and home-care assistance.

15.3.2 Medical Diagnosis

Early diagnosis of diseases in the 5G era will be facilitated through the use of wearables, implantable medical devices, healthcare monitoring devices, mobile robotic home assistants as well as mobile-based diagnostic tools, as discussed earlier. In addition, two approaches are particularly helpful in medical diagnosis, notably, medical imaging and laboratory tests, which have opportunities to advance further under 5G.

During the previous decades, medical imaging (e.g., X-rays, magnetic resonance imaging [MRI] and computerized tomography [CT] scan) has significantly influenced medical practice whereby improving diagnosis of diseases and treatment (Comaniciu, et al., 2016). Through medical imaging, visual representations of body parts (e.g. organs, bones and tissues) are created, and these are utilized for analysis, diagnosis of medical issues and medical intervention. The 5G mobile networks enable rapid sharing of diagnostic details and medical imaging whereby creating avenues for developments in remote and cooperative diagnosis. This will enable medical practitioners to send or receive high-quality medical images quickly when seeking expertise and second thoughts from doctors practicing in a different geographical location. Similarly, patients from rural areas or in developing countries will be able to better communicate their medical diagnoses as well as relevant images to specialists based in abroad whereby reducing the need for travel in order to access quality medical support. Through remote collaborative medical diagnosis, medical practitioners are able to benefit from enhanced knowledge as well as increased innovation in practice, and patients are able to benefit from improved quality of diagnosis and treatment (Morley & Cashell, 2017). In addition, a mixed approach involving medical practitioners and machines can reliably improve diagnostic performance. The increase in the use of medical imaging within the 5G era provides opportunities for the application of AI techniques to screen and analyze medical images, while also comparing with previous cases in order to improve detection of medical issues.

Laboratory services are also essential in medical diagnosis whereby providing vital information to medical practitioners following medical tests (e.g., blood test and urine test). With improved connectivity in the 5G era, enhanced interconnectivity between laboratories in different geographical areas can enhance access to a huge number of lab test reports. Value could be generated from these reports through application of data analytics and AI. For instance, following a lab test of a patient, results could be compared with the database of reports to identify associations and enhance interpretation of results. The availability of large amount of data is considered beneficial to machine learning algorithms since more training data improve accuracy of results following application of these algorithms (Holzinger, et al., 2015). In addition, platforms and applications for patients to analyze their lab reports could emerge whereby enabling patients to upload their lab reports and obtain support to interpret values as well as medical advices from practitioners.

15.3.3 Medical Treatment

Data-driven technologies form the core of the digital health revolution, and improved connectivity under 5G era will generate an increased volume and variety of data that can be leveraged to better personalize treatments provided to patients. Exploitation of these data can help to determine the most appropriate or effective treatment to be provided to a patient while considering the various factors related to the patient (e.g. blood group, blood pressure, etc.) and the medical condition being treated. Even if a treatment course has started, AI techniques could be applied to data obtained from health monitoring devices and wearables to further assess the performance of the medication being taken in order to prevent any side effects or to even change medications at an early stage. In other words, the application of data analytics and AI techniques (e.g., machine learning, deep learning and cognitive computing) can find associations between factors investigated while better representing and enabling interpretation of complex medical data. In addition to determining the performance of the treatment to be provided to patients, AI techniques (e.g., linear regression and artificial neural network) can also be used to improve estimations of treatment expenses, thereby supporting decision-making by patients and medical practitioners. Also, with advances in AI and robotics, the use of robotic prosthetics and bionic body parts is expected to revolutionize healthcare where neutrally controlled organs can replace and restore different capabilities of patients who have nonfunctioning or amputated body parts.

Moreover, medications are essential during treatment, and adherence could be followed more effectively by doctors in the 5G era. In order to better guide prescriptions, clinical research databases (e.g., ClinicalTrials.gov) could be utilized by medical practitioners to verify and obtain answers pertaining to drug interactions, risk factors and indicator thresholds, among other factors. Furthermore, the prescription of smart pills or ingestibles enables medical practitioners to remotely monitor whether patients are taking medications correctly, with the proper dose, frequency and timing. When such smart pills are ingested by the patient, essential information about medication adherence is transmitted wirelessly and is accessible by the patient, caregiver or even family members on their smartphones. In case the patient forgets to take medications, such smart systems also enable notifications to be sent. In addition, while various brands of drugs are available on the market, prescription of genuine ones is important. Counterfeit medicines are extremely hazardous for the human organism. In the 5G era, opportunities arise for blockchain-based applications for medical practitioners to check for counterfeit illegal medicines through the medical logistic chains. Such applications could be utilized during medical consultation when doctors prescribe drugs to patients.

Additionally, continuous professional development is recommended for doctors in order to enable competent practice, and this could be achieved by periodically undertaking a range of accredited educational activities to further build key skills and acquire new knowledge on innovative diagnostic techniques and treatments, among others. In addition to online and mobile learning, two cutting-edge technologies that have huge potential in medical training under 5G include AR and VR (Jun & Kim, 2017). AR has shown prospects to assist in medical training and surgical procedures whereby enabling computer-generated images or 3D models to be overlaid on real-time patient body parts, models or even images of body parts in order to visualize unapparent anatomical details while obtaining key information. In addition to enable visualization of human anatomical structure, AR has been utilized to teach about 3D lung dynamics and training laparoscopy skills to medical doctors. While AR is commonly used to educate medical professionals whereby simulating critical situations within a safe environment (Botden & Jakimowicz, 2009), its use in medical education can considerably increase attractiveness and engagement during learning of different concepts whereby complementing books with 3D models through which learners can interact with. On the other hand, virtual reality enables user immersion into a computer-generated virtual world where the real-world environment surrounding the user is replaced by a virtual one. Together with the integration of haptics and novel interaction technologies, VR has been utilized in medical education whereby simulating dental, bone, eye and minimally invasive surgeries in the past (Ruthenbeck & Reynolds, 2015). As AR and VR have started to benefit from data streaming technologies and wireless networks during recent years, bandwidth and latency have been key barriers that inhibit high fidelity telepresence and collaborative applications that implement these technologies (Orlosky, et al., 2017). As such, with massive improvements in bandwidth and latency made possible through 5G, fully immersive AR and VR applications for medical training will emerge that enable 3D models, videos and information to be retrieved in a timely manner from remote servers for real-time display.

15.3.4 Surgical Services

In the 5G era, key developments in surgical services relate to surgical telementoring and remote surgery. In surgical telementoring, real-time guidance and assistance on surgical procedures are provided by an expert surgeon to another practitioner located at a remote location. In this telemedicine concept, assistance provided by the expert surgeon can vary from verbal feedback given when watching a real-time video stream of surgical procedures to even taking control over the procedure through some robotic device such as a robotic arm. In addition to providing care to patients located remotely, surgical telementoring can help professional development of junior practitioners while also fostering networking and collaboration. In the past, surgical telementoring systems were hindered by low transmission speed and latency, where adverse consequences on surgical performance were raised (Fabrlzio, et al., 2000). Even with 4G, a lag time of 0.27 seconds was noted, which hampers some real-time features in surgical telementoring (Houser, 2019) as a shorter delay of 0.25 is recommended for more interactive mentoring scenarios within such systems (Lema, et al., 2017). The latency period of just 0.01 seconds in 5G has established a key milestone in telementoring. As such, 5G provides opportunities for a vast number of surgeries to be telementored while also advancing robotic telementoring platforms.

On the other hand, in remote surgery or telesurgery, the entire surgical procedure is controlled by a surgeon from a remote location, and the procedure normally involves utilization of robotic systems. Since its first application in medicine in 1985, robots have slowly integrated into operation rooms of many hospitals around the world (Beasley, 2012). Among the surgical robots, the da Vinci Surgical System is considered as the most popular one where over 4000 units have been installed around the world whereby facilitating more than 5 million surgeries using a minimally invasive approach. Throughout the surgical procedure tele-operated by this medical robotic system, a surgeon sits at a special console within the same room as the patient and is able to view the operative field through 3D small cameras while at the same time manipulating surgical instruments attached to robotic arms to perform the surgery. Robotically assisted surgery has been successful in different medical disciplines including laparoscopy, orthopedic surgery and neurosurgery. The use of robotic systems within surgery brings different advantages including improved precision, flexibility and control, in addition to remote access to surgical sites (Burgner-Kahrs, et al., 2015). Although during the past decade, there has been some use of surgical robots, 5G provides opportunities for more substantial growth of such robotic systems in order to enhance the benefits of minimally invasive surgery, while also creating avenues for their applications in new types of surgical procedures (Beasley, 2012). For instance, use of surgical robots can be extended such that expert surgeons having specialized robotic consoles within their offices can conduct surgical procedures on patients located in a remote location (e.g., operation room in a different building or even hundreds of miles away within a different hospital). In the past, such types of operations were limited due to the maximum delay of 150 ms that remote telesurgery can tolerate (Lema, et al., 2017). However, the ultralow communication delay within 5G addresses this concern where commands from robotic arms and systems can be transmitted smoothly. Furthermore, due to high bandwidth possible with 5G, surgeons operating surgical robots will be able to benefit from high-definition view of the operative field in real time. These advances in telesurgeries will not only improve access to expert surgeons through reduced traveling needs for performing surgeries but also provide surgical services to patients located within rural areas or in developing countries whereby improving collaboration between surgeons.

Furthermore, the surgical world could also benefit from the application of innovative technologies including AR to either train doctors on surgical procedures or augment body parts of patients during operations by overlaying computer-generated information (e.g., text, images, models and videos of body parts). This technology has been successfully implemented to assist with laparoscopic surgical procedures (Fuchs, et al., 1998), breast-conservative cancer surgery (Sato, et al., 1998) and liver surgery (Hansen, et al., 2010), among others. With 5G, further opportunities will arise pertaining to development and deployment of AR applications to be used during medical education or even within the distinct phases of any surgical procedure (referred as perioperative period) that consists of the preoperative, intraoperative and postoperative phases. For instance, in the preoperative phase, this technology could be used to improve planning of surgical procedures toward diminishing adverse clinical happenings that can affect postoperative outcomes. Even during the intraoperative stage, 3D models can be overlaid on body parts needing surgical procedures to enable surgeons virtually explore, simulate and apply corrective procedures toward guiding operations. Finally, in the postoperative phase, outcomes could be compared with initial stages toward generating reports and even assessing outcomes of surgical procedures.

15.3.5 Emergency Services

In emergency cases, race against time is primordial to provide adequate prehospital care and transport the patient to the hospital in a timely manner in addition to ensuring the emergency team is prepared for reception and treatment of the patient. In relation to emergency services, widespread 5G deployment is expected to provide various opportunities and benefits (Zhang & Pickwell-Macpherson, 2019). To start with, innovative architectures for emergency healthcare systems involving 5G are expected to emerge that leverage the benefits provided by 5G while interconnecting various components (e.g., stakeholders and other enabling technologies). Instead of waiting for an emergency to occur, advances in health monitoring and diagnostics discussed earlier can help early detection of medical issues so that care could be provided in a timely manner. In other words, information retrieved from wearables and monitoring devices attached to patients can be connected to AI-driven early warning systems that can notify the medical team or the nearest medical center in a timely manner so that emergency actions could follow. In telehomecare situations, doctors can even provide medical assistance (e.g., advice or control wearables) while waiting for ambulance and emergency team to reach the site (Oleshchuk & Fensli, 2011). Furthermore, widespread rollout of 5G will foster developments in intelligent transportation systems, and this can improve services provided by emergency vehicles (e.g., transporting patients to hospital quicker). Even within ambulances, 5G provides opportunities for additional features including high-definition video communication with medical experts, improved connection with hospital and emergency departments, collection and transmission of observational data within the vehicle to improve diagnosis and even treatment. With early transmission of medical reports from the ambulance itself, the emergency team at the hospital can be better prepared (e.g., preparing operation room or medical devices needed) so as to provide immediate medical care, while also reducing chances for the situation to worsen.

15.3.6 Preventive Care

Preventive care is another essential service within healthcare systems and principally involves advising people to prevent potential medical issues in addition to screenings at different levels in order to prevent health-related problems. Motivated by a data-driven healthcare, the application of sophisticated AI techniques and data analytics is expected to provide incentives in preventive healthcare and early diagnosis of diseases with lesser available symptoms (Ukil, et al., 2016). Due to its essential role in predicting, preventing and managing undesired health conditions, data are a critical resource in the healthcare sector, and historically, this sector has been generating huge volume of data. Since data are widely considered as a valuable asset, generating meaningful information from data sourcing from different origins is even more valuable. The rising flood of data to be generated with 5G provides opportunities to big data scientists to determine correlations, comprehend patterns and analyze trends toward early identification of diseases at different levels (e.g., patient level and national level) in order to improve decision-making, enhance care, save lives and lower costs. At the patient level, predictive modeling techniques could be applied to patient-related data obtained from different sources including wearables and monitoring devices to improve diagnosis and guide prescription of medicines (Clifton, et al., 2013). For example, with GoCap, diabetic patients can automatically log insulin doses, and medical practitioners can view the shared logbook in real time and the data enable early identification of problems and to tweak dosages if needed. Data analytics can even help medical practitioners improve their knowledge and enhance decision-making in response to medical emergency following data mining techniques applied to a vast amount of biomedical literature available from books, journals and medical reports, among other documents (Ukil, et al., 2016). Even at a broader level (e.g., in a particular region, country, continent or even globally), data analytics and AI techniques provide opportunities to better profile diseases and identify predictive events (outbreaks, epidemics and pandemics). For example, big data analytics has become an important tool in the fight against influenza where respiratory illnesses related to seasonal flu kill between 291,000 and 646,000 people yearly around the world (WHO, 2018). With over 700,000 flu reports received by the Centers for Disease Control and Prevention (CDC) on a weekly basis that document details on sickness, treatment provided and outcome of treatment, the CDC has created an application called FluView that provides real-time view of how this disease is spreading spatially and temporally. In the same application, information on locations of patients combatting flu as well as remedial information to caregivers (e.g., vaccines and antiviral medications) are provided. Likewise, in the 5G era, the explosion in data due to enhanced connectivity provides opportunities for better-informed decision-making for a range of medical issues following application of the discussed techniques.

15.4 Challenges of Smart Healthcare and Prospective Solutions

Even though 5G will bring a massive boost in the digital transformation of healthcare, various challenges can hamper delivery and consumption of the smart and innovative healthcare services discussed earlier. Rather, smart healthcare in the 5G era can experience the phenomenon where potential problems can grow proportionally to the benefits provided by this powerful communication technology.

15.4.1 5G Challenges and Deployment Issues

Since 5G is a complementary technology and is expected to work together with other communication technologies, smart healthcare solutions will be adversely impacted by their associated limitations as well as deployment issues of 5G. Key 5G issues that need further attention include deployment in dense heterogeneous networks, multiple access techniques as well as full-duplex transmission (Li, et al., 2018). In order to deploy 5G wireless networks, significant investments will be needed to install the required infrastructure (e.g., deploying thousands of new cell sites) as well as to upgrade required software, which could be a challenging process for wireless providers facing financial difficulties. For healthcare solutions to take full advantages of 5G, ubiquitous and reliable coverage is needed. Failure to install a fully reliable infrastructure by service providers would raise important concerns. Moreover, to take advantage of the technical features of 5G, existing devices utilized for healthcare applications (e.g., smartphones, medical imaging machines and surgical robots) that are not compliant with 5G will need to be replaced if upgrades are not possible, and this would be costly for healthcare institutions. As key solution, effective deployment strategies backed up with cost–benefit analyses need to be devised by key stakeholders to better plan key phases until 5G is fully deployed.

15.4.2 Issues with Smart Devices

As 5G has been designed for mass connectivity to support billions of connected devices, various healthcare-related devices such as wearables and sensors are expected to proliferate at a remarkable rate on the market (Dunn, et al., 2018). It is expected that solution providers and manufacturers of such healthcare devices remain competitive on the market by continuously rethinking their strategies and adopting innovation models. Since these devices are fabricated by different manufacturers, interoperability is a major challenge for systems involving heterogeneous devices that can hinder connection and communication between such devices (Akpakwu, et al., 2017). Heterogeneous devices can also generate ambiguous data, thus making it difficult for humans, machines and software agents to process, interpret and integrate in health-related information systems. In case of inaccuracies leading to failure for certain devices to communicate potential medical issues as it should have been, problems regarding liability and responsibility can also arise in case of lack of policies.

Since interoperability issues within heterogeneous systems are principally due to lack of universal standards, standardization bodies should come up with standards that manufacturers of healthcare devices should adhere to (Ahad, et al., 2019). Until such solutions are devised, manufacturers and developers of smart health devices and systems need to identify interoperability issues at various levels (e.g., in terms of application, devices or communication) and appropriate conversions conducted wherever possible. As a prospective solution to work with and integrate data from heterogeneous devices, manufacturers can provide semantic annotation of the data whereby describing key aspects such as what the data means, their origin, properties and what key fields represent, among others. As for accuracy of results provided, further testing strategies could be adopted by manufacturers to test their products under various conditions. Also, medical practitioners should recommend only devices known to provide results with the highest accuracy. Upon implementation of such standards, suppliers and buyers of these healthcare devices need to ensure that only devices that adhere to these standards are procured and deployed. Platforms that verify adherence to standards as well as authenticity of devices would be helpful in the process. Moreover, regulatory bodies could come up with stronger laws and policies that prevent manufacturing and adoption of counterfeit and substandard medical devices.

15.4.3 Software/Application-Related Issues

With the benefits provided by 5G, there will be a considerable increase in the number of health-related software (e.g., diagnostic tools and health information systems) as well as mobile applications for use by doctors and patients. The availability of a multitude of tools is challenging in the way that users can be confused about which tool to use where some published applications can also be from untrusted sources. Also, software that have reached the end of their lifetime and that have limited compatibility with devices should also be progressively discontinued in order to avoid associated vulnerabilities.

A growth in the use of automated and intelligent smart healthcare systems is envisaged in the 5G era, and proper solutions need to be established to ensure fully resilient and reliable software are being utilized. For instance, in order to reduce time to market or with lack of expertise in the domain, software solutions might not be sufficiently tested, and if deployed, the presence of bugs can adversely impact various aspects ranging from reduced accuracy of results to critical failures. In order to address such issues, effective testing strategies can be implemented in addition to comprehensive fault management systems for detection and control of malfunctions in systems such as health information systems (Shladover, 2018). Also, similar to medical devices, software systems within hospitals are typically installed from various vendors giving rise to compatibility issues that can hamper installation, operation and data management (Huyck, et al., 2015). In order to avoid such pitfalls, right from the beginning, medical providers need to ensure that software components are properly designed for smooth integration with existing solutions.

15.4.4 Data-Related Challenges

With smart healthcare solutions in the 5G era, there will be rapid generation of a huge volume of digital data that also need to be stored. This will make healthcare a key user of big data. With the various aspects related to data in smart healthcare, management and governance are among key problems to be dealt with (Kaisler, et al., 2013), and for these, further partnerships will need to be brokered to ensure data-related services are effectively dealt with. In addition, associated challenges will be reflected in terms of the different Vs of big data, notably volume, variety, velocity, veracity and value.

Under 5G, voluminous healthcare data will be generated from a variety of sources such as wearables, mobile applications and medical consultation. Storage of an ever-growing amount of data is critical for decision-making pertaining to services such as early disease diagnosis and preventive care. For this, effective data storage strategies are needed that administrate various aspects such as what data are stored, how they are stored, where they are stored, their lifetime and who can access those, among others. Likewise, if proper solutions are not planned beforehand, transportation of data from the storage point to the processing point can be a costly process. This could be avoided by processing the data at the storage location itself and to only transfer results to wherever needed, while also maintaining integrity and provenance of data.

With the integration of multiple health sources toward a connected healthcare, a variety of data (structured, semistructured and unstructured data) will have to be manipulated. Since the most significant volume of data within healthcare is unstructured, fragmented and rarely standardized (Kruse, et al., 2016), aggregation, storage, processing and analysis will be more challenging. Storage of varying types of data in formats that are incompatible with applications and cutting-edge technologies can lead to interoperability issues, where further efforts are needed for cleansing and processing. In order to address the data variety–related issues, metadata protocols as well as semantic models of data integration can be considered (Andreu-Perez, et al., 2015). Another potential solution is standardization of data, which has its own challenges.

Other key characteristics of big healthcare data include velocity and veracity where velocity relates to the continuous generation of data at a fast speed and veracity is about the conformity to facts and accuracy. Key causes for high velocity in healthcare include high-frequency data sources in the form of pervasive sensors as well as social media, in addition to continuous streaming data (e.g., video communication between medical practitioners for telementoring or remote surgeries). For addressing velocity-related issues, potential solutions include lightweight processing models and high-performance computing. On the other hand, with a large amount of healthcare data available, the accuracy of data reported can be compromised in case of measurement imprecisions, confounding factors as well as inference certitude of outputs. Such inaccuracies in healthcare can lead to imprecisions when diagnosing and providing prescriptions and treatments to patients and can have adverse consequences. In order to address the issue of veracity, causality and uncertainty quantification techniques could be applied to the data. Similarly, self-reported data can be collected in a consistent manner whereby reducing mix-ups (Kruse, et al., 2016).

In order to generate value from the stored data, timely processing and analytics of big data are essential. However, processing of voluminous healthcare data that are stored at various locations can have associated performance issues and introduce delays, thus hampering real-time healthcare and emergency response services. For alleviating performance issues, high-performance or processing intensive systems are recommended, although such systems can be costly to procure and operate. Indexes could also be built right from the beginning while collecting and storing data in order to reduce processing time. Similarly, optimized algorithms and techniques such as abstraction into smaller data sets can improve execution time for analytical processing. On the other hand, for big data analytics tools to be successful in smart healthcare, methodological issues still warrant further attention and are hindered by data quality, inconsistencies as well as legal hurdles. In addition, application of big data analytics in smart healthcare necessitate appropriate skill sets and resources, where healthcare professionals working with data will have to improve their knowledge and skills to keep up to date with the fast evolving technologies and techniques. Also, healthcare organizations implementing big data analytics need to be aware of and comply with the various legal issues and standards (Kruse, et al., 2016).

15.4.5 Security and Privacy

Security and privacy are major design requirements of smart healthcare solutions under 5G, and these will not be without associated challenges (Akpakwu, et al., 2017). Several aspects need to be considered when ensuring security of such solutions in the 5G era to ensure key security objectives including confidentiality, integrity, availability, access control and nonrepudiation are catered for. Violation of any of these security objectives can lead to serious consequences whereby even leading to life-threatening situations for patients (Islam, et al., 2015). Furthermore, the amount of medical information gathered on patients including medical diagnosis history, treatments and medical images, among others, will increase over time. Due to the sensitive nature of such information, privacy is one of the major challenges smart health solutions can face in the 5G era. Patients need assurances that their health data will not be compromised purposefully by medical service providers and malicious users. Similarly, users of smart health solutions need to know about their associated positive or negative influences whereby ensuring transparency of underlying invisible processes (e.g., operations, data collection and processing as well as any monitoring activities) (Jones, et al., 2015). Likewise, confidentiality of data transmitted between devices to relevant servers should be protected to prevent adversaries from eavesdropping and/or tampering with the medical data. In certain circumstances, ensuring confidentiality of transmitted data will be challenging especially for devices with limited processing capabilities such as sensors, which cannot process strong cryptographic algorithms.

A portion of the security solutions utilized within 4G will progress directly into 5G, whereas others will need reengineering (Akpakwu, et al., 2017). Moreover, security frameworks and architectures compliant with 5G wireless networks are needed that account for security management involving heterogeneous devices (Fang, et al., 2017). Similarly, to ensure ethical considerations are respected in the development of 5G-enabled intelligent systems and environments, frameworks such as eFRIEND could be used by system designers and developers (Jones, et al., 2015). In addition, cutting-edge technologies and data analytics can also improve efficiency and effectiveness in detecting frauds (Kruse, et al., 2016). On the other hand, to warrant privacy of data and communication among various entities in the 5G era, regulations are needed that govern various aspects such as security of data access, location privacy as well as mutual authentication between patients and medical practitioners. Designers of smart health solutions need to emphasize and ensure that all data are safeguarded. In addition, various other approaches such as privacy-aware routing mechanisms, privacy by design and regular security and privacy assessment could be implemented as prospective solutions (Liyanage, et al., 2018).

15.4.6 Issues with Cutting-Edge Technologies

In the 5G epoch, different cutting-edge technologies such as IoT, AI, AR and VR, robotics, among others, will help to provide smart and innovative healthcare services while also facilitating the coming of higher-level human–computer interactions in healthcare. However, some challenges and limitations of each affiliated technology can adversely influence advances of smart healthcare solutions. A few of these challenges along with potential solutions are summarized in Table 15.1.

15.4.7 Acceptance and Adoption

Acceptance and adoption of smart health solutions by end users (patients and doctors) are crucial and can be challenging at the same time. While technology acceptance is an attitude toward a technology, adoption is the process during which a user becomes aware of a

TABLE 15.1

Challenges and Potential Solutions with Cutting-Edge Technologies

Cutting-Edge Technology	Challenge	Potential Solution
Robotics	Robot ethics (delegation of sensitive tasks, blame for a failure and unemployment)	Regulations, legislation developments and strategic planning
	Issues pertaining to implementation of robotic systems (e.g., computer vision related issues and effective detection of failed hardware parts)	Further technological developments in robotics by R&D community and industry, among others.
Mobile healthcare applications	Smartphone platform variability (variability and compatibility issues among platforms and mobile operating systems)	Cross-platform application design and development
Internet of things	Working with devices having limited resources and processing capabilities (Li, et al., 2018)	Lightweight algorithms and solutions
Augmented reality	Object recognition and tracking issues (Chen, et al., 2017): • low sensitivity and tracking accuracy • challenges with varying object sizes, smooth surfaces and variability of lighting conditions • mismatch of virtual and physical distances	Improvements in AR development tools and libraries
	Performance-related issues (overheating of device and latency issues with object detection and overlays)	• Enhancements in object detection and overlay methodologies • Code optimization
Virtual reality	Limitations of technological solutions for multiuser settings (Wiederhold & Riva, 2019)	Development of new tools to cater for multiuser collaborative virtual environments
Artificial intelligence techniques (machine learning, deep learning, etc.)	Accuracy and validity of records in data set can influence training process and outputs produced	Enhanced evaluation metrics to assess accuracy and validity of training and outputs produced
	Processing requirements (duration and speed) for real-time and emergency healthcare services	Use of high-performance systems and optimized algorithms
Blockchain	Scalability issues – blockchain systems become heavy with an increasing number of transactions	• Storage optimization of blockchain • Redesigning blockchain
	Selfish mining – nodes with high computing power could reverse the blockchain and the transaction that took place	Strategies to stop selfish mining

technology, embraces it and utilizes the technology. Since adoption and acceptance of such innovative healthcare solutions depend on various factors such as ease of use, attitudes, convenience, perceived usefulness, cost and demographic factors, among others, failure to consider such factors by device designers may adversely impact acceptance and adoption. Ultimately, user engagement is considered as a critical factor when monitoring and treating medical issues, and some solutions such as wearables were claimed to limit user engagement and interaction capabilities, which also are key factors that affect user acceptance (Baig, et al., 2017). In addition, innovations in medical solutions having substantial complexity necessitate more technical skills for operation. If appropriate training is not provided, users may struggle when using such solutions whereby leading to various issues, ranging from inducing errors to even causing damage to the solution if it is a piece of hardware (e.g., medical robot). Over time, a group of users can become overly dependent on smart healthcare solutions, and to cater for their needs, designers and developers of such solutions need to consider important aspects such as quality of service, robustness, reliability and dependability.

15.5 Conclusion

5G has the potential to bring a massive boost in the digital transformation of healthcare whereby enabling innovations within the various healthcare services such as medical diagnosis, treatment and surgery. This transformation will be facilitated by an assortment of cutting-edge technologies such as robotics, IoT, AI and big data, which will be propelled to the next level under 5G. The convergence of these technologies can bring boundless opportunities in terms of smart healthcare solutions toward reducing pressures faced by medical practitioners and providers due to the large number of medical issues that need attention. For instance, delivery of medical care has prospects to be extended from the traditional settings (e.g., in hospital or by private doctors) toward personalized healthcare accessible remotely to patients with the rapid expansion of telemedicine and wearable technologies, among others. In this chapter, the key opportunities that 5G is expected to bring in healthcare services such as medical diagnosis, in-home medical care, treatment surgery, emergency services and preventive care are discussed. On the other hand, various challenges can negatively impact delivery and consumption of the smart and innovative healthcare services in the 5G era, where potential problems can grow proportionally to the benefits provided by this powerful communication technology. These include 5G's own limitations and deployment issues, challenges associated to applications, medical devices, data and cutting-edge technologies as well as security and adoption-related hurdles. These challenges are presented as part of this chapter in addition to potential solutions for overcoming these challenges.

References

Agiwal, M., Roy, A. & Saxena, N., 2016. Next generation 5G wireless networks: a comprehensive survey. *IEEE Communications Surveys & Tutorials*, 18(3), pp. 1617–1655.

Agyapong, P. et al., 2014. Design considerations for a 5G network architecture. *IEEE Communications Magazine*, 52(11), pp. 65–75.

Ahad, A., Tahir, M. & Yau, K., 2019. 5G-based smart healthcare network: architecture, taxonomy, challenges and future research directions. *IEEE Access*, 7, pp. 100747–100762.

Akpakwu, G., Silva, B., Hancke, G. & Abu-Mahfouz, A., 2017. A survey on 5G networks for the Internet of Things: communication technologies and challenges. *IEEE Access*, 6, pp. 3619–3647.

Al-Turjman, F., Zahmatkesh, H. & Mostarda, L., 2019. Quantifying uncertainty in Internet of medical things and big-data services using intelligence and deep learning. *IEEE Access*, 7, pp. 115749–115759.

Andreu-Perez, J., Leff, D., Ip, H. & Yang, G., 2015. From wearable sensors to smart implants–toward pervasive and personalized healthcare. *IEEE Transactions on Biomedical Engineering*, 62(12), pp. 2750–2762.

Andrews, J. et al., 2014. What will 5G be? *IEEE Journal on Selected Areas in Communications*, 32(6), pp. 1065–1082.

Anwar, S. & Prasad, R., 2018. Framework for future telemedicine planning and infrastructure using 5G technology. *Wireless Personal Communications*, 100(1), pp. 193–208.

Baig, M. et al., 2017. A systematic review of wearable patient monitoring systems–current challenges and opportunities for clinical adoption. *Journal of Medical Systems*, 41(7), p. 115.

Baker, S., Xiang, W. & Atkinson, I., 2017. Internet of things for smart healthcare: technologies, challenges, and opportunities. *IEEE Access*, 5, pp. 26521–26544.

Banerjee, A. & Gupta, S., 2015. Analysis of smart mobile applications for healthcare under dynamic context changes. *IEEE Transactions on Mobile Computing*, 14(5), pp. 904–919.

Beard, J. et al., 2015. *World report on ageing and health 2015*. Geneva, Switzerland: World Health Organization.

Beasley, R., 2012. Medical robots: current systems and research directions. *Journal of Robotics*, 2012, pp. 1–14.

Botden, S. & Jakimowicz, J., 2009. What is going on in augmented reality simulation in laparoscopic surgery? *Surgical Endoscopy*, 23(8), p. 1693.

Burgner-Kahrs, J., Rucker, D. & Choset, H., 2015. Continuum robots for medical applications: a survey. *IEEE Transactions on Robotics*, 31(6), pp. 1261–1280.

Chen, L., Day, T.W., Tang, W. & John, N.W., 2017. Recent developments and future challenges in medical mixed reality. In *2017 IEEE International Symposium on Mixed and Augmented Reality (ISMAR)*. IEEE, pp. 123–135.

Clifton, L. et al., 2013. Predictive monitoring of mobile patients by combining clinical observations with data from wearable sensors. *IEEE Journal of Biomedical and Health Informatics*, 18(3), pp. 722–730.

Comaniciu, D., Engel, K., Georgescu, B. & Mansi, T., 2016. Shaping the future through innovations: from medical imaging to precision medicine. *Medical Image Analysis*, 33, pp. 19–26.

Demirkan, H., 2013. A smart healthcare systems framework. *IT Professional*, 15(5), pp. 38–45.

Dunn, J., Runge, R. & Snyder, M., 2018. Wearables and the medical revolution. *Personalized Medicine*, 15(5), pp. 429–448.

Fabrizio, M. et al., 2000. Effect of time delay on surgical performance during telesurgical manipulation. *Journal of Endourology*, 14(2), pp. 133–138.

Fang, D., Qian, Y. & Hu, R., 2017. Security for 5G mobile wireless networks. *IEEE Access*, 6, pp. 4850–4874.

Fox, S. & Duggan, M., 2012. Mobile Health 2012 [Online]. Available at: http://www.pewinternet.org/Reports/2012/Mobile-Health.aspx

Fuchs, H. et al., 1998. *Augmented reality visualization for laparoscopic surgery*. Berlin, Heidelberg: Springer, pp. 934–943.

GSMA Intelligence, 2014. *Understanding 5G: Perspectives on Future Technological Advancements in Mobile*. https://www.gsma.com/futurenetworks/wp-content/uploads/2015/01/Understanding-5G-Perspectives-on-future-technological-advancements-in-mobile.pdf

Hansen, C. et al., 2010. Illustrative visualization of 3D planning models for augmented reality in liver surgery. *International Journal of Computer Assisted Radiology and Surgery*, 5(2), pp. 133–141.

Holzinger, A., Röcker, C. & Ziefle, M., 2015. From smart health to smart hospitals. In: Holzinger, A., Röcker, C., Ziefle, M. (eds) Smart Health. Lecture Notes in Computer Science, vol 8700. Cham: Springer, pp. 1–20.

Houser, K., 2019. The Next G [Online]. Available at: https://futurism.com/the-byte/5g-powered-surgery-worlds-first [Accessed 9 Aug 2019].

Huyck, C., Augusto, J., Gao, X. & Botía, J., 2015. *Advancing ambient assisted living with caution.* s.l., Cham: Springer, pp. 19–32.

Islam, S. et al., 2015. The internet of things for health care: a comprehensive survey. *IEEE Access*, 3, pp. 678–708.

Jones, R. & Katzis, K., 2018. 5G and wireless body area networks. In *2018 IEEE Wireless Communications and Networking Conference Workshops (WCNCW)*, IEEE, pp. 373–378.

Jones, S., Hara, S. & Augusto, J., 2015. eFRIEND: an ethical framework for intelligent environments development. *Ethics and Information Technology*, 17(1), pp. 11–25.

Jun, S. & Kim, J., 2017. 5G will popularize virtual and augmented reality: KT's trials for world's first 5G olympics in Pyeongchang. s.l., ACM, p. 4.

Kaisler, S., Armour, F., Espinosa, J. & Money, W., 2013. Big data: issues and challenges moving forward. In: 2013 46th Hawaii International Conference on System Sciences. IEEE, pp. 995–1004.

Kruse, C., Goswamy, R., Raval, Y. & Marawi, S., 2016. Challenges and opportunities of big data in health care: a systematic review. *JMIR Medical Informatics*, 4(4), p. 38.

Latif, S., Qadir, J., Farooq, S. & Imran, M., 2017. How 5G wireless (and concomitant technologies) will revolutionize healthcare? *Future Internet*, 9(4), p. 93.

Lema, M. et al., 2017. Business case and technology analysis for 5G low latency applications. *IEEE Access*, 5, pp. 5917–5935.

Li, S., Da Xu, L. & Zhao, S., 2018. 5G Internet of Things: a survey. *Journal of Industrial Information Integration*, 10, pp. 1–9.

Liyanage, M., Salo, J., Braeken, A., Kumar, T., Seneviratne, S. & Ylianttila, M. 2018. 5G privacy: scenarios and solutions. In *2018 IEEE 5G World Forum (5GWF)*, IEEE, pp. 197–203.

Milenković, A., Otto, C. & Jovanov, E., 2006. Wireless sensor networks for personal health monitoring: issues and an implementation. *Computer Communications*, 29(13–14), pp. 2521–2533.

Morley, L. & Cashell, A., 2017. Collaboration in health care. *Journal of Medical Imaging and Radiation Sciences*, 48(2), pp. 207–216.

Oleshchuk, V. & Fensli, R., 2011. Remote patient monitoring within a future 5G infrastructure. *Wireless Personal Communications*, 57(3), pp. 431–439.

Orlosky, J., Kiyokawa, K. & Takemura, H., 2017. Virtual and augmented reality on the 5G highway. *Journal of Information Processing*, 58(2), pp. 133–141.

Ruthenbeck, G. & Reynolds, K., 2015. Virtual reality for medical training: the state-of-the-art. *Journal of Simulation*, 9(1), pp. 16–26.

Sato, Y. et al., 1998. Image guidance of breast cancer surgery using 3-D ultrasound images and augmented reality visualization. *IEEE Transactions on Medical Imaging*, 17(5), pp. 681–693.

Scheffler, R. et al., 2016. *Health workforce requirements for universal health coverage and the sustainable development goals.* Geneva, Switzerland: World Health Organization.

Shladover, S., 2018. Connected and automated vehicle systems: introduction and overview. *Journal of Intelligent Transportation Systems*, 22(3), pp. 190–200.

Tehrani, M., Uysal, M. & Yanikomeroglu, H., 2014. Device-to-device communication in 5G cellular networks: challenges, solutions, and future directions. *IEEE Communications Magazine*, 52(5), pp. 86–92.

Ukil, A., Bandyoapdhyay, S., Puri, C. & Pal, A., 2016. *IoT healthcare analytics: the importance of anomaly detection.* IEEE: Crans-Montana, Switzerland, pp. 994–997.

Vaidya, A., 2017. Global telemedicine market to experience 16.5% annual growth rate through 2023 [Online]. Available at: https://www.beckershospitalreview.com/telehealth/global-telemedicine-market-to-experience-16-5-annual-growth-rate-through-2023.html [Accessed 18 Jul 2019].

West, D., 2016. How 5G technology enables the health internet of things. *Brookings Center for Technology Innovation*, 3, pp. 1–20.

WHO, 2018. Influenza (Seasonal) [Online]. Available at: https://www.who.int/news-room/fact-sheets/detail/influenza-(seasonal) [Accessed 6 Aug 2019].

Wiederhold, B. & Riva, G., 2019. Virtual reality therapy: emerging topics and future challenges. *Cyberpsychology, Behavior, and Social Networking*, 22(1), pp. 3–6.

Yang, Z. et al., 2016. An IoT-cloud based wearable ECG monitoring system for smart healthcare. *Journal of Medical Systems*, 40(12), p. 286.

Zhang, Y. & Pickwell-Macpherson, E., 2019. 5G-based mHealth bringing healthcare convergence to reality. *IEEE Reviews in Biomedical Engineering*, 12, pp. 2–3.

16

5G Edge-Based Video Surveillance in Smart Cities

Aleksandar Sugaris

ICT College

CONTENTS

16.1 Introduction

This chapter explores deployment possibilities of cloud- and edge-based Internet protocol (IP) video surveillance in smart city (SC) environment. IP video surveillance solution becomes one of the multimedia services integrated in IP networks. As video surveillance can significantly enhance public safety and property security, this chapter discusses the solutions that prevent IP surveillance networks from failing to deliver video and lose key images within seconds. By its nature, video surveillance is mission-critical (MC) IP-based application. A platform for MC communications and services has been a priority of 3GPP in developing technical specification and service requirements for the 5G system. This chapter analyzes how IP surveillance – where machines stream video all the time with high bit rate (fulfilling 4K and 8K video resolution security requirements) and pan–tilt–zoom (PTZ) interactions – fits in to the 5G SC system migrating to a video surveillance as a service (VSaaS) cloud model. Additionally, in the SC scenario with the large number of cameras (massive IoT), along with the introduction of central cloud, this chapter evaluates mobile edge computing to support video analytics in artificial intelligence (AI) manner, enabling much more powerful processing and analytical capabilities that overcome limitations of central cloud-only solution. Moreover, this chapter emphasizes that the imperatives of AI and the introduction of 5G delivering lower latency with higher bandwidth force network providers to apply an end-to-end AI processing chain, bringing the fusion of cloud and edge, and ultimately smarter cities.

Over the past few years, traditional communications industry approach is in change as over-the-top (OTT) competition has arrived. OTT services have been increasing their dominance not only in multimedia streaming applications but also in core communication services such as voice and messaging. In order to recover from this competition and before moving in the new revenue deployment scenarios, telecom companies are now modernizing their operations mostly by investing in network throughput improvements, such as 5G and fiber upgrades. These upgrades open strategic opportunity in deploying sophisticated applications enabled by ubiquitous media networks. Deployment scenarios fall under the massive machine-type of communications 5G requirement category, and they are also strongly related to ultrareliable and low-latency communications and enhanced mobile broadband (eMBB) 5G requirements [1]. MC applications are among those sophisticated deployments enabled by 5G networks, such as video surveillance as a real-time video MC service (MC Video) [2]. The MC Video service supports video media communications between several users with the permissions to gain access to the system (i.e., MC Video user access to MC Video server) [3]. Furthermore, machine-to-machine (M2M) communications and Internet of things (IoT) networking are typical applications for opening wide customer bases such as SCs, utilities, security industry specialists, smart homes and so on where telecom companies can develop business models for managing and maintaining connected services and monetizing 5G opportunities. For entering to those new markets, telecom companies make either acquisitions or partnerships with appropriate adjacent industry leader or start-up company specialized, for instance, in SC applications.

IoT is a distributed system that creates the value from collected data and gives various physical sensors/devices the opportunity to exchange the information based on which the decisions are made. IoT systems are researched in many disciplines, but rapid growth is enabled by advances in digital technologies, such as decreasing cost of miniaturized electronics components, ubiquitous wireless connectivity, expansion of storage capacity and computational power of processing systems and invention of advanced software applications for analytics. There is an increased interest in applying IoT technologies in public safety [4], and 4G to 5G evolution challenges are researched in order to support public safety networks [5]. As a part of IoT ecosystem, M2M communications grow rapidly, and IoT technology connects digital devices with several sensing, actuation and computing capabilities with the Internet, thus offering services in the SC context [6].

We are regularly witnessing large investments in SC projects. Despite that, there is no universally accepted definition of SC. Different projects develop different smart parts of urban areas – essentially various types of data collection sensors to provide the information for further analysis. Regardless of SC type, all of them have three technology components – IoT sensor, Internet connectivity and computing resources for data analysis. What kind of technology is used depends on the SC objectives within the urban area, for instance, safety, security, energy, healthcare, transportation, government and so on. This chapter concentrates mainly on safety and security of SC aspects that improve everyday life in normal as well as emergency situations. Generally, in this SC scenario, there are three technology components: connected video surveillance camera as IoT sensor, 5G network as Internet connectivity and cloud/edge servers as computing resources for data analysis.

Following the aforementioned information and communications industry trends, this chapter provides the reader with IP video surveillance deployment possibilities in 5G SC environment where video surveillance in IP system is treated like a MC multimedia service for which it is ensured quality of service (QoS) throughout the network. This chapter is about new trends in IP video surveillance and open research challenges that arise from the implementation of cloud/edge computing, video analytics and AI within IP surveillance 5G communications as one of the top 5G use cases (part of massive IoT and SC).

The next subsection provides basics of IP video surveillance as a MC multimedia SC service – from technological, architectural and network point of view to the protocols and tools for ensuring QoS of this multimedia service. The following subsection focuses on VSaaS cloud solutions and mobile edge computing in support of video surveillance analytics and surveillance processing tasks. The fourth subsection deals with fitting massive IoT video surveillance concept to 5G SC system including cloud/edge AI application. This chapter is closed by concluding remarks and reference list.

16.2 Mission-Critical Multimedia Smart City Service

Recently, the city administrations around the world and their technology partners started to establish a unified service and video surveillance technology platform for collaboration and information sharing. By using that system, city staff can monitor large areas and zoom in on small details doing that from a central location, see in bright light and in the dark night and dispatch personnel to the right place at the right time.

Video surveillance system and its equipment are most effective when integrated with other security hardware/sensors forming a coherent security system. Synergy obtains when video system is combined with intrusion and motion alarm sensors, electronic access control, fire alarms, communications network and personnel [7]. The integration of those sensors and video subsystem improves the whole security system and makes it more intelligent in many different scenarios. For instance, if the intruder violates a barrier, the intrusion detection system is able to determine that a person – not a bird or leaf – pass through the barrier. Video provides this information. This chapter focuses on video surveillance component of the SC. Figure 16.1 depicts typical digital video surveillance multisite application.

Digital video security is made possible due to rapid improvements of broadband telecommunication technologies and video compression algorithms. Once the network capacities reached a level to constantly handle heavy loads of data, coupled with increased efficiency of video compression standards, and processing power, it was possible to deliver high-bandwidth-demanded video streams (today 4K resolution is demanded). What video surveillance makes a smart component of SC is video management system (VMS) running on high computing power server. VMS solutions with video analytics and AI, such as license plate recognition, behavior analysis, tripwire detection, loitering detection, people counting, abandoned object detection or vehicle tracking, take data from smart sensors and by analyzing them enable various types of optimizations.

16.2.1 Digital Video Technology

About 15 years since the introduction of IP cameras, 21% of systems have exclusively analog cameras, and 36% of systems are mixture of IP and analog (IFSEC trend report, 2017.) [8]. Combining that, we get that analog cameras are still installed in a majority (57%) of

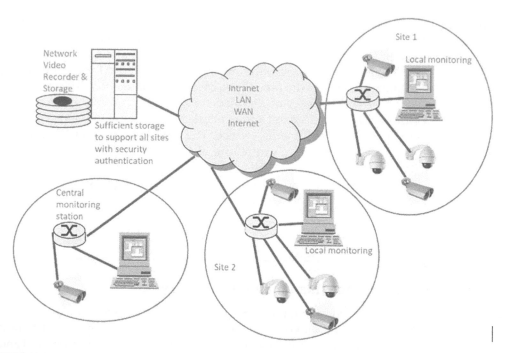

FIGURE 16.1
Digital video surveillance in multisite application. LAN, local area network; WAN, wide area network.

systems (43% of systems are IP-only). Revenues from global network camera sales overtook those of analog cameras in 2014 [8].

There are couple of key compression techniques when video signal is compressed in IP camera (or in encoder for IP network adaptation purposes).

One of the key compression techniques is discrete cosine transform (DCT) followed by a quantization [9]. Generally, an image (frame) within video sequence is divided in 16×16 pixel macroblocks (used for motion-compensated compression), and they are divided in four 8×8 pixel blocks (used for DCT compression). Images of video stream are encoded into intraframes (I-frames), forward predicted frames (P-frames) and bidirectional predicted frames (B-frames). The I-frame is encoded as a single image with no reference to any past or future frames, which makes I-frame a key frame (a reference point for future frames). The data for each frame and 8×8 block are first DCT-transformed and then quantized. The first coefficient in the top left corner represents null horizontal and vertical frequency, and the bottom right coefficient represents the highest spatial frequency component in the two directions. The example with typical DCT coefficient values is shown in Figure 16.2a. Depending on the number of details contained in the original 8×8 block, the high-frequency coefficients will be bigger or smaller, but in a vast majority of images, the amplitude decreases quickly with the frequency, due to the smaller energy of high spatial frequencies. The DCT thus has for compression very useful property of concentrating the energy of the block on a low number of coefficients situated in the top left corner of the DCT-transformed block; thus there is only few nonzero values to record/transmit compared with 64 nonzero pixel values in the original 8×8 block. Quantization is a complex process of, simply said, ignoring less important bits of the image. All DCT coefficients in 8×8 block are divided by suitable quantization factors defined in quantization tables. The compression standards define quantization tables in accordance of the

591	106	-18	28	-34	14	18	3
35	0	0	0	0	0	0	0
-1	0	0	0	0	0	0	0
3	0	0	0	0	0	0	0
-1	0	0	0	0	0	0	0
0	0	0	0	0	0	0	0
-1	0	0	0	0	0	0	0
0	0	0	0	0	0	0	0

16	11	24	47	99	99	99	99
18	21	26	66	99	99	99	99
24	26	56	99	99	99	99	99
47	66	99	99	99	99	99	99
99	99	99	99	99	99	99	99
99	99	99	99	99	99	99	99
99	99	99	99	99	99	99	99
99	99	99	99	99	99	99	99

FIGURE 16.2
(a) Two-dimensional DCT coefficients – typical values. (b) Quantization table – example. DCT, discrete cosine transform.

psychophysiological specificities of human vision (reduced sensitivity to high spatial frequencies). An accuracy decreases with the increase of spatial frequencies, and DCT coefficients are divided by larger quantization factor, as it is shown in Figure 16.2b. That way the dynamic range of 8×8 block as well as the quantity of information required to encode a block is reduced. The quantization also usually results in a great number of zero values within 8×8 block. Quantization process ensures low bit rate without perceptible degradation of the picture quality, and it is one of the main tools for controlling the data rate of the compressed video stream.

If video surveillance video stream is encoded as constant bit rate (CBR), then the quantization level is adjusted automatically by the compression algorithm so the bit rate is constant. The quantization level is directly configurable if video stream is encoded as variable bit rate (VBR). In video surveillance encoder quantization settings in VBR case, the compression level is increased, and the image quality is lowered if on the reference scale the quantization level is lowered [10].

The other key compression technique is motion compensation applied to macroblocks for P-frames and B-frames. Motion compensation technique use the strong correlation between successive frames in order to maximally reduce the amount of information required to transmit or store them. It consists of predicting most of the pictures of a sequence from preceding and even subsequent pictures. It is done with a minimum of additional information representing only the differences between pictures, and that so-called prediction error is also encoded using DCT and quantization. A P-frame is encoded as a 16×16 area relative to the closest preceding I-frame or P-frame, based on movement in the frame. If there is an object displacement in the frame, a motion vector is transmitted. If there are multiple areas of motion, then more motion vectors are added in relation to those positions. The compression rate of P-frames is higher than that of I-frames. B-frames are coded by bidirectional interpolation between the I or P frame which precedes and follows them. The future reference frame is the closest I-frame or P-frame. The difference between a B- and a P-frame is the use of reference areas in a future frame, where the two 16×16 macroblock areas are then averaged. B-frames offer the highest compression rate compared with I-frames and P-frames. The arrangement of different types of frames is shown in Figure 16.3a.

FIGURE 16.3
(a) I-, B- and P-frames in motion compensation process. (b) Group of pictures example.

Video frames in video surveillance applications are normally transmitted from a few frames per second up to 20–30 frames per second. Each frame is compressed and converted in bits and bytes for transmission over IP networks. Frames may be compressed independently of other frames, such as I-frames, or compressed based on the differences, such as P- or B-frames. These frames form a group of pictures (GoP), composed of an I-frame and variable number of P- and B-frames, which only encode the motion vector changes from the previous I-frame, like in Figure 16.3b. It is possible to encode I-only, I- and P-, or I-, P- and B-frames. The choice of frames and GoP structure affect strongly the overall compression rate and quality of video, as well as random access resolution, which is very important for video surveillance. Two parameters, M and N, describe the order of I-, P- and B-frames. M is the distance (in number of frames) between two successive P-frames, and N is the distance between two successive I-frames, defining a GoP. The parameters generally used are $M=3$ and $N=12$ as shown in Figure 16.3b, obtaining a satisfactory video quality with an acceptable random access time of less than half of the second.

At present, digital video surveillance encoding engines support mainly the following compression algorithms: MJPEG and H.264 (MPEG-4 AVC). Those algorithms are based on previously briefly explained key compression techniques. They each have their own application areas with advantages and disadvantages [11].

MJPEG (Motion Joint Photographic Experts Group) is the digital video version of JPEG standard, based on I-frame compression. It forms the video stream from the sequence of JPEG-encoded images without predictions so it produces high-quality video, but bit rates and required memory space of MJPEG are much larger than the other two algorithms. The advantages are convenience for detailed video analysis, easy search of the material and robust transmission as every frame is a key frame.

The Moving Picture Experts Group (MPEG) was formed to create a standard in digital video. There is a whole family of standards based on MPEG compression techniques. Although the methods of compression are similar to JPEG, MPEG provides motion compensation, and the bandwidth and storage space is drastically reduced compared with MJPEG. The prevailing compression standard at the moment in video surveillance is H.264. Due to many improvements in the algorithm (more efficient prediction methods) compared with older standards, H.264 can encode and decode video at a higher quality with the same bit rate or using half the bandwidth of its codec predecessors with the same video quality. It requires high level of processing power because there are more mathematical computation requirements necessary to achieve the higher-quality image. H.264 can provide better quality video for video analytics, and due to the required high level of processing power on both sides, it can easily support the most demanding video analytics software application and alarm signals processing of any kind. Recently, H.265 or high-efficiency video coding (HEVC) – the latest MPEG standard, entered in video surveillance industry mainly to support higher-resolution video (4K and 8K) that is very important in security applications. If compared with H.264, HEVC achieves the same perceptual quality of video with only half (or even less than half) of the bit rate used by H.264 [12]. HEVC extends the compression concepts implemented in H.264 and achieves improvement in many areas [13], like in intrapicture prediction, like in motion compensation precision and motion vector prediction, or extended macroblock sizes, and many more.

16.2.2 Internet of Things Camera Sensor

Video surveillance camera converts the visual light image (focused by the camera lens) into an electrical time-varying video signal for later presentation on a monitor display and/or permanent recording on a video recorder. The lens collects the reflected light from the scene and focuses it onto the camera image sensor. The sensor converts the light image into a time-varying electronic signal. The camera electronics processes the information from the sensor and forms the video signal. If it is IP camera, video signal is digitized, compressed and packed into IP stream over Ethernet interface. IP packets are sent to a viewing monitor by two-wire unshielded twisted-pair, wireless or other IP transmission system.

Digital video surveillance systems use CIF, VGA and megapixel (MP) resolution camera sensors. In Table 16.1, typical digital resolutions in video surveillance systems are presented. 4K is an ultrahigh-definition standard [14] adopted by International Telecommunication Union (ITU) and across the electronics industry that provides 16:9 aspect ratio with approximately 4000 (4K) pixels in horizontal resolution. It can be seen from Table 16.1 that 4K resolution is double that of full HD (FHD) both horizontally and vertically, and it results in four times total resolution. There are many megapixel formats available on the video surveillance market, but 4K as a standard ensures easy integration into the whole system. Ultra HD (UHD) resolutions are very important for surveillance applications, as by using UHD, we can more precisely extract image detail form of a live or recorded video image. Identifying a face requires 250 ppm (pixels per meter) horizontally [15]. An FHD sensor has 1920 pixels horizontally, and a 4K sensor has 3840. It implies that with 4K, a person can potentially be identified further away, or alternatively a wider view angle can be used to cover a larger area, leading to a fewer number of surveillance cameras required in the system. The drawback is high bit rate (around 20 Mbps [12]) required if 4K video is encoded by H.264, and this is where H.265 comes in.

IP network infrastructure is the ideal solution for remote monitoring since it allows for control PTZ over a single network that is accessible to users anywhere in the world. The control and management are available in most cases by a standard web browser and Internet access; in some cases, a special software application needs to be installed on viewing device (PC/laptop/smart phone). Fixed cameras are not equipped with PTZ control, and they appear in box and so-called bullet shape. PTZ control enables motorized camera to physically move its pointing axis horizontally (pan/360°) and vertically (tilt/180°) and also to zoom in or zoom out objects in the scene. Today, PTZ control is usually provided with dome cameras.

TABLE 16.1

Digital Resolutions in Video Surveillance Systems

CIF	352×288 (PAL)	HD	1280×720 (0.9 MP)
4CIF	704×576 (PAL)	Full HD	1920×1080 (2.07 MP)
VGA	640×480	Ultra HD 4K	3840×2160 (8.3 MP)
SXGA	1280×1024 (1.3 MP)	Ultra HD 8K	7680×4320 (33.2 MP)
WUXGA	1920×1200 (2.3 MP)		

16.2.3 Network Aspects

Network aspects of a video surveillance as an MC multimedia SC service are of the highest importance; especially it is important that the insurance of required QoS as surveillance shares the network resources with other services. In IP network best effort environment, special care must be taken on ensuring QoS for video surveillance.

IP cameras and encoders communicate with the servers and monitoring system end devices in two different ways on transport layer 4 [16] – using transmission control protocol (TCP and HTTP) or UDP (user datagram protocol). MJPEG/H.264 can use both transport layer protocols, but typically MJPEG is transported through TCP, and H.264 video is typically transmitted over UDP and real-time transport protocol (RTP). For slightly congested network or networks with inherent low packet loss such as a wireless network, the retransmission of TCP can be beneficial. Video displaying at the receiving end may appear to stall or be choppy when packets are retransmitted, but with the use of MJPEG, each image stands alone so the images that are displayed are typically of good quality. RTP/UDP transport is most suitable for networks with bandwidth that is guaranteed through QoS mechanisms and very little packet loss. H.264 over RTP/UDP is intolerant to packet loss, as there will be visible artifacts and degradation of quality in the decoded images if there is packet loss in the network. Some IP cameras and encoders also provide H.264 transport over TCP. TCP encapsulation can be beneficial for networks with low-level packet loss because due to heavy packet loss and retransmissions, video would be practically stopped. TCP may be also useful for fixed cameras and may not be a desirable configuration for use with a PTZ-controlled camera, as TCP induces a latency in the transport due to the required packet acknowledgments. Besides that, H.264 over TCP is useful with streams that are only being recorded and not viewed live [16].

If IP camera uses RTP/UDP, the communication is like in the case of multimedia RTP streaming (Figure 16.4a). Real-time streaming protocol (RTSP) is an application layer control protocol for RTP. Figure 16.4b shows combined work of RTSP and HTTP for RTP streaming service.

QoS refers to the ability of the network to provide special service to a set of users or applications or both. Proper design of QoS in an IP video surveillance environment is crucial. Video signal shares network resources with other services. Video transport requires unique network infrastructure characteristics to ensure that the service is usable, reliable

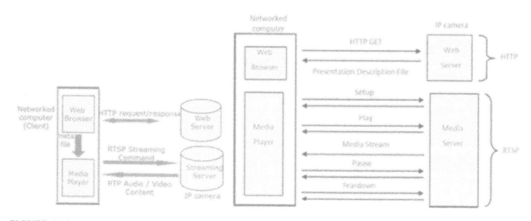

FIGURE 16.4
(a) IP surveillance over RTP streaming. (b) RTSP control communication. IP, Internet protocol; RTP, real-time transport protocol; RTSP, real-time streaming protocol.

and available to media servers and end users. Normally, local area network (LAN) connections provide higher bandwidths than wide area network (WAN) connections, and communication over WAN brings more challenge. If remote viewer accesses IP camera (or stream-recorded video) over WAN network, routers need QoS planning and configuration. In order to support QoS, network architectural components have to introduce several important improvements and settings in standard best effort behavior. In order to ensure end-to-end QoS functionality, all network switches and routers must include support for QoS. There are several different ways (models) of implementing QoS. Video surveillance system products most frequently use the differentiated services (DiffServ) model. Besides that, the cameras, servers, and encoders can be deployed on separate virtual LANs (VLANs) to provide isolation at layer 2 and transported over the WAN with layer 3 isolation over a multiprotocol label switching (MPLS). Additionally, the network connectivity between the remote and central surveillance location could either be over a private WAN service such as mentioned MPLS, or over the public Internet, typically over a secure virtual private network service such as IPsec VPN or HTTPS/SSL (secure socket layer).

DiffServ layer 3 (L3) can be used for both traffic marking and classification. DiffServ-supported networks mark their traffic so the router knows which service to apply to the packet. By setting a field called the differentiated services code point (DSCP), the marking is done in the IP header, and the intelligence of a DiffServ network is set up in the routers. Each DSCP value represents a QoS class, also known as a behavior aggregate. A particular DSCP value is mapped to a particular routing behavior. This behavior is known as a per-hop behavior (PHB) and is implemented in the router using different queuing disciplines. Network elements that are DiffServ-compliant must conform and implement the specifications of the PHB. The classes usually are CS0 (000000 or 0) – CS7 (111000 or 56); higher classes provide increasingly better service treatment. For instance, video surveillance live stream standard value for DSCP is 32 decimal (or 100000 binary). However, users can select different value for live surveillance, DSCP 40 as suggested in Ref. [10]. Less strict DSCP classification is used if we deal with the transmission of recorded video signal that should be replayed – because recorded video typically uses TCP protocol.

Besides marking and classification, QoS takes care about scheduling and priority queuing. The basic QoS queuing tools in routers for video surveillance are LLQ PQ (low-latency queuing – priority queue) and class-based weighted fair queuing (CBWFQ). In WFQ, each class may receive a differential amount of service in any interval of time. Each class i is assigned a weight, wi. During any interval of time during which there are class i packets to send, class i will be guaranteed to receive a fraction of bandwidth equal $wi/\sum(wj)$.

CBWFQ defines traffic classes to be assigned to each queue with minimum bandwidth guarantees provided to prevent starvation. LLQ by using a strict priority queue provides low-delay guarantees to certain traffic types (e.g., video surveillance). Any traffic in the PQ is prioritized and scheduled over all other traffic on the interface. That is, it provides a minimum bandwidth but does not exceed that if there is congestion. LLQ's can set multiple priority queues, with different rates for different traffic types.

LLQ PQ is the primary QoS tool for live video surveillance multimedia traffic on a converged network, while for a recorded video surveillance stream, QoS configuration with CBWFQ is normally used because usually recorded video uses TCP and the large decoder buffer prevents the requirement for a strict LLQ PQ priority queue. It should be noted that the application of mentioned QoS rules is not a substitute for proper bandwidth level calculation and network provisioning of link speeds to handle traffic loads. On the other hand, planning a large bandwidth on a link for a specific application (video surveillance) is not a substitute for configuring QoS on the network elements.

16.3 Cloud and Edge Solutions

In a SC concept, a remote viewer connects over WAN to the surveillance cameras and/or to network video recorder (NVR). An NVR is basically a standard networked server with a VMS software application that controls the flow of video surveillance streams. Normally, today, NVRs include VMS in client/server architecture. Modern VMS solutions are software based where customers select independently hardware (server machine and storage) and software (VMS application that would be later installed on server machine) components. Servers, storage and workstation technologies are key components of an IP-based physical security solution, which ensure MC SC video surveillance.

16.3.1 Cloud-Based Internet Protocol Video Surveillance

The term VSaaS refers to hosted IP video surveillance based on cloud technologies. By using cloud solutions, video surveillance becomes a service that normally includes remote viewing, management alerts, video recording, storage and network security aspects. VSaaS is enabled by the availability of larger network bandwidth and cloud technology advances. Global data center traffic will more than triple from 2016 to reach 20.6 ZB annually by 2021 [17]. In an SC deployment scenario, a large number of high-resolution video surveillance cameras integrated in management system produce large amounts of storage capacity and network bandwidth. Cloud-based surveillance systems are used for big data processing and storage, which is the reason for surveillance video processing optimization [18]. This makes VMS and the current infrastructure (for supporting a high-resolution video surveillance) very expensive. Planning, implementation and maintenance in this scenario is time consuming and also complicated to deal with. Those commercial driven reasons for cloud-based video surveillance imposed some technoeconomic analysis [19].

If you are deploying a traditional surveillance system, the video processing and recording/management occur on a system installed at a local site with the option for remote viewing. The video may be accessed via WAN connection for live viewing or for storage retrieval. If we are deploying a true cloud solution, the video processing/management is performed by the cloud (Figure 16.5). The system may have a local onsite device (gateway) to communicate with the cameras and the cloud. Secure remote video access and cloud storage are parts of this solution. From the technology prospective, the traditional surveillance systems have a shorter time to obsolescence. They may start with robust features, having options for feature upgrades at some future time by downloading firmware updates, but it is limited for real technology update. Programming interfaces in that case are closed. On the other hand, VSaaS is designed with the possibility for rapid technology evolution. With VSaaS solution, the provider sends automatic technology online updates to customer local equipment, which evolves continuously for innovations. In that case, programming interfaces for analytics and applications are open and publicly published.

Cloud computing provides convenient, ubiquitous, on-demand network access to a shared configurable computing resources [20]. Cloud computing model incudes three different types of service models for delivering the essential characteristics of cloud

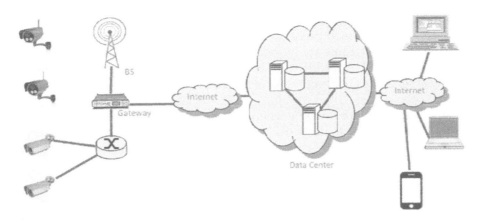

FIGURE 16.5
Cloud-based multimedia surveillance system. BS, base station.

computing, software as a service (SaaS), platform as a service (PaaS) and infrastructure as a service (IaaS). The most frequently used service model is SaaS (by 2021, 75% of the total cloud workloads will be SaaS [17]) where the user does not have the access to the underlying cloud infrastructure (e.g., operating system, hard disc drive, cloud network), but the service provides applications running on the cloud infrastructure. SaaS is standard hosted VSaaS, typically for the system of up to 15 cameras per location, and due to limited bandwidth resources, usually the camera streams video with resolution less than megapixel, lower than 15 frames per second and with efficient compression H.264 standard. These limitations will disappear with using 5G networks for accessing IP cameras. The PaaS service model provides a platform to create custom applications by using the PaaS-provided programming languages, libraries and tools. IaaS service model provides the consumer additional options on demand [20]. If more control of the application is required, customer can on demand use hardware and virtual machines (VMs), so on cloud infrastructure, they can install operating system, programming tools and application software.

There are three main cloud deployment models, public, private and hybrid. Public clouds are the most common way of deploying cloud computing (by 2021, 73% of the total cloud workloads will be public cloud data centers, up from 58% in 2016 [17]). In that case, all hardware/software and other cloud infrastructure is owned and managed by the cloud provider, and it is available to the customers so they share the same cloud resources with other organizations or cloud tenants. A private cloud consists of computing resources used exclusively by one organization and may exist on premise or off premise. Private clouds are often used by government institution or, any other organizations with MC operations seeking enhanced control over their environment, as it is the case with SC. Hybrid clouds combine private clouds with public clouds, so organizations can use the advantages of both. In a hybrid cloud, data and applications can move between private and public clouds, which are bound together by standardized or proprietary technology that enables data and application portability. Video surveillance systems consume high amounts of data. It is possible that organization's private cloud may be overloaded during peak hours so the information could be offloaded on the public cloud, but this solution leads to high financial cost and may be risky due to potential data leakages. Although it is possible to protect the sensitive streams by encrypting them, such techniques are still too expensive for large-scale video data. This is where hybrid cloud can be efficient solution as an organization's private data center is seamlessly integrated with the elastic public cloud. A hybrid cloud allows an organization to strategically push certain computation to the public cloud, potentially solving mentioned problem. There are special solutions for video surveillance in which partial video streams are pushed to public cloud while keeping sensitive content in the private cloud [21]. A middleware is desired that unifies private and public clouds and effectively schedules the processing of mixed-sensitivity video streams on the hybrid cloud.

Different approaches are possible when it comes to the design problems and choices relevant to a cloud-based multimedia surveillance system [22]. A surveillance system framework for the cloud core components propose publish–subscribe broker of the media streams; multimedia surveillance service directory registration; cloud manager for overall management between users and other cloud components; monitoring and metering responsible for performance monitoring/usage tracking of cloud VM and resource allocation manager for configuration of VM dynamically according to the current demand. The services provided by cloud could be media processing service (e.g., face recognition, motion detection and event detection), storage service, big data analytics service and so on.

16.3.2 Edge Solutions

Edge video surveillance is the application of edge computing where the video analysis and the most of the computing are done at the edge instead of the cloud (see Figure 16.6 where the common architecture of edge computing is shown).

In SC surveillance system, the distributed IP cameras (IoT devices) capture videos and send a large volume of data to the remote servers of the city surveillance system to potentially congest backbone network. Besides that, the problem is the lack of nearby computing power for real-time video analysis and decision-making. In that case, pure cloud video surveillance solution introduces significant challenges for the SC applications that require low backbone bandwidth consumption and latency in tracing criminal, recognizing moving cars and other quick reaction video analytics, meaning that a delay in delivering packets to and from the cloud is crucial to SC applications. Latencies are different for each country and region, based on urban and rural deployment of fixed and mobile broadband technology, proximity cloud data centers and the quality of customer premises equipment. Global average latency is 31 ms for fixed networks, and global average

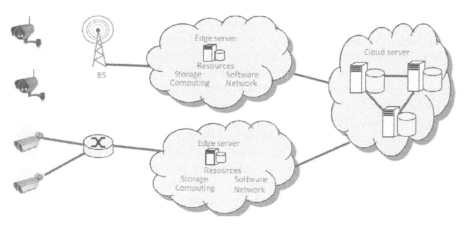

FIGURE 16.6
Edge computing–based video analytics. BS, base station.

latency is 55 ms for mobile networks [17]. In the case of video surveillance traffic, the time that is required to encode, transmit, buffer and decode the video to be displayed at the receiving end is the total end-to-end delay. This delay is critical in a live PTZ operation where the commands from the VMS must result in a rapid change in the displayed video. As a guideline, the total one-way end-to-end delay, both video and command, in a PTZ environment should not exceed 150 ms when UDP is the transport protocol in order to provide an acceptable system reaction to viewing clients. For TCP, the round-trip time (RTT) should not exceed 50 ms [10]. For mobile network with cloud, global average latency (55 ms) is higher than this recommendation, so there is a need to shorten the transmission path and use edge solution. Of course, the particular network latency will drive the appropriate solution. The delay is less critical in a fixed-camera operation, but for quick reaction video analytics, RTT should be even less than 50 ms to ensure enough time for the sophisticated cloud/edge video analytics, so there is a need for 5G bandwidth and latency in the access network.

Edge computing is an architecture that distributes resources and services of computing, control and storage, anywhere along the route from cloud to edges of the network. Edge computing concept reduces the amount of data transferred from an edge to cloud, and most of the computations and storage are performed closer to the IoT terminals, either within the edge or near the edge. That way, it is possible to use video resolutions higher than in the case of core cloud solution, and we can move to 4K and 8K resolution, if edge computing is used. Thus, the main idea of edge computing is to have computing facilities between the IP camera and the current cloud. Bringing computing and storage to the edge of the network reduces the latency and jitter [23], which are very important communication characteristics for real-time applications such as video surveillance. IP cameras are connected to an edge server, which itself is connected to the cloud through the rest of the network. The edge server is at a fixed physical location and has relatively high computational power, but it is less powerful than a conventional data center used in the cloud. There is a clear distinction between the device (camera) level and the edge (server) level.

One of the key enabling edge technologies is effective orchestration. Orchestration is defined as a group of operations that the cloud providers undertake to either manually or automatically select, deploy, monitor and control the configuration of hardware and software resources for application delivery [24]. Despite that intention, many current orchestration approaches, either open-source or commercial cloud platforms, are not for edge IoT applications. They are manual or simple algorithms that are prone to errors when there are needs to orchestrate very sophisticated cloud/edge services. On top of that, most of the orchestration procedures are application specific, and using such algorithm for edge orchestration has limitations. For those reasons, there is a high demand for automated tools to turn the application requirements into orchestration schemes optimizing the resource allocation and provisioning for the IoT applications [24]. In edge computing, the orchestrator needs to offload IoT data and computation to the edge cloud. It dynamically allocates appropriate resources to carry out these offloaded tasks matching the surveillance demands.

The edge video surveillance solutions are already developed. One of them is elastic real-time surveillance system developed on the top of a distributed edge platform where edge server resources are dynamically allocated and adjusted based on the current workload of the surveillance application [25]. That system integrates network functions virtualization and software-defined networking (SDN) which ensures easy operation on the hardware resource virtualization and the programmable virtual networks including entities configuration. For a specific surveillance task, launched in the distributed edge servers, a group of VMs (configured, monitored and managed by the SDN controller) work together. The first step is surveillance application request collection (e.g., video resolution, processing frequency, task location and storage space). SDN controller translates request to system behaviors and data flow configurations. Then, resource orchestration is performed to satisfy the requirements of computing power for specific surveillance tasks including IP addresses, bandwidth levels and so on. VMs are launched on distributed compute nodes to flexibly utilize the hardware resources. The parameters for VMs are retrieved from the orchestration file to specify CPU capacity, memory space and IP addresses. Routers follow the SDN routing strategies, and the source data are uploaded and forwarded to the optimal edge node [25].

Despite the fact that by implementing edge computing, the latency is reduced, compared with cloud computing, the resources in edge computing are still limited which can have negative repercussion on task scheduling, surveillance application performance and the overall QoS. Edge computing resource management is a challenge when the application requirements are of video surveillance complexity level because there are many available resources and choosing the suitable resource to meet the user's requirements is the main task. Edge computing decentralizes cloud computing in terms of computation and storage. The purpose of edge computing is data preprocessing, to reduce the delay and serve the applications with real-time response. The resources near the user support the real-time communication with IP cameras. Resources include computing resources, storage resources, network resources and software resources. The resource prediction methods enable the adequate resource for users by analyzing the load of the resource itself. The edge servers first analyze the requested service QoS for end users, and the service is separated into a number of tasks that are processed locally as far as possible. Based on the estimation results, each task chooses the best matching resource from the resource pool. Finally, the selected resource is allocated and scheduled to meet the QoS requirements [26]. There are also other recent solutions [27] that allocate resources for latency-sensitive services (such as surveillance face recognition) over edge computing networks with caching. This solution jointly considers computation offloading, content caching and resource allocation as an integrated model, everything in order to minimize the total latency of the computation tasks. Besides that, the increase of computation capacity at edge devices contributes to a new challenge such as edge learning, which is intended to be solved in the fields of wireless communication and machine learning. To minimize the power consumption and the task execution latency, there is another proposed solution for resource allocation for edge computing in IoT networks via reinforcement learning [28]. Edge resources are typically resource constrained compared with the cloud, and as resource management is important challenge, there are many more solutions identified and classified in different architectures, infrastructures and algorithms [29].

The distributed edge system brings design challenges such as specification for distributed edge control, distributed or centralized resource assignment strategies, orchestration of computing tasks and network routing, traffic load balancing and so on. In order to provide a system design parameters understanding (the resources proportion in local cloud vs. data center, relative latency of core and edge clouds) and determine their impact on response time and service bit rate, a model is developed for performance analysis of an edge cloud system designed to support latency-sensitive applications in city urban areas [30], which is a category where an SC video surveillance service belongs to. By using the model, we can determine application response time (sum of network delay, queuing and processing time), as a function of offered load for different values of edge and core computing resources, and network bandwidth parameters.

16.4 5G Surveillance Intelligence

5G connectivity is an important part of the IoT in particular and in a SC concept in general. Wireless technologies provide flexibility and efficiency to many systems. For the various IoT applications (including video surveillance with a camera as IoT sensor), wireless connectivity is especially beneficial for sensor-to-edge communication but possibly also for edge-to-edge and edge-to-cloud interconnections. 5G as wireless metropolitan area networks connectivity solution (or wireless WAN as a superset of both) envisages higher capacity and much lower latency than the current cellular systems [31].

16.4.1 Characteristics and Applications

There are many purposes of video surveillance in SC, including (but not limited to) monitoring busy public places, transportation centers, intersections, business centers, institutions and residential areas, strategic infrastructure (energy grid, telecom data centers, pumping stations) and crime areas. High demand for video surveillance is driven by the innovations in video camera technology that make it genuine smart IoT sensor (besides features explained earlier, there is a novel application of body-worn and in-vehicle cameras) and the cloud solutions that support data collection/analytics. Even if 4K resolution is applied, single wireless camera does not use much of the bandwidth (especially with H.265), but along with the introduction of multiple high-resolution camera per site in SC environment, the current network access infrastructure could not respond in an adequate manner so 5G technology is needed.

Industry initiatives produced a set of 5G requirements; among others, the bit rate per connection should increase up to 100 times compared with 4G, meaning peak data rate 1–10 Gbps to end points in the field (downlink) [32]. The latest 3GPP technical specifications [1,33] introduce the new (higher than for 4G) eMBB service requirements. They are specified for various deployment scenarios and different service areas (local and wide areas, indoor/outdoor, office/home, urban and rural areas,), as well as special deployments (massive gatherings, broadcast and high-speed vehicles). For instance, one of the highest data rates is 1 Gbps for the downlink and half of that value for the uplink. Note that for video surveillance application, IP cameras are connected in uplink (the connection from the user device to the server). To support required capacity, 5G system architecture defines QoS model based on QoS flows that require guaranteed flow bit rate (GBR QoS flows) and that do not require guaranteed flow bit rate (non-GBR QoS flows), for both uplink and downlink [34]. To identify a QoS flow in the 5G system, a QoS flow ID (QFI) is used. The QoS flow is the finest granularity of QoS differentiation in the 5G sessions. The system defines a flow-based QoS with or without QoS-dedicated signaling. The standardized packet marking is applied for the option without any specific flow QoS signaling, which informs the QoS enforcement functions what QoS to provide. QoS support for finer granularity and more flexibility is offered by the option with QoS-dedicated negotiation. Also, a new reflective QoS type is introduced in which the user equipment (IP camera) requests for the uplink traffic the same QoS rules as the ones it received for the downlink, thus reducing control plane signaling and ensuring symmetric QoS differentiation over downlink and uplink [1]. Besides the mechanisms provided by 5G standard, there are other options for the self-optimization video streaming use case that goes beyond traditional QoS approaches to network management. If video is generated by SC application, like an M2M application for automated video surveillance, network management system monitors the video and network state metrics that must reflect the information content of a video stream and its capability in performing surveillance application tasks (such as automated face recognition or motion detection), which means that it must support appropriate decision-making rather than perceptual quality from a consumer's perspective. If the network is congested, smart algorithm runs on media-aware network elements used in the core or at the edge to adapt video stream in a manner that by dropping irrelevant video parts optimizes this quality utility function rather than more traditional perceptual quality [35].

Besides bit rate per connection, another very important parameter is among the 5G technology requirements – the latency that is essentially end-to-end round-trip delay. The initial requirement for 5G latency was sub-1 ms [32]. Delivering 1 ms latency over a large network is very difficult, and we can see this condition relaxed. The overall service latency depends on many different delays in the system [33], such as on the radio interface, within the 5G transmission system, transmission to a server which may be outside the 5G system and data processing. Some of these delays depend directly on the 5G system itself, whereas others delays can be reduced by edge servers outside of the 5G system to allow local hosting of the services. In 3GPP specifications, end-to-end latency is defined as the time that takes from the moment an information is transmitted by the source to the moment it is successfully received at the destination [33]. Practically, it is the delay allowed for the 5G system to deliver the service in case the end-to-end latency is completely allocated to the 5G system from the user equipment to the interface to data network (Internet). In 5G system, the packet delay budget (PDB) defines an upper limit for the end-to-end latency. Low-latency eMBB (e.g., augmented reality) could be the closest service to MC SC video surveillance with quick reaction video analytics, with standardized 5G QoS characteristics of PDB less than 10 ms [34]. If in that case edge computing is used, RTT is two times 10 ms plus video analytics processing time that in total should be less than 50 ms (previously defined PTZ surveillance RTT limit for TCP connections). If there is no need for PTZ and quick video analytics, it is possible to use mobile core cloud solution (Figure 16.5), now already a mature concept [36]. Edge computing in mobile network (Figure 16.6) is strongly supported by the information and communications technology industry [37]. It has been evolving from mobile edge computing (MEC) industry initiative to multiaccess edge computing (MEC, again) framework to be used by the European Telecommunications Standards Institute (ETSI) to coordinate and promote multivendor proof-of-concept (PoC) projects and MEC deployment trial (MDT) projects illustrating key aspects of MEC technology [38]. In order to demonstrate the viability of a new technology during its early days (and prestandardization phase), PoCs are used as an important tool. MDTs come after PoC to demonstrate the viability of MEC in a commercial deployment and to provide feedback to the standardization efforts where general principles for MEC service application programming interfaces (APIs) are introduced [39]. Video surveillance is part of ETSI specification as use case in multiaccess edge computing (MEC) environment, named camera as a service [40].

5G evaluations for IoT and video surveillance SC scenario are on the way [41,42]. The goals of that kind of experimentations are mainly to test the 5G technology by using real network deployments and to evaluate innovative services enabled by the 5G network. In smart video surveillance system evaluation [41], the solution is capable of performing automatic analysis of video streams, and the use case is part of the eMBB usage scenario. It is considered the possibility of using 4K resolution enabled by the high 5G capacity, which enhances face recognition and person reidentification performance and also reduces a number of cameras for the installation to fully capture a scene. Within that same project, traffic control should be performed by using 4K high-resolution cameras in order to recognize license plates at higher distance or to improve recognition rates at nearer distance. This evaluation project for some specific event recognitions uses higher frame rates than for previous purposes as events are dynamic, so higher bit rates are used. Video streams are processed by GPU (graphics processing unit) computing platforms that are designed for massive parallel computing environments and based on deep learning techniques. Use cases of this experimentations take advantage of 5G high bit rates, and also the reduced latency is beneficial by allowing to use PTZ cameras that are controlled from the control room [41]. There are evaluations of high-precision 4K surveillance cameras as IoT terminals for enabling safe workplace environments [42]. In this project, the 5G system transmits images from construction equipment at the work site to a database and uses AI to assess the interactions and movements of workers in the database. A system makes possible to avoid crashes by informing truck and crane drivers and notifying automatic driving construction equipment of that information. By using AI within the range where signal-to-noise ratio does not drop below 15dB due to radio interference, simulation results showed a workers' recognition rate of 90% [42].

16.4.2 Video Analytics and Artificial Intelligence Surveillance

The human evaluation and assessment of surveillance data make SC video surveillance not so efficient. AI (combined video analytics and machine learning) makes the concept of video surveillance automated in real time where the data are fully utilized. This means that intelligent surveillance systems are capable to identify special events of interest and alert appropriate personnel. Currently, many different video analytics applications are available (face recognition, people holding weapons, etc.), and some others are under way (suspicious loitering, repeated visits in a single day and other unusual behaviors). Face recognition is one of the leading applications of IoT analytics based on video data, and usually, it is in line with the following process. The video analytics application analyzes individual frames to locate and capture a human face as a sample and stores it in customer database. People are detected through cameras and matched against the database. The system alerts if a face is recognized or not. For a recognized face, the system depicts additional registered information and an image of that person. The application is usually based on 3D structures. The unique facial characteristics of the captured sample are evaluated, and a unique data set is extracted from the 3D structure of the detected face. The extracted data set for the sample is matched with the models available in the database of registered faces. Time to results really depends on processing power and the number of entries into the database. Normally, it could be about 20–30 ms for a thousand of registered faces analyzed.

Video analytics application software is installed either on the camera where the camera processes the video (and create metadata), or it is installed on the server where video streams are sent to and processed on the server, independently from the cameras. Camera-based video analytics is software built into physical camera hardware where all analytics and computing occur at the source of the video data, with only relevant video streams leaving the camera [43]. If the analytics is processed on the camera, the main advantage is reduced bandwidth usage because the camera filters the information, thus avoiding to transfer all video data to the servers (only storing the video that the analytics solution has classified as relevant). If the intention is to keep all videos for a certain period of time, then as earlier emphasized, there is a camera with internal SD memory card option with, for instance, 20 days recording time. The negative side of camera-based analytics is that it is built directly into hardware that means it is less flexible than other options and also there is a processing power limitation. Server-based analytics can have different implementation options but, compared with camera-based solution, generally servers have more processing power and are able to process more video and more advanced analytics applications. In a security platform with unified analytics, it is possible to configure the analytics from the same interface as VMS. In larger deployments, this solution provides the same user experience across all types of video analytics, which simplifies configuration and operation. Because VMSs are built for video management, their analytics capabilities are less powerful than other server-based options. Another server-based solution is video analytics software installed on the separate server machine, often designed to perform specific types of analysis. Features of cloud provider software server–based analytics solution highly depend on cloud computing model, Saas/PaaS/IaaS. Finally, there is AI server-based solution that allows the platform to learn what data are important and what are not, improving application accuracy (machine learning).

Figure 16.7 shows MEC server–based distributed analytics solution [44]. The video management application transcodes and stores captured video streams from cameras received on the 5G uplink. The video analytics application receives the video stream from video management application or directly from the IP camera and processes the video data to detect and notify specific configurable events, e.g., lost child, abandoned luggage and so on. The application sends low bandwidth video metadata to the central SC operation for database searches [37]. If needed, alarm analytics operations are then generated automatically. The alerts can be viewed either via the management application clients or in the analytics application viewer clients.

Novel current solutions spread the required compute resources between the cameras, edge servers, specific video analytics applications and cloud server [44]. Video analytics application can run locally at the edge and/or on the cloud, with the ability to send alarms via WAN. Cloud server can also host alerts (Figure 16.8). The genuine AI autonomous

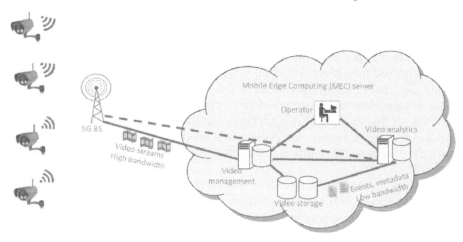

FIGURE 16.7
MEC distributed analytics solution.

surveillance system imposes the combination of cloud and edge computing and moreover forces an end-to-end AI chain to bring processing based on requirements and equipment. The autonomous system needs an intelligently distributed architecture in which captured data and reference data sets can be shared and synchronized in real time.

The need for AI is not only because it makes surveillance system more efficient but also because it protects data centers from overloading. AI system separates the video that needs to be stored from the rest that should not even be captured in the first place. Without AI, the system experiences difficulties to find anything. With AI, the data that are captured and shared have more value. With advanced SC surveillance and end-to-end AI scenario, regular analysis can remain at the edge due to all positive reasons explained earlier, but there is occasional requirement to transmit video (for further analysis and higher-level decision-making) to operators in control room, so the cloud computing enables them to see critical visual data – exactly what they need (Figure 16.8).

In short, a trend is to distribute AI across the network and the processing power and storage capacity to be shared, and this trend leads to the cloud and edge fusion, rather than reliance on one over the other. In video surveillance, the first phase of AI was filtering data, forming metadata descriptors and using non–real-time algorithms to classify and match objects. Currently, AI algorithms are more advanced, matching faces and vehicles with automating low-level decisions and detecting and tracking objects with deep learning techniques in video [43,44]; it is possible to detect threats by learning what such threats look like. By data mining and deep learning, AI system can even predict events [44]. Deep learning is AI form of machine learning and demonstrates the ability to learn without human input. It is modeled on the human brain – deeply layered, tightly packed neural networks. Deep learning algorithms can constantly calibrate weighting assigned to various inputs to better understand their environment. Although only 6% of end users applied deep learning algorithms, this could change dramatically since half of the rest expect to introduce it within the next 5 years [45]. The next AI phase is autonomous surveillance, bringing the integration of different automated processes, where the process is fully automated and self-improving, and direct human intervention is not needed beyond setting an objective.

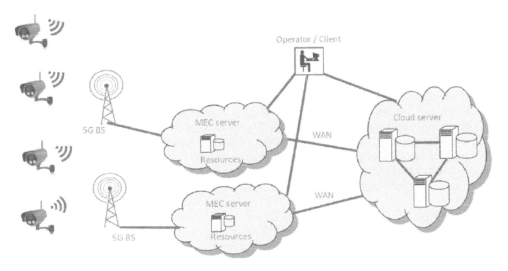

FIGURE 16.8
AI cloud/edge server–based video analytics. AI, artificial intelligence.

16.5 Conclusions

This chapter analyzed IP surveillance in the 5G SC scenario with the large number of cameras (massive IoT).

First, the overview of digital video technologies basics is given, emphasizing key elements for surveillance implementation including various compression algorithms' advantages and disadvantages for video surveillance application. IP camera is defined as IoT sensor with the important features as a sensing device. Special care is devoted to network aspect of IP video where video surveillance is treated as a MC multimedia service for which it is ensured QoS throughout the network by implementing appropriate HTTP/TCP or RTP/UDP streaming service DiffServ marking and classification, for both surveillance live video and recorded video, as well as scheduling and priority queuing with LLQ PQ and CBWFQ.

SC cloud-based IP surveillance is evaluated with different cloud computing models and public/private/hybrid deployment models including VSaaS capabilities. Due to potential limitations in network capacity and overall service latency with the pure cloud solution, mobile edge computing SC possibilities are analyzed for supporting the video surveillance analytics, with the overview of edge server resource management solutions. Moreover, this chapter emphasizes the imperatives of the introduction of 5G delivering lower latency with higher bandwidth including the use of 5G flow models bringing required sub-50 ms RTT latency for PTZ control and video analytics processing.

Finally, there is a machine learning AI need for smarter cities. The distributed 5G AI solutions are presented, including the one with deep learning techniques, bringing processing power and storage sharing based on particular requirements and equipment.

References

1. 3GPP TR 21.915 V2.0.0: 3rd Generation Partnership Project; Technical Specification Group Services and System Aspects; Release 15 *Description*; Summary of Rel-15 Work Items (Release 15), September 2019.
2. 3GPP TS 22.280 V17.1.0: 3rd Generation Partnership Project; Technical Specification Group Services and System Aspects; *Mission Critical Services Common Requirements* (MCCoRe); Stage 1 (Release 17), September 2019.
3. 3GPP TS 23.281 V17.0.0: 3rd Generation Partnership Project; Technical Specification Group Services and System Aspects; *Functional Architecture and Information Flows to Support Mission Critical Video* (MCVideo); Stage 2 (Release 17), September 2019.
4. P. Fraga-Lamas et al., A review on Internet of Things for defense and public safety, *Sensors*, Volume 16, Issue 10, 44 pages, October 2016.
5. R. Fantacci et al., Public safety networks evolution toward broadband: Sharing infrastructures and spectrum with commercial systems, *IEEE Communications Magazine*, Volume 54, Issue 4, pp. 24–30, 2016.
6. Y. Mehmood et al., Internet-of-Things-based smart cities: Recent advances and challenges, *IEEE Communications Magazine*, Volume 55, Issue 9, pp. 16–24, 2017.
7. H. Kruegle, *CCTV Surveillance Video Practices and Technology*, Second Edition, Elsevier, Burlington, MA, 2007.
8. IFSEC, The video surveillance report 2017, IFSEC trend report, *IFSEC Global*, 2017.

9. W. Fischer, *Digital Video and Audio Broadcasting Technology, A Practical Engineering Guide*, Third Edition, Springer, Berlin/Heidelberg, Germany, 2010.

10. *Cisco Video Surveillance Solution: Reference Network Design Guide*, Release 7.0, Cisco Systems Inc., San Jose, CA, June 2013.

11. A. Caputo, *Digital Video Surveillance and Security*, Elsevier, Burlington, MA, 2010.

12. T. K. Tan et al., Video quality evaluation methodology and verification testing of HEVC compression performance, *IEEE Transactions on Circuits and Systems for Video Technology*, Volume 26, Issue 1, pp. 76–90, January 2016.

13. K. R. Rao et al., *Video Coding Standards*, Springer, Dordrecht, Netherlands, 2014.

14. Recommendation ITU-R BT.2020–2: *Parameter values for ultra-high definition television systems for production and international programme exchange*, October 2015.

15. European Standard BS EN 62676-4: *Video surveillance systems for use in security applications - Application guidelines*, 2015.

16. J. King, *Cisco IP Video Surveillance Design Guide*, Cisco Systems Inc., San Jose, CA, August 2009

17. Cisco Global Cloud Index: *Forecast and Methodology* 2016–2021, updated 2018, https://www.cisco.com/c/en/us/solutions/collateral/service-provider/global-cloud-index-gci/white-paper-c11-738085.html, accessed 14-August-2019.

18. L. Tian et al., Video big data in smart city: Background construction and optimization for surveillance video processing, *Future Generation Computer Systems*, Volume 86, pp. 1371–1382, September 2018.

19. D. J. Neal, S. Rahman, Video surveillance in the cloud? *International Journal on Cryptography and Information Security (IJCIS)*, Volume 2, Issue 3, 19 pages, 2012.

20. P. Mell, T. Grance, The NIST definition of cloud computing recommendations of the National Institute of Standards and Technology, *NIST Special Publication*, Volume 145, pp. 1–7, 2011.

21. C. Zhang, E. Chang, Processing of mixed-sensitivity video surveillance streams on hybrid clouds, In *Proceedings of IEEE 7th International Conference on Cloud Computing*, USA, pp. 9–16, 2014.

22. M. A. Hossain, Framework for a cloud-based multimedia surveillance system, *International Journal of Distributed Sensor Networks*, Volume 10, Issue 5, 11 pages, 2014.

23. K. Tocze, S. Nadjim-Tehrani, A taxonomy for management and optimization of multiple resources in edge computing, *Wireless Communications and Mobile Computing*, Volume 2018, Article ID 7476201, 23 pages, 2018.

24. J. Pan, J. McElhannon, Future edge cloud and edge computing for Internet of Things applications, *IEEE Internet of Things Journal*, Volume 5, Issue 1, pp. 439–449, February 2018.

25. J. Wang et al., Elastic urban video surveillance system using edge computing, In *Proceedings of Smart IoT'17*, San Jose/Silicon Valley, USA, 6 pages, October 2017.

26. G. Li et al., Method of resource estimation based on QoS in edge computing, *Wireless Communications and Mobile Computing*, Volume 2018, Article ID 7308913, 9 pages, 2018.

27. J. Zhang et al., Joint resource allocation for latency-sensitive services over mobile edge computing networks with caching, *IEEE Internet of Things Journal*, Volume 6, Issue 3, pp. 4283–4294, June 2019.

28. X. Liu et al., Resource allocation for edge computing in IoT networks via reinforcement learning, In *Proceedings of IEEE International Conference on Communications*, Shanghai, China, 6 pages, May 2019.

29. C.-H. Hong, B. Varghese, Resource management in Fog/Edge computing: A survey on architectures, infrastructure, and algorithms, *ACM Computing Surveys (CSUR) Journal*, Volume 52, Issue 5, Article No. 97, September 2019.

30. S. Maheshwari et al., Scalability and performance evaluation of edge cloud systems for latency constrained applications, In *Proceedings of 2018 Third ACM/IEEE Symposium on Edge Computing*, USA, pp. 286–299, October 2018.

31. OpenFog Architecture Working Group, *OpenFog Reference Architecture for Fog Computing*, OpenFog Consortium Publication, USA, 2017.

32. D. Warren et al., *Understanding 5G: Perspectives on Future Technological Advancements in Mobile,* GSMA Intelligence Analysis Report, December 2014.
33. 3GPP TS 22.261 V16.8.0: 3rd Generation Partnership Project; Technical Specification Group Services and System Aspects; *Service Requirements for the 5G System;* Stage 1 (Release 16), June 2019.
34. ETSI TS 123 501 V15.5.0: 5G; *System Architecture for the 5G System* (3GPP TS 23.501 version 15.5.0 Release 15), April 2019.
35. J. Nightingale et al., QoE-driven, energy-aware video adaptation in 5G networks: The SELFNET self-optimisation use case, *International Journal of Distributed Sensor Networks,* Volume 12, Issue 1, 15 pages, 2016.
36. F. Liu et al., Gearing resource-poor mobile devices with powerful clouds: Architectures, challenges, and applications, *IEEE Wireless Communications,* Volume 20, Issue 3, pp. 14–22, June 2013.
37. M. Patel et al., Mobile-Edge Computing (MEC) industry initiative, *Introductory Technical White Paper,* September 2014.
38. ETSI GS MEC 005 V2.1.1: Multi-access Edge Computing (MEC); *Proof of Concept Framework,* July 2019.
39. ETSI GS MEC 009 V2.1.1: Multi-access Edge Computing (MEC); *General Principles for MEC Service APIs,* January 2019.
40. ETSI GS MEC 002 V2.1.1: Multi-access Edge Computing (MEC); *Phase 2: Use Cases and Requirements,* October 2018.
41. F. Nizzi et al., Evaluation of IoT and video surveillance applications in a 5G smart city: The Italian 5G experimentation in Prato, In *Proceedings of 2018 AEIT International Annual Conference,* 110th Edition, Italy, 6 pages, October 2018.
42. D. Nozaki et al., AI management system to prevent accidents in construction zones using 4K cameras based on 5G network, In *Proceedings of the 21st International Symposium on Wireless Personal Multimedia Communications (WPMC- 2018),* Thailand, pp. 462–466, November 2018.
43. J. Barthélemy et al., Edge-computing video analytics for real-time traffic monitoring in a smart city, *Sensors,* Volume 19, Issue 9, 23 pages, May 2019.
44. J. Chen et al., Distributed deep learning model for intelligent video surveillance systems with edge computing, *IEEE Transactions on Industrial Informatics,* https://doi.org/10.1109/TII.2019.2909473, arXiv:1904.06400v1, 9 pages, April 2019.
45. IFSEC, The video surveillance report 2019, IFSEC trend report, *IFSEC Global,* 2019.

17

Challenges of Implementing Internet of Things (IoT) in 5G Mobile Technologies

V. Hurbungs, V. Bassoo, V. Ramnarain-Seetohul, Tulsi Pawan Fowdur, and Y. Beeharry
University of Mauritius

CONTENTS

17.1 Introduction

The next evolution in mobile communications is 5G, and it is expected to be deployed by 2020 [1]. It will represent a massive change from the current 4G/LTE networks. 5G is set to transform our society into a completely mobile and connected one. It will provide a platform to devise and expedite the development of a wide range of applications, some of which are currently unimaginable, which will substantially improve our quality of life and well-being. Moreover, the need for a better network is being felt around the world as it is predicted that our existing cellular networks will not be able to sustain the ever-increasing growth in the number of smart devices and exponential data demands [2].

The 5G network is being designed to provide end-to-end latency of less than 1 ms to sustain a throughput of 10 Gbps per connection [3]. It also supports networks of 1 million connections per square kilometer to improve mobility up to 500 km/h and provides a flexible network architecture that will allocate network resources and capabilities on demand and in real time [3]. NB-IoT and LTE-M are low power wide area technologies that will be part of 5G network and support applications that will require longer battery lifetime and global coverage [4]. It is expected that 5G will support a massive number of connected devices. With the gain in popularity of Internet of things (IoT), connected devices are increasingly impacting on how we live and how we work. The cost of broadband connection is decreasing worldwide, and devices have wireless connectivity and inbuilt sensors that can capture, for example, temperature, proximity, light intensity and heart rate, among others.

IoT brings a lot of opportunities, and today, we can think of smart buildings, smart agriculture, smart cities, smart products or smart businesses. Additionally, the new capabilities of the 5G network will give rise to a wide range of emerging use cases. The ultrareliable and low latency feature of 5G will work well for applications in the areas of autonomous driving, healthcare and emergency services. The ability of 5G to support a massive number of devices will enable the development of smart cities. Moreover, the enhanced broadband experience that is offered by 5G will enrich entertainment experiences with technologies such as augmented and virtual reality and provide services such as mobile office [3,5].

This chapter addresses challenges of implementing IoT in 5G. The second section of this chapter presents emerging concepts related to the physical layer; millimeter wave communication (mmWave) technologies, massive MIMO systems and small cell concepts are discussed. Section 17.3 presents challenges in the area of IoT middleware. The last section of this chapter addresses important security aspects related to the implementation of IoT in 5G mobile technologies.

17.2 Millimeter Wave Communication Connectivity

With the rise in the number of wireless products, the lower end of the radio spectrum is significantly clogged. The majority of the cellular communication devices operate in the spectrum range of the 300 MHz to 3 GHz, whereas WLAN devices operate in the bands of 900, 2.4 and 5.8 MHz. Some of the other services that use the ultrahigh frequency band and below are AM radio broadcasting, military communications and television broadcasting [6]. The 30–300 GHz spectrum is sparsely used by the military and is therefore underutilized. The band is referred to as the mmWave. As shown in Figure 17.1, around 252 GHz of the spectrum is considered suitable for mobile broadband use. Frequencies around the 60 GHz band are referred to as oxygen absorption band and can undergo an attenuation of 15 dB/km. Frequencies between 164–200 GHz are referred to as the water vapor absorption band and can suffer from attenuation of up to tens of dBs per kilometer depending on the humidity in the air [7]. Excluding these two bands and assuming that 40% of the remaining band can be made accessible, a spectrum of approximately 100 GHz of mmWave is therefore available for mobile broadband communication [8]. This band is essential to support IoT applications in 5G networks.

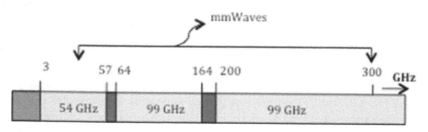

FIGURE 17.1
mmWave band. mmWave, millimeter wave.

17.2.1 Characteristics of Millimeter Wave Communication

mmWave is a technology that appears to be a prominent enabler of 5G. It offers numerous advantages. Larger bandwidths can be made available, and therefore, it can deliver the high data rates of around 10 Gbps for uses such as real-time gaming, high-quality video streaming or advanced IoT applications [9].

The radio frequency components of the wireless communication systems of an mmWave device are small in size (around 40 nm) due to the extremely short wavelengths [9]. All the different components including complex antenna arrays can therefore be packed on ICs (integrated circuits) using semiconductor materials [10]. It eliminates the need for coaxial cables, printed circuit boards and transmission lines that are currently being used to connect antennas to transceiver circuits [6]. The miniaturization of the systems opens up the use of mmWave frequencies for many applications. The narrow beam width of the small antennas allows for better spatial resolution. It also helps to achieve greater privacy and security, as the signals are constrained to only a limited area [11].

However, mmWaves have many unfavorable propagation features. It experiences higher propagation loss compared with other communication systems operating at lower carrier frequencies. It suffers from serious atmospheric absorption and is attenuated by rain, therefore restricting the mmWave communication range. The signal is also vulnerable to blockages [12]. Building materials such as bricks can mitigate signals by 40–80 dB, and even the human body can cause an attenuation of around 20–35 dB [13].

Most of the unfavorable elements of mmWaves can be curbed by new design features. Small cell sizes can help address the atmospheric absorption and rain attenuation problems [12]. Directional antennas with narrow beam can mitigate multipath reflection components [6].

17.2.2 Millimeter Wave Massive Multiple-Input Multiple-Output in Ultradense Networks

Massive MIMO is another technology that is set to become a main enabler for 5G, and it is expected to support the proliferation of IoT devices. As shown in Figure 17.2, massive MIMO uses a much larger number of antennas (100 or more) as compared with the conventional MIMO. Massive MIMO operating at mmWave frequencies is more advantageous, as it provides very high data rates due to the larger mmWave bandwidth and delivers higher gains because of the use of a large number of MIMO antenna arrays. Therefore, it drastically increases the capacity of networks by vigorous spatial multiplexing due to the sheer number of antenna elements and improves energy efficiency by using steerable antennas and beamforming to direct energy to desired receivers, at the same time mitigating interference [8,12].

Since the size of mmWave frequency antennas is significantly smaller, it is consequently "easier" to design massive MIMO systems as it is feasible to place a greater number of antenna elements in a small area at the base stations and even in the user equipment. Moreover, the radiation power of the antennas is reduced due to the small antenna size, therefore limiting interference. The reduction in power and size ensures lower costs [8].

The mmWave massive MIMO system becomes more interesting with the advent of small cell concepts in ultradense networks [14]. The small cells are anticipated to work well with massive mmWave MIMO antenna system, which inherently has low radiation power, therefore limiting interference with neighboring cells. Due to the size of the cells (up to a 100 m) [14], line-of-sight transmission is possible, and by employing mmWave massive

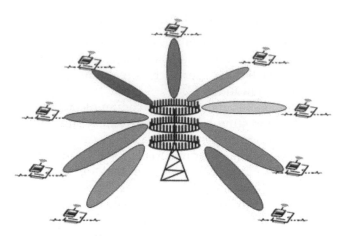

FIGURE 17.2
Massive MIMO system. MIMO, multiple-input multiple output.

MIMO systems, wireless links with astounding data rates in the order of 100 Gbps can be expected. However, the complexity of such a system is very high [6].

17.3 Internet of Things Middleware

Middleware is usually an abstraction layer located between the application layer and the operating system. Middleware is mostly used in distributed systems to hide complexity and heterogeneity of hardware, operating systems and protocols. IoT can be considered as a complex distributed system with heterogeneous objects, embedded with electronics, software, sensors and actuators, which can connect to exchange data [15,16]. IoT introduces significant new challenges for the middleware layer given the number of connected objects, the volume of data, heterogeneity of components and continuous connectivity between devices. A key technology in the success of IoT systems is middleware, which acts as the intermediary between IoT devices and applications as shown in Figure 17.3.

A number of middleware are emerging to simplify the integration of sensors, devices, networks, applications and services. Table 17.1 highlights some examples of open-source IoT middleware [18].

IoT provides a lot of new possibilities in terms of information and communication technology. However, given its heterogeneous nature, it imposes several challenges on the middleware layer. The tasks of an IoT middleware are more challenging as the IoT ecosystem must deal with scalability, reliability, availability, security and privacy, among others. Moreover, an IoT middleware must deal with programming abstraction, interoperability, adaptability, context awareness, autonomy and distributiveness [19].

The arrival of 5G technology in the near future imposes additional challenges for the IoT middleware layer. Since 5G aims at providing higher bandwidth, reliability, mobility, faster speeds, low latency and better user experience, 5G systems are expected to provide enhanced device and network capabilities [20]. The IoT middleware layer should therefore be able to support 5G-enabled applications and services.

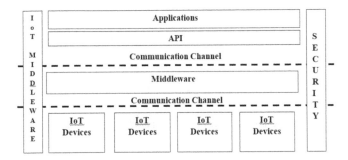

FIGURE 17.3
IoT middleware [17]. API, application programming interface; IoT, Internet of things.

TABLE 17.1

Open-Source IoT Middleware

Middleware	Description
AllJoyn	• Collaborative open-source software framework • Allows devices and applications to discover and communicate with each other • Includes security mechanisms such as encryption and authentication • Operating system: Windows, Linux, OS X, Android, iOS, Arduino
Kaa	• Building complete end-to-end IoT solutions, connected applications and smart products • Communication and monitoring between IoT devices and back-end infrastructure • Can be deployed on Amazon's cloud • Operating system: Linux
Mango	• High-performance NoSQL database • Supports data acquisition, real-time monitoring and security • Operating system: Windows, Linux, OS X
Nimbits	• Data logging service and rule engine platform for connecting people, sensors and software to the cloud • Offer a public cloud through a Nimbits Server
OpenIoT	• Blueprint middleware infrastructure for implementing/integrating IoT solutions • Collect and process data from any IoT device • Stream, analyze and visualize data to the cloud • Operating system: Windows, Linux, OS X

IoT, Internet of things.

17.3.1 5G-Based Challenges for Internet of Things Middleware

According to a study by Bandyopadhyay et al. [21], the required functional blocks of an IoT middleware are illustrated in Figure 17.4. The need for these functional blocks arises from the need to enforce a common standard among devices, bind heterogeneous components together and integrate different layers and application programming interface (API) with the aim of providing transparent access to applications and services. Other layers such as network, storage and applications are built on top of the functional blocks. 5G will impact on these functional blocks since a larger number of heterogeneous IoT devices and applications will interact together, leading to higher data traffic demands. 5G networks will also handle millions or billions of devices, generating context-specific data, which need to be discovered. Since 5G is also about connectivity and people-oriented services, device and traffic data over 5G networks need to be secured and made available only to authorized

IoT Middleware

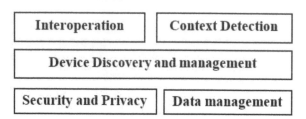

FIGURE 17.4

5G research and development span from simulation and design verification, to device acceptance and ultimately to manufacturing and deployment. IoT, Internet of things.

devices, services and applications. The major challenges of 5G-based IoT middleware are discussed in the next subsections with respect to the functional blocks.

Interoperation. Interoperation is one of the fundamental enablers of IoT. The lack of communication between devices may result in sensor data not being used efficiently. Therefore, the middleware layer has a crucial task to ensure that IoT devices are able to talk to each other and exchange data. A major challenge is that there should be proper standards defining IoT device requirements, architecture, API specifications and communication layers. In a 5G environment where many devices are expected to collaborate together, it is therefore important that all devices can communicate and interact with each other, and sensor data can be shared across multiple devices in the IoT ecosystem. 5G will require abstractions at several layers, namely hardware, software, data and network. The 5G-enabled IoT middleware must therefore implement a common protocol for a large number of components with different configurations and be able to efficiently manage new resources that may be discovered in the future [20]. Instead of creating vertical layers of integrated IoT products, developers should promote frameworks that cut across hardware and software horizontally to support an interoperable framework [22]. Devices, sensors and applications should therefore not require customized middleware that does not allow data sharing between applications. Another direction is to move toward open-source frameworks to solve interoperability [23]. This will avoid companies from developing solutions independently of each other with different platforms and frameworks. Interoperation is closely related to scalability in the context of 5G since the IoT middleware must also be able to adapt itself to both small-scale and large-scale deployment of IoT devices.

Context detection. IoT middleware must be context aware in order to operate in smart environments where connected devices perform context-triggered actions for the end users. Since one of the objectives of 5G is to provide a high quality of experience to the user, context awareness will enhance existing applications and provide more user-centered services [20]. According to Bandyopadhyay et al. [21], context awareness can be categorized as context detection and context processing. Context detection is responsible for characterizing the IoT environment, whereas context processing prepares the data for further analysis. The work in Ref. [24] defines two types of context, viewed from the operational and conceptual perspectives, which are of importance to the IoT paradigm: primary and secondary context. Location, identity, time and activity identification among others represent the primary context [25], whereas secondary context refers to any information that can be derived using primary context [24]. Examples of primary context are as follows: GPS location (longitude and latitude), user identity based on RFID tag, time and sensor data (e.g., temperature, humidity) [24]. Examples of secondary context are as follows: compute distance of two sensors computed using GPS values, determine the user activity based on mobile phone sensors such as GPS, gyroscope and accelerometer and identify a face of a person using facial recognition system [24]. Since 5G is also characterized by providing any service anytime, anywhere and with anyone, context management therefore becomes a prime objective of IoT middleware. The latter must be able to define a context categorization scheme (primary or secondary) that can be used to classify data which may be considered as primary in one situation and as secondary in another situation. The quality and accuracy of the acquired data will vary from different categorization techniques, and this becomes even more challenging with the large volume of data generated by sensors in IoT environments. The IoT middleware must therefore be able to handle context information generated from devices, applications, users, environment and 5G networks. New mechanisms or abstractions should be devised to generate and share context information.

Device discovery and management. IoT devices need to make themselves visible in a 5G environment where they need to find each other with the required profile and characteristics. Discovery protocols should be dynamic to accommodate for a wide range of network topologies. Standard discovery interfaces and semantic annotations at the middleware layer are therefore essential to ensure interoperability among devices in 5G networks [20]. Once discovery is done, communication is established, and devices can share data and cooperate together to provide novel services compared with traditional networks. In addition to communication protocols, network topology will play an important role in IoT device discovery and management. In Ref. [26], the authors propose an IoT registry infrastructure after evaluating three network topologies using parameters such as discovery time, server-side infrastructure, reliability and network traffic. Device discovery is also closely related to security since devices generating sensitive data should not be visible or accessible by unauthorized objects. There should be proper discovery mechanisms setup specifically for 5G environments to ensure proper visibility and access control since 5G will be characterized by millions of devices that are close to people. It is equally important to know which device is on a network; is it a valid device or a hacker trying to spy on network communication? Traditional network access controls may not be appropriate since not all IoT devices are secure. There should be a service to display a picture of the devices in range and determine which ones are trusted or untrusted.

Security and privacy. 5G creates a new set of challenges to security and privacy for connected devices. Since 5G networks will connect more people, new approaches to security at the middleware layer are necessary. In Ref. [27], the authors present a matrix model for evaluating threats to IoT systems in terms of hardware/device, network and cloud/server. Existing middleware protocols may not be appropriate for connecting low-power devices. Privacy becomes an important requirement for personal data collected by sensors, which are transferred to the cloud via the middleware layer. With the arrival of 5G technology, IoT middleware will have to cater for new types of security threats. The study in Ref. [17] recommends that the IoT middleware security architecture should be based on lightweight approaches since 5G applications and devices will operate in constrained environments with low memory, computing power and network requirements. The middleware layer must therefore authenticate applications and connected devices, manage access rights to resources, data and services, protect context data from unauthorized access and ensure integrity of sensor data.

Data management. Middleware usually enables communication and manages data in distributed applications. The challenge of IoT middleware is now to manage large volumes of data generated by 5G networks. Therefore, it is important to establish new methods to acquire, store and transfer data between connected devices. Mobile devices will form part of the 5G landscape; when they move, the IoT middleware layer should still be able to manage these data. In an IoT environment, the life cycle of data starts from data acquisition, aggregation, transfer, processing and finally storage [28]. According to Ref. [28], there is a need for data-centric IoT middleware that will handle data between the devices and the data stores. In addition, the IoT middleware should also make provision for large data volume transfer from the 5G network to data stores. The study in Ref. [29] highlights the need for green data management systems for IoT to minimize energy costs. One of the objectives of 5G is to reduce energy consumption of connected devices. To reduce energy consumption by IoT middleware, communication overheads and storage mechanisms should be optimized.

17.3.2 Perspectives of Internet of Things Middleware toward 5G

The middleware layer has evolved considerably from facilitating interoperability in distributed systems to connecting devices in an IoT environment in order to provide distribution, heterogeneity, mobility, interoperability and data transparency, among others. With the arrival of 5G networks, IoT middleware should now be designed using the service-oriented architecture (SOA) instead of a vertical architecture, which caters only for specific areas [20]. SOA is essentially a collection of services that communicate with each other across different platforms, using standard communication protocols and interfaces, and therefore promotes code reuse, interaction, scalability and reduced costs, among others [30].

SOA-based IoT middleware perfectly fits the requirements of 5G. New types of applications and services can be created without worrying about the technologies on which platforms are built. Instead, these applications can rely on the interoperability that SOA provides. In addition, SOA can help scale 5G applications at the same time decreasing the costs of developing business service solutions. The adoption of SOA principles will allow the decomposition of complex 5G networks into a set of well-defined modular components that offer their own functionality as services as shown in Figure 17.5.

Each layer is characterized by different technologies and protocols that interact together to provide a common service. Tables 17.2 and 17.3 summarize some of the requirements imposed by 5G systems on technology, services and applications [19,20].

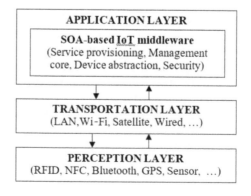

FIGURE 17.5
5G research and development span from simulation and design verification, to device acceptance and ultimately to manufacturing and deployment. GPD, global positioning system; IoT, Internet of things; LAN, local area network; NFC, near-field communication; RFID, radio frequency identification; SOA, service-oriented architecture.

TABLE 17.2

Technological Requirements of 5G Systems

Requirements	Explanation
Massive system capacity	More IoT devices and applications Higher data traffic
Very low cost and energy consumption for mobile devices	Low cost devices, extending battery life
Ultrahigh reliability and availability for mobile connectivity	Reliability and availability for mission-critical applications
Powerful nodes at the edge of the network	Process raw data at the IoT device, offload data processing from the core of the network to IoT devices
Higher ubiquitous data rates for real-life condition situations	More bandwidth and high-speed network connectivity
Virtualized network technology support	Multiple virtual core networks
Very low latency for next-generation networks	1 ms or less for latency-critical applications

IoT, Internet of things.

TABLE 17.3

5G-Based IoT Services and Applications Requirements

Requirements	Explanation
Everything on cloud	Cloud computing technologies Very high data rates
Massive applications with very large number of IoT devices	Alternative connectivity such as Wi-Fi, Bluetooth
User experience	Enriched context information, high-definition videos, augmented reality Fast data rates
Uninterrupted and reliable operation of critical applications	High availability and reliability of mobile network technology
Ubiquitous connectivity of smart objects and applications	Intelligent on-demand connection

IoT, Internet of things.

The purpose of each layer is summarized as follows [20]:

- Perception layer: Sensing the environment and real-time data collection
- Transportation layer: Delivery of collected data to applications and servers
- Applications layer: Responsible for device discovery, device management, data filtering and data aggregation, among others.

17.4 Security and Internet of Things

Nowadays, with the expansion of the vertical industries including IoT, a high-speed pervasive network access is essential [31]. This can be achieved with the 5G broadband technology [32]. 5G will link vital infrastructure that will necessitate further security to guarantee safety of devices and society [33]. Consequently, the security must be reviewed to preserve the privacy, confidentiality and availability of data. The 5G network is very broad, and it is impossible to secure the whole of it with one security solution. A holistic approach should be applied to secure data. The 5G network should have necessary in-built security capabilities, and additional layer of protection could be applied to services that need more safeguarding [31].

17.4.1 Challenges of Ensuring Secure Communications

The 5G network will have to address new security issues and challenges as the security features provided in current networks may not be appropriate and adequate [34]. The 5G network will be used intensively by vertical industries, and consequently, a variety of new services are going to emerge. At this point, security requirements could differ considerably among the services [34]. For example, mobile IoT devices need lightweight security compared with high-speed mobile services, which require effective security. The traditional network-based security approaches may not be adequate to develop security mechanisms for IoT devices over 5G links.

Availability. Availability can be termed as the extent to which a service is reachable to an authorized user [35]. By 2020, the 5G network is believed to endorse 50 billion connected devices and 212 billion connected sensors as well as allow access to 44 zettabytes (ZB) of data [32]. These will include smartphones, tablets, smart watches, cars, machinery, appliances, remote sensing and monitoring devices, among others. Consequently, all these will produce an enormous amount of data. The availability of these data is important in order to provide enhanced services to users in an IoT network [31]. Availability is an important metric to guarantee the reliable communications in 5G [36]. Denial of service (DoS) attack is among one of the major attacks, which prevents the right person form getting access to data at the right time. With several unprotected IoT nodes, it therefore becomes a challenge for 5G networks to prevent DoS attacks [35].

Authentication. One of the main issues with the interconnection of many devices is the availability of a huge amount of data, which must be protected at both the communication and processing levels. However, since IoT devices are characterized by low processing capabilities, it is inconvenient to process data on the devices. The inability of these devices to encrypt and decrypt data makes authentication more difficult, and this becomes a major challenge to preserve the integrity of data [36]. This can in turn give rise to malicious attacks; intruders can invade the network and steal personal information due to insufficient security. Thus, to prevent unauthorized access to connected devices, there is a need for more rigorous authentication mechanisms [34]. In addition, a well-established security framework is required to help users manage the accessibility of their devices. However, the problem in the 5G network is that services will degrade as more security is enforced [37]. For instance, low latency is an essential requirement in the 5G network, and if huge amount of data are encrypted and decrypted, this can degrade the services, thereby violating latency and efficiency requirement of the 5G network [35]. A trade-off is therefore important, as both security and enhanced services are vital.

Privacy. Privacy is also a concern in 5G networks, as they are open network platforms. Privacy leakage can have serious impact [34] since mobile networks transfer data that contain confidential information, for example, location of users [36]. An intruder may deduce the location of the IoT devices based on communication patterns [36]. Furthermore, to provide differentiated quality of service, networks may need to inspect the kind of service that is being used, and this process may lead to privacy violation. Although there are efficient privacy-preserving algorithms that exist [36], it might be difficult to implement these algorithms in 5G networks given the limited capabilities of IoT devices [34,36]. The solution is to devise lightweight security mechanisms for resource-constrained devices so as to ensure privacy of data. Lightweight cryptography consists of different types of algorithms [38], and the elliptic curve cryptography is one such example that can be used to preserve privacy [39].

Integrity. The aim of the 5G network is to provide connectivity at any time, at any place in order to support applications and services which are closely related to our daily life. Hence, integrity of data is an important requirement in these applications. For instance, data can get tampered when they are at rest and in motion [39]. As some IoT networks are opened, harmful devices can connect and compromise the network by committing different attacks or by introducing viruses or trojans. The following types of attacks usually affect the integrity of messages: cloning, message modification, message insertion and replay, among others [40]. The use of cryptographic protocols can help to accomplish the security goals of IoT. However, they are difficult to implement on the resource-constrained IoT devices [41]. Furthermore, data can get tampered by insider malicious attacks, and it is hard to detect these attacks as the insider attackers may have valid identities. Hence, maintaining data integrity can be a very challenging task in the 5G network [35].

17.4.2 Security Solutions to Ensure Communications

5G networks will be service oriented. This indicates that security will be required for the different services that emanate from the various angles of services [31]. However, all the 5G services will need some standard security aspects such ass authentication, preserving integrity and privacy and protecting confidentiality of data. Despite the differences among network technologies, devices and services, there is a need for a set of common security standard [31]. Furthermore, the next-generation access networks would also comprise of different network systems (e.g., 5G, LTE) and technologies. There should be flexibility in the management of security for different network technologies [31].

Security is more efficient when it is added during the design process. The security mechanisms must be able to evolve to accommodate for protocol changes and addition of new algorithms [42]. Network slicing can be part of the solution to manage the different requirements of the different applications and user groups. By slicing the network into isolated subnetworks, this can produce smaller networks with distinct set of traffic characteristics. The security requirements may be restricted to single slices instead of the whole network [43]. Figure 17.6 shows how a network can be divided, and each subnetwork will have distinct end-to-end security requirements.

The advantage that end-to-end security provides is flexible data protection for distinct services. The security needs differ considerably within different types of services. In addition, with end-to-end data protection, there is no need to encrypt and decrypt data repeatedly at intermediate nodes. As compared with hop-by-hop data protection, end-to-end data protection needs a smaller number of encryption and decryption and less delay in data processing and provides greater communication efficiency [31].

The security infrastructure for the 5G network will have to tackle security needs from service scenarios and customers. Hence, security strategies should be arranged on the basis of service scenarios and then put into operation to relevant network slices and nodes. The security requirements of different services can then be achieved with greater flexibility [31].

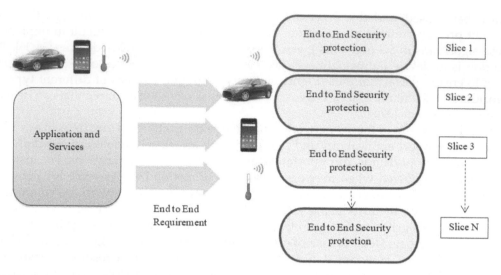

FIGURE 17.6
Network and subnetwork security requirements [34].

Authentication. Authentication is very important in 5G networks where low-latency and efficient authentication schemes are required [35]. Thus, it is a great challenge for network and service providers to make access and authentication easier and less expensive. Three models have been proposed to handle different authentication needs [31]:

1. Network authentication only: In this model, service providers will pay the network for service authentication, and users will get access to the various services after they have been authenticated once. Users will require authentication each time they access a different service.

2. Authentication by service providers only: In this model, authentication done by the vertical industries will be taken into account by networks, and devices will not need radio network access authentication, which in turn will help in decreasing operation costs.

3. Authentication by both networks and service providers: The network access will be controlled by the networks, and service access will be handled by service providers.

Privacy. 5G networks will need improved procedures for transparency, openness, controlling access and accountability. It is very important to consider strict privacy rules and regulations during the standardization of 5G networks. The regulatory procedure can be grouped into three categories, namely [33]:

- Government level: The government will need to formulate privacy regulations for each country.

- Industry level: Different businesses and consortia such as 3GPP, ETSI and ONF will have to work together to outline the best methods and operations to preserve privacy.

- Consumer level: Privacy will depend on consumers' specifications. For instance, to protect privacy for location of customers, techniques based on anonymity can be employed to hide the real identity of the subscriber with pseudonyms. In addition, messages can be encrypted before sending to location-based service providers. Methods such as obfuscation can further be used in protecting location privacy. Additionally, location cloaking–based algorithms can be helpful to manage location privacy attacks [33].

Confidentiality. One of the most effective and simple solutions to protect data is to use encryption techniques such as public key infrastructure (PKI) and elliptic curve cryptography (ECC). However, these methods require a lot of computational power and battery usage. As IoT devices are resource constrained, these methods may not be utilized in 5G networks. A lightweight encryption algorithm and key management protocols are therefore required for IoT devices [44]. Moreover, in 5G networks where latency is an important factor, the computational overhead for encryption, hashing and secure protocols can affect battery-powered devices. One solution is precomputation to save time and power, but protocols need to be reviewed to make this possible [42].

17.5 Conclusion

The 5G network represents a massive change from the current 4G/LTE network infrastructure and services. It has the potential of offering very high-speed connection and low latency and supports more bandwidth, because of its built-in computing ability to manage data effectively. It will also be an intelligent and service-oriented network, which will integrate data analytics as part of the delivered services. The 5G network will support IoT environments by connecting more and more devices that will change our daily lifestyle. Emerging concepts related to the physical layer such as mmWave technologies, massive MIMO systems and small cell can provide support for 5G networks. IoT middleware is an important area of research since 5G networks aim at simplifying the integration of sensors, devices, networks, applications and services. 5G networks will have to address new security issues and challenges with the interconnection of IoT devices to ensure privacy.

References

1. IEEE 5G, "IEEE 5G and Beyond Technology Roadmap," IEEE, White Paper, 2018.
2. M. Agiwal, A. Roy, and N. Saxena, "Next Generation 5G Wireless Networks: A Comprehensive Survey," *IEEE Communications Surveys & Tutorials*, vol. 18, no. 3, pp. 1617–1655, 2016.
3. NGMN Alliance, "NGMN 5G Initiative White Paper," NGMN Alliance, Reading, UK, White Paper, 2015.
4. Z. Bojkovic and D. Milovanovic, "5G connectivity Technologies for the IoT: Research and Development Challenges," in *ELECOM 2018*, 2018, pp. 351–361.
5. Intel, *Intel 5G Network Cloud Client*, Intel, 2016.
6. T. Rappaport, R. Heath Jr, R. Daniels, and J. Murdock, *Millimeter Wave Wireless Communications*, 1st ed., T Rappaport, Ed.: Pearson Education, 2015.
7. Z. Pi and F. Khan, "An Introduction to Millimeter-Wave Mobile Broadband Systems," *IEEE Communications Magazine*, vol. 49, no. 6, pp. 101–107, June 2011.

8. S. A. Busari, K. M. Huq, S. Mumtaz, Linglong Dai, and J. Rodriguez, "Millimeter-Wave Massive MIMO Communication for Future Wireless Systems: A Survey," *IEEE Communications Surveys and Tutorials*, vol. 20, no. 2, pp. 836–869.

9. S. Garg, "Is Millimeter Wave Technology Future of Wireless Communications?" *Microwave Journal*, July 2017. https://www.microwavejournal.com/blogs/25-5g/post/28775-is-millimeter-wave-technology-future-of-wireless-communications

10. A. Kingatua. (February 2017) allaboutcircuits. [Online]. https://www.allaboutcircuits.com/news/the-role-of-millimeter-waves-in-ever-expanding-wireless-applications/

11. Y. Yu, "Millimeter-wave wireless communication," in Integrated 60GHz RF Beamforming in CMOS, Yikun YuPeter, G.M. BaltusArthur, H.M. van Roermund, pp. 7–18, 2011, Springer, Dordrecht.

12. Y. Niu, Y. Li, D Jin, L. Su, and A. Vasilakos, "A Survey of Millimeter Wave Communications (mmWave) for 5G: Opportunities And Challenges," *Wireless Networks*, vol. 21, no. 8, pp. 2657–2676, November 2015.

13. S. Rangan, T Rappaport, and E Erkip, "Millimeter-Wave Cellular Wireless Networks: Potentials and Challenges," *Proceedings of the IEEE*, vol. 102, no. 3, pp. 366–385, March 2014.

14. M. Kamel, W. Hamouda, and A. Youssef, "Ultra-Dense Networks: A Survey," *IEEE Communications Surveys and Tutorials*, vol. 14, no. 4, pp. 2522–2545, 2016.

15. J. Morgan. (2014) A Simple Explanation of 'The Internet of Things'. [Online]. https://www.forbes.com/sites/jacobmorgan/2014/05/13/simple-explanation-internet-things-that-anyone-can-understand/#739669af1d09

16. Technopedia. (2018) Middleware. [Online]. https://www.techopedia.com/definition/450/middleware

17. R. Tiburski, L. Amaral, and F. Hessel, "Security challenges in 5G-based IoT middleware systems," in Internet of Things (IoT) in 5G Mobile Technologies, George Mastorakis, Jordi Mongay Batalla, Constandinos Mavromoustakis, Vol. 8, pp. 399–418, 2016, Springer, Cham.

18. C. Harvey. (2015) 6 Open Source Middleware Tools for the Internet of Things - Datamation. [Online]. https://www.datamation.com/mobile-wireless/slideshows/6-open-source-middleware-tools-for-the-internet-of-things.html

19. M. Razzaque, M. Milojevic-Jevric, A. Palade, and S. Clarke, "Middleware for Internet of Things: A Survey," *IEEE Internet of Things Journal*, vol. 3, no. 1, pp. 70–95, 2016.

20. L. A. Amaral et al., "Middleware technology for IoT systems: Challenges and perspectives toward 5G," in *Internet of Things (IoT) in 5G Mobile Technologies*, George Mastorakis, Jordi Mongay Batalla, Constandinos Mavromoustakis, Vol. 8, pp. 333–367, 2016, Springer, Cham.

21. S. Bandyopadhyay, M. Sengupta, S. Maiti, and S. Dutta, "Role of Middleware for Internet of Things: A Study," *International Journal of Computer Science & Engineering Survey*, vol. 2, no. 3, pp. 94–105, 2011.

22. Telcordia Technologies. (2017) Overcoming Interoperability Challenges in the Internet of Things. [Online]. Available at: https://iconectiv.com/sites/default/files/documents/iconectiv%20white%20paper%20IoT%20challenges.pdf

23. G. Eastwood. (2017) IoT's Interoperability Challenge. [Online]. https://www.networkworld.com/article/3205207/internet-of-things/iots-interoperability-challenge.html

24. C. Perera, A. Zaslavsky, P. Christen, and D. Georgakopoulos, "Context Aware Computing for the Internet of Things: A Survey," *IEEE Communications Surveys & Tutorials*, vol. 16, no. 1, pp. 414–454, 2014.

25. K. Ahmad. (August 2017) Linkedin. [Online]. https://www.linkedin.com/pulse/relation-between-contextual-computing-iot-kabir-ahmad/

26. P. Ccori, L. De Biase, M. Zuffo, and F. da Silva, "Device discovery strategies for the IoT," in *IEEE International Symposium on Consumer Electronics*, pp. 97–98, 2016, Sao Paulo.

27. P. Fremantle and P. Scott, "A Survey of Secure Middleware for the Internet of Things," *PeerJ Computer Science*, vol. 3, p. 114, 2017.

28. M. Abu-Elkheir, M. Hayajneh, and N. Ali, "Data Management for the Internet of Things: Design Primitives and Solution," *Sensors*, vol. 13, no. 11, pp. 15582–15612, 2013.

29. N. Ali and M. Abu-Elkheir, "Data Management for the Internet of Things: Green Directions," in *2012 IEEE Globecom Workshops*, pp. 386–390, 2012, Anaheim, CA.
30. S. Watts. (2017, May) Bmc blogs. [Online]. https://www.bmc.com/blogs/service-oriented-architecture-overview/
31. Huawei, *5G Scenarios and Security Design*, Huawei, Shenzhen, 2016.
32. M. West Darrell, *How 5G Technology Enables the Health Internet of Things*, vol. 3, pp. 1–20, 2016.
33. A. Ijaz et al., "5G security: Analysis of threats and solutions," in *IEEE Conference on Standards for Communications and Networking (CSCN)]*, pp. 193–199, 2017, Helsinki.
34. Huawei, *5G Security: Forward Thinking*, Huawei White Paper, 2015.
35. D. Fang, Y. Qian, and R. Qingyang Hu, "Security for 5G Mobile Wireless Networks," *IEEE Access*, vol. 6, pp. 4850–4874, 2018.
36. A. Alrawais, A. Alhothaily, C. Hu, and X. Cheng, "Fog Computing for the Internet of Things: Security and Privacy Issues," *IEEE Internet Computing*, vol. 21, no. 2, pp. 34–42, 2017.
37. X. Huang, P. Craig, H. Lin, and Z. Yan, "SecIoT: A Security Framework for the Internet of Things," *Security and Communications Networks*, vol. 9, pp. 3083–3094, 2016.
38. A. Sedrati and A. Mezrioui, "A Survey of Security Challenges in Internet of Things," *Advances in Science, Technology and Engineering Systems*, vol. 3, no. 1, pp. 274–280, 2018.
39. D. M. Mena, I. Papapanagiotou, and B. Yang, "Internet of Things: Survey on Security and Privacy," *Information Security Journal: A Global Perspective*, vol. 27, no. 3, pp. 162–182, 2017.
40. M. A. Ferrag, L. Maglaras, A. Argyriou, D. Kosmanos, and H. Janicke, "Security for 4G and 5G Cellular Networks: A Survey of Existing Authentication and Privacy-Preserving Schemes," *Journal of Network and Computer Applications*, vol. 101, pp. 55–82, 2018.
41. Y. Fu, Z. Yan, J. Cao, O. Koné, and X. Cao, "An Automata Based Intrusion Detection Method for Internet of Things," *Mobile Information Systems*, vol. 2017, pp. 1–13, 2017.
42. 5G Innovation Center, University of Surrey, *5G Security Overview*, 2017.
43. Ericsson, "5G Security, Ericsson, White Paper, UEN 284 23-3269," June 2017.
44. S. K. Lee, M. Bae, and H. Kim, "Future of IoT Networks: A Survey," *Applied Sciences*, 16, vol. 7, no. 2, pp. 1072, October 2017.

Index

Note: *Italic* page numbers refer to tables, **bold** page numbers refer to figures.